KNOWING NATURE

KNOWING NATURE

CONVERSATIONS AT THE INTERSECTION OF POLITICAL ECOLOGY AND SCIENCE STUDIES

EDITED BY
MARA J. GOLDMAN,
PAUL NADASDY, AND
MATTHEW D. TURNER

The UNIVERSITY of
CHICAGO PRESS
Chicago & London

Mara J. Goldman is assistant professor of geography at the University of Colorado–Boulder. Paul Nadasdy is associate professor of anthropology and American Indian studies at the University of Wisconsin–Madison and author of Hunters and Bureaucrats: Power, Knowledge, and Aboriginal-State Relations in the Southwest Yukon. *Matthew D. Turner is professor of geography at the University of Wisconsin–Madison.*

The University of Chicago Press, Chicago 60637
The University of Chicago Press, Ltd., London

Printed in the United States of America

20 19 18 17 16 15 14 13 12 11 1 2 3 4 5

ISBN-13: 978-0-226-30140-2 (cloth)
ISBN-10: 0-226-30140-0 (cloth)
ISBN-13: 978-0-226-30141-9 (paper)
ISBN-10: 0-226-30141-9 (paper)

Library of Congress Cataloging-in-Publication Data

Knowing nature: conversations at the intersection of political ecology and science studies / edited by Mara J. Goldman, Paul Nadasdy, and Matthew D. Turner
p. cm.
Includes bibliographical references and index.
ISBN-13: 978-0-226-30140-2 (cloth: alk. paper)
ISBN-10: 0-226-30140-0 (cloth: alk. paper)
ISBN-13: 978-0-226-30141-9 (pbk.: alk. paper)
ISBN-10: 0-226-30141-9 (pbk.: alk. paper) 1. Political ecology. 2. Environmental sciences. I. Goldman, Mara. II. Nadasdy, Paul. III. Turner, Matt (Matthew D.)
JA75.8K66 2011
333.7—dc22 2010024824

♾ The paper used in this publication meets the minimum requirements of the American National Standard for Information Sciences—Permanence of Paper for Printed Library Materials, ANSI Z39.48-1992.

CONTENTS

MARA J. GOLDMAN

AND MATTHEW D. TURNER

INTRODUCTION

Knowing nature is a complex, multiple, and highly political process. This is clearly illustrated by looking at the knowledge and management of a piece of land, seemingly isolated but impacting and impacted by decision-making processes, politics, and technology around the world.

A barren stretch of ground in the Sahelian region of West Africa holds diverse meanings to different people and institutions. Livestock herders value it for its proximity to a water point and for the grass it will grow once the rains come. To a group of remote-sensing analysts working in the northeastern United States, the stretch of ground is represented on their computer screens as a cluster of twelve pixels. They are working on an FAO-funded project to identify the extent of land degradation in the region. The twelve pixels show higher than expected reflectivity, symptomatic of degradation, with the cluster's proximity to surrounding villages suggestive of human-induced degradation. Their findings, covering many thousands of such pixels, were published and then used by a group of British climate scientists to argue at an international conference that the prolonged dry period in the Sahel is produced by the cumulative effect of many such ("degradation") clusters across the region resulting in greater regional reflectivity, surface cooling, and lowered potential for convective thunderstorms. This hypothesis points to local mismanagement as the cause behind persistent drought in the region, in contrast to alternative hypotheses that point to inherent climate variability or greenhouse-gas warming (and fossil-fuel consumption in industrialized nations).

The barren stretch is of interest to other actors as well. A government official sees it as a place with good potential to establish a mango plantation because it is low-lying and has received animal manure every year. He states that herders have degraded the land because they put down few permanent ties and therefore have no incentive to manage the land properly. He proposes that the management of the land needs to be changed. He hopes to persuade the development agency official who is running a small-scale irrigation program to fund his plantation. The official's program staff have held meetings with local villages—finding that irrigation of such low-lying areas would support villagers' interest in gardening projects. Yet other groups are promoting not "development," but conservation. An international environmental group, working with the government's forestry department, is seeking to protect important biodiversity in the area from grazing livestock. Their claims are supported by a report prepared by a panel of experts identifying the zone as underrepresented in terms of biodiversity protection. Moreover, a consultant hired by the forestry department found that the area was not meeting its potential in terms of vegetative productivity and recommended reducing stocking rates of domestic livestock.

Despite its isolation and limited resource potential, the stretch of land in question has been variously interpreted and enrolled by a range of actors who use arguments about its "barrenness" to support their arguments for its and the region's future. One could analyze this situation as simply a politics about access to and control over land—the herders need land for fodder and access to water; conservationists are seeking to protect biodiversity (on the land) by removing the damaging herders and their livestock; and development workers are seeking to increase the land's productivity through irrigation, thereby supporting agriculturalists' claims to the land for gardens or mango orchards. It is hard to ignore these politics—there will likely be clear losers and winners associated with different futures. But can we understand these politics simply by understanding the divergent material interests of herders, farmers, conservationists, and government officials?

Actually, the conduct and outcome of these politics depends heavily on the knowledge claims about past and future changes to the land. Like many environmental questions, there is significant uncertainty about the land's past and future.[1] As a result, different knowledge claims about what has created the land's "barrenness" and what will produce a productive future necessarily rely on invocations of a mix of observations and truth claims not only about the stretch of land in question but about global climate change, regional trends in land degradation, productive potential of the region's soils, and even the cultural characteristics of local land users. These environmental politics *are* a "politics of knowledge." What counts as valid understandings of the environment shapes contestations and outcomes.

These politics influence what could be seen as the production, application, and circulation of environmental knowledge. How applied research agendas are established, funded, pursued, and evaluated shapes the production of environmental knowledge. Management approaches (i.e., the "application" of environmental knowledge) are constructed from a mix of common understandings about human societies and the environment, scientific findings and technologies, standard (accepted) management approaches, political and economic prerogatives, and location-specific understandings. The example above shows how difficult it is to separate the politics of production, application, and circulation of environmental knowledges. The production of environmental scientific knowledge is shaped by management goals and directives as well as widely circulated ideas about society and environment. Clearly, the politics surrounding environmental management is not simply the playing out of material interests but is animated by competing knowledge claims about the environment. The outcome of these struggles shape the kind of "environments" that are produced and promoted, and whose purpose stretches of ground, water, and air serve.

Environmental knowledge production is framed, funded, and publicized in widely different social arenas. The livestock herders, who arguably know the stretch of land in question best, are not even aware that their eventual exclusion from their pasture is because they have been identified as agents of its demise by research conducted halfway around the globe by scientists who have never set foot on the pasture or, in some cases, the region. The mix of complex systems, limited research funding, and crisis mentality leads to the common practice of findings/impressions/common knowledges developed in one locality strongly influencing environmental research and management in "analogous systems" around the world (i.e., the circulation of knowledge). Research conducted on the grazing impacts of Maasai-managed livestock in Kenya influences how Fulβe pastoralism is viewed in Mali, West Africa. How are different socioecological systems determined to be sufficiently analogous? How do certain ideas become dominant by resonating within multidisciplinary forums? How do development and conservation goals influence applied research? How do different environmental knowledges, developed through different types of engagement with the material world, relate to each other? And what are the impacts on the people and environments involved (i.e., through the application of knowledge)?

These and other questions, which transcend the conventionally understood divisions of production, application, and circulation of knowledge, need to be addressed for us to understand the politics surrounding most environmental questions. How do we make sense of these politics? Conventional treatments of environmental science, policy, and management implicitly or explicitly treat production, application, and circulation of scientific knowledge as loosely articulated spheres of activity and discourse. Environmental scientists, working in relative isolation in universities and government labs, are seen as providing information to the public and government about changes to the environment and the role of humans leading to these changes. Different government agencies and civil-society groups then utilize this information to wage debates, make decisions, and inform resource management programs. The public or the government may misuse scientific information for certain purposes, but the production of scientific knowledge is seen as separate from management and policy. As shown in the example above, the reality is much more complex. A conventional political-economic analysis of the struggles surrounding the pasture would seek to explain the outcome as simply the result of divergent powers and interests, ignoring the broader politics that shapes the production and circulation of the competing knowledge claims that are enrolled by actors and help shape the outcome of these struggles. A "common property" analyst may use this case to identify the need to develop local political institutions to make more effective decisions about commonly held resources such as this "barren" stretch of land. A conventional sociology-of-science approach may analyze the remote-sensing analyst's categorization scheme and how it masks underlying assumptions, reflects her competition for funding opportunities, and depends on its resonance with common (mis)understandings in the field to gain broader acceptance. Unfortunately, all such accounts (made in isolation from each other) would mischaracterize the environmental politics surrounding this stretch of land and thereby would likely misdiagnosis the "problem." Clearly, broader and more integrated analyses of the "knowledge politics" surrounding the use of natural resources are needed.

In this volume, we seek to illuminate the politics that run through the conventional trichotomy of production, application, and circulation of environmental knowledge. Rather than treating environmental knowledge as a disembodied product that circulates in certain networks and then is applied, contributors to this volume focus on the practices of multiple actors involved in the inseparable nexus of production, application, and circulation of environmental knowledge. Environmental knowledge is seen as embodied in local contexts and contested in debates not only about science policy or resource politics but about expertise—about how knowledge claims regarding the environment are generated, packaged, promoted, and accepted by the diversity of actors involved in environmental management, conservation, and development.

Contemporary approaches to the study of the politics surrounding environmental scientific knowledge and practice are many, including sociology's analysis of the "risk society" (e.g., Beck 1992), historical treatments of environmental conservation and the meanings attached to "nature" (e.g., Cronon 1996; Grove 1996; Worster 1993), electoral and institutional politics as they relate to the environment (e.g., Dowie 1995), and environmental politics as social movements (Gottlieb 1993; Merchant 1992). Political ecology and science and technology studies (STS) are relatively new fields of interdisciplinary academic inquiry[2] that show great promise in reaching beyond conventional boundaries to analyze a broader environmental politics. Work within political ecology has focused primarily on the politics that surround environmental change, conservation interventions, and natural resource economies (see figure I.1). To understand these politics, scholars working in the field of political ecology have increasingly felt it

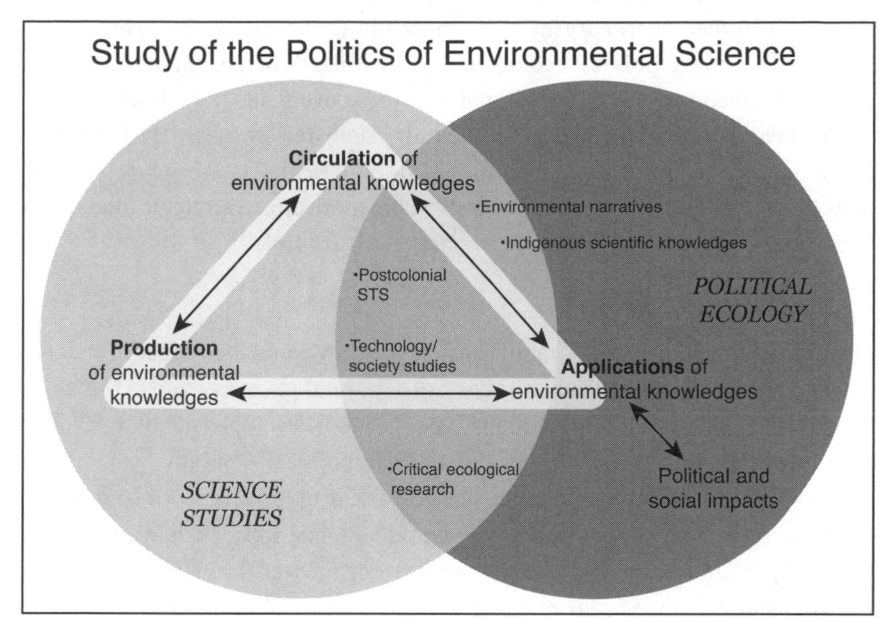

FIGURE I.1 Research emphases of political ecology and STS in relationship to the nexus of knowledge production, application, and circulation.

necessary to engage with the different ways in which "nature" is perceived, studied, and presented by different social groups—from local resource users to scientific "experts" whose assumptions, visions, and management techniques can be imposed on others. Work within STS has been extremely valuable in uncovering how "expert" or scientific knowledge is constructed, used, and transmitted (i.e., how it travels/circulates). STS scholars have shown how scientific practice and knowledge production are shaped not only by the scientist-subject relationship but by a broader set of social relations (see figure I.1).

While political ecology and STS have emerged and developed separately, they have also begun to merge, overlap, and borrow from each other. Political ecologists have increasingly sought to better understand the politics of environmental knowledge by reaching out to incorporate the concepts and tools developed in STS. Some STS scholars have likewise sought to engage more seriously with the politics surrounding the application of environmental scientific knowledge by blurring the assumed boundaries between environmental scientific practice (production) and management (application). Whether identifying themselves as political ecologists, STS scholars, or neither, all contributors to this volume are actively working on theoretically informed empirical work that seeks to cross these conventional boundaries.

Given that these works are informed most fully (but not exclusively) by new engagements between political ecology and STS scholarship, it is important that we provide the reader an understanding of how these two fields have developed and where areas of fruitful exchange lie. We only briefly review these developments. More comprehensive reviews are available for both fields.[3] Our brief reviews will be followed by a fuller discussion of areas where more direct, strategic interaction between these two fields is likely to be particularly fruitful.

POLITICAL ECOLOGY

Political ecology is a cross-disciplinary field of study concerned broadly with the politics surrounding (1) the use and control of natural resources and (2) environmental change and its representations. Simply put, it developed out of the uneasy marriage between cultural ecology and agrarian political economy—a marriage that emerged from an interest in the political and economic roots of land degradation in rural areas of the "developing world" (Blaikie 1985; Blaikie and Brookfield 1987; Watts 1983). Early contributions engaged critically with dominant environmental analyses (those strongly shaped by neo-Malthusian and "tragedy of the commons" concepts) and sought to combine a detailed understanding of rural producers' use of the natural resources as conditioned by their "access to resources," which is shaped by changes in the biophysical environment and the broader political economy. The explanatory focus was the dialectical relationship between social and environmental change with a particular emphasis on the connection between poverty and environmental mismanagement. Since the early 1990s, there has been a movement away from the structural roots of political ecology (Escobar 1999) along with a diversification of political-ecological scholarship. The rapid growth of political ecology scholarship is fueled by its attraction to scholars as a framework for drawing connections between social and ecological change; the environment and social justice; global and local change; as well as the construction of dominant views of "nature" or "the environment."

Political ecologists' treatment of environmental politics has always been broadly defined as a focus on the power relations and political-economic processes mediating knowledge about, and access to, natural resources. This is reflected in the large scope of work covered by political ecology researchers, including work on indigenous scientific knowledge, social natures, social movements, conservation and development politics, and critical analysis of dominant environmental narratives. Political ecologists use a mix of methods to examine the many different ways that nature can be "known" and managed, including critical social theory, historical analysis, ethnographic techniques, and ecological analysis. Despite the diversity of topics and approaches under its umbrella, political ecology generally differs from other fields that address a broadly defined environmental politics (i.e., land-change science, resilience theory, environmental history, cultural ecology, environmental sociology, ecological anthropology, political science, etc.) by variously combining some aspect of the following three features in its analysis (discussed in detail below):

1. A commitment to incorporating understandings of the biophysical processes that underlie environmental change and the availability of natural resources
2. An emphasis on understanding environmental politics as geographically and historically situated (i.e., the "case study" approach)
3. Strong commitments to social justice

COMMITMENT TO INCORPORATING UNDERSTANDINGS OF BIOPHYSICAL PROCESSES

Rather than treating "nature" as a backdrop to social conflict, biophysical processes and ecological response to human resource use are seen as part of the analysis of environmental politics. This has been a major focus since the initial framing of political ecology (Blaikie 1985; Blaikie and Brookfield 1987), which was largely concerned with the material effects of the decisions made by local land users (devegetation, soil erosion, etc.) as shaped not only by local land endowments but by broader political-economic change. A major contribution of early political ecology was the recognition that the prospect for land degradation was not predictable simply from the level of human demand on natural resources (human population, livestock population, consumption, wealth, etc.). Ecological response to resource extraction exhibits spatial heterogeneity, nonlinearities, and threshold effects, which require the political ecologist to engage seriously with ecological dynamics in order to understand society-nature relations. Moreover, the nature, rate, and spatial extent of ecological change are seen to matter in the unfolding politics surrounding the environment. Despite its acknowledgment of the importance of understanding the physical processes leading to environmental change and social conflict, political ecology has been critiqued for "missing" the ecology (P. Walker 2005). In other words, political ecologists have generally treated resource-related conflict as inherently social. Changing material (ecological) conditions are seen as influencing environmental politics only through the divergent meanings attached to change by individuals and groups with divergent powers (Carney 1992; D. Moore 1993; Peluso and Watts 2001; Schroeder 1999b).

Ecological processes are shaped by geographic and historical context. The reestablishment of a forest, the response of grasslands to grazing, and the maintenance of biodiversity all are strongly dependent on the ecological history and the surrounding landscape matrix of any site. Political ecologists, by seeking to understand local ecological dynamics in areas where prior ecological research is limited, have often discovered evidence that contradicts common understandings of ecological processes held by resource managers and therefore find themselves engaging critically with dominant discourses about agroecological change (Bassett and Zuéli 2000; Braun 2002; Carney 2001; Fairhead and Leach 1996; Ferguson 1990; Forsyth 1998; Leach and Mearns 1996; Neumann 1998; Prudham 2005; Robbins 1998; Rocheleau 1991). As a result, political ecologists have been drawn outside of their traditional scope of inquiry, the politics surrounding the application of scientific knowledge, toward gaining a better understanding of scientific knowledge production (see figure I.1). For the most part, these critical engagements have remained somewhat coarse—focused more on science-based arguments by policy makers, conservationists, and developers and less on environmental scientific practices themselves (for exceptions see Bassett and Zuéli 2000; Forsyth 2003; Langston 1996; Robbins 2001; Robertson 2006; Sayre 2008; Taylor 2005; Turner 1998, 2003; Zimmerer 1996).

EMPHASIS ON UNDERSTANDING ENVIRONMENTAL POLITICS WITHIN PARTICULAR PLACES AND HISTORICAL MOMENTS

Political ecology's emphasis on analysis within particular geographic and historical contexts stems from its intellectual heritage (cultural ecology) and its methodological commitments to make connections between social and ecological change explicit within particular places. Along with this commitment to context is the ambition to relate local processes to broader social and ecological changes. The common emphasis of seeking causal connections between global or regional changes to local resource-use decisions is one of the most appealing and ambitious characteristics of political ecology. Such causal connections are difficult to make and fraught with theoretical problems. While the role of markets as intermediate networks has been a major emphasis in political ecology work, there is a growing interest in the global circulation of knowledge and its impact on local society-environment relations (e.g., Escobar 1999; Forsyth and Walker 2008; Fairhead and Leach 1996; Goldman 2005; Neumann 1998). This reflects in part the proliferation of institutions involved in environmental governance at local levels across the world. Claims made by academics, World Bank officials, or members of international environmental organizations about the effects of resource use on environmental health may strongly influence government and NGO actions and, in so doing, shape local communities' access to natural resources. Scholars taking a political ecology or related approach have also contributed to nuanced understandings of the changing role of state-society interactions impacting environmental knowledge politics (e.g., Agrawal 2005; Drayton 2000; Grove 1996; Heynen 2007; Li 2007; D. Moore 2005; Mitchell 2002; J. Scott 1998). Therefore, political ecologists have increasingly been concerned not only with how ideas about development and conservation develop, but how scientific understandings of the environment are packaged, stabilized, and circulated within and across scientific and resource management communities—and with what effects.

STRONG COMMITMENTS TO SOCIAL JUSTICE

Political ecology first developed from cases in rural areas of the developing world where state-sponsored conservation, resource extraction, or development often pitted local communities against outside interests. The links between political-economic subordination, vulnerability, and environmental mismanagement have remained major foci of political-ecological analysis. Moreover, there has long been an emphasis on treating social movements as motivated not solely by "environmental" concerns but by social justice (Bryant 1996; Escobar 1998; Guha 2000; Peet and Watts 1996; Peluso 1993; Rangan 2000; Rocheleau, Thomas-Slayter, and Wangari 1996). Accordingly, environmental justice figures prominently in first-world political ecology as well (e.g., Keil 2003; McCarthy 2002; Swyngedouw and Heynen 2003; Wainwright and Robertson 2003). Political ecology shares with cultural ecology an attention to the politics surrounding what is often referred to as "indigenous knowledge" (Agrawal 1995; Murdoch and Clark 1994; Richards 1985). The work of cultural and political ecologists has demonstrated how subordinate groups are disempowered by the characterization of their understandings of the natural world as nonscientific, tradition-bound, overly risk-adverse, shaped by superstition, or simply biased. Political and cultural ecologists have sought to illuminate, translate, and in some cases champion local understandings of the environment (Carney 1991; Escobar 1998; Fairhead and Scoones 2005; D. Moore 1993; Peluso 1992; Sundberg 2003; Zerner 2000). A new interest in political ecology is analyzing the engagement of Western scientific with other environmental knowledges (Goldman 2007; Hayden 2003; Akhil Gupta 1998; Nadasdy 1999).[4]

REACHING OUT TOWARD STS SCHOLARSHIP

The three overlapping areas of concern sketched above are shared by the diverse treatments of environmental politics by scholarship labeled "political ecology." Political ecologists have increasingly been pulled into the arenas of environmental science production and circulation from their core focus of analyzing society-environment relations in particular places and historical moments.

The commitment that, to understand society-environment relations, one must understand biophysical processes *in place* has led political ecologists to perform or collaborate with those performing ecological research. In so doing, political ecologists have found themselves in social, discursive, and methodological positions that fall between the multiple knowledges about the local environment held by local people, state-sponsored managers, international NGOs, and the broader environmental scientific community. Such knowledge interfaces are power-laden and have potential for significant material impact on the environment (human and nonhuman). Moreover, new trends in international conservation and development emphasizing participation and decentralization of resource management authority have made the politics surrounding divergent knowledge claims more obvious (e.g., Agrawal and Ribot 1999; Brosius, Tsing, and Zerner 1998; Escobar 1998; Goldman 2003; Nadasdy 1999; Sundberg 2003). Understandably, political ecologists' social justice commitments have drawn them to analyze these interfaces.

These trends have led to two impulses to de-essentialize environmental knowledges through more in-depth, critical engagements with knowledge production and circulation. The first impulse is to engage critically with multiple knowledge claims in order to come closer to underlying truths or to seek cross-knowledge accommodations (critical realism). The second is the recognition that to fully understand environmental politics, a focus on the exercise of power at the places and moments where competing knowledge claims are debated publicly (application) is insufficient. Social power that shapes on-the-ground impacts operates in the realms of knowledge production and circulation as well (often far from the place of application). Therefore, to fully analyze environmental politics, political ecologists need to not see divergent knowledge claims as the starting point for politics but instead seek to understand how these knowledge claims are constructed and travel to the places of interest.

The traditional tools of political ecologists are best suited for analyzing how the applications of knowledge, markets, capital, labor, and political power in particular contexts affect local society-environment relations. These tools are most appropriate for understanding the construction of landscapes and the circulation of commodities and wealth through markets, and power and authority through governance structures. They are less insightful for understanding the production and circulation of scientific knowledge. Political ecologists have therefore increasingly looked toward STS scholarship for ways to de-essentialize environmental knowledges through the adoption of tools and concepts that allow more productive analysis of the production and circulation of these knowledges. The goal of "reaching out" in these ways is not to replace but to enrich and broaden political ecology's historical emphasis on place-based analysis of political economy and ecology. These new areas of inquiry will be discussed more fully after a brief review of STS scholarship as it relates to environmental politics.

SCIENCE AND TECHNOLOGY STUDIES

STS is a growing cross-disciplinary field of study concerned with the production of scientific knowledge and technologies within a social (cultural, political) context (Law 2004, 12). Previously referred to as science, technology, and society studies (Law 1991), and often referenced simply as "science studies," STS draws from and incorporates work within the fields of the sociology of scientific knowledge, the history of science, the philosophy of science, science-policy studies (see Forsyth 2003), and feminist science studies (Haraway 1991; Harding 1986; Keller 1999), to analyze the various ways in which science "works" to produce (and circulate) knowledge about the world.

Work within STS has highlighted the mutually constituted nature of scientific knowledge and society (social relations, institutions, power dynamics, and politics). Scientific knowledge, once recognized as an objective reflection of reality (nature), is exposed in STS work as the outcome of messy and situated practices: practices that are shaped by particular historical, socioeconomic, political, and cultural contexts. In other words, "scientific knowledge and technologies do not evolve in a vacuum. Rather they participate in the social world, being shaped *by* it, and simultaneously *shaping* it" (Law 2004, 12, emphasis in original). This has been highlighted in different ways by different spheres of STS. Feminist historians and philosophers of science have exposed how specific historical contexts have shaped particular moments of scientific knowledge production as linked to imperialism, racism, and sexism (Haraway 1989; Harding 1986, 1988). Laboratory ethnographies (or "lab-studies") have uncovered the messy, culturally specific, day-to-day practice (and negotiations) involved in the creation of scientific "facts" in the lab (Latour and Woolgar 1979; M. Lynch 1982, 1985; Pickering 1992); as related to epistemic cultures (Knorr-Cetina 1999); as a part of cultural style and accepted norms (Traweek 1988); and through long-distance networks of people, animals, objects, and institutions (Callon 1986; Latour 1987, 1988, 1996).

STS has provided different theoretical approaches and analytical tools, often accompanied with catchy terminology, to understand the production of scientific knowledge and the location of "expertise." Such contributions include (but are not limited to) actor-network theory (ANT), hybrids (cyborgs), boundary objects, standardized packages, black-boxing, natureculture, work objects, bandwagons, and co-production. ANT is perhaps the most well-known and well-utilized approach/language taken up from STS (Callon 1986; Latour 1987, 2005; Law and

Hassard 1999), particularly by political ecologists (e.g., Birkenholtz 2009; Gareau 2005; Kirsch and Mitchell 2004; Robbins 2007; Swyngedouw 1999; Perkins 2007; Prudham 2005). ANT provides a framework for analyzing the production of knowledge as occurring through relational networks, where objects (e.g., people, animals, microbes, tools, institutions) contribute equally as agents (actants) to the configurations and reconfigurations of the network itself. One could, as such, follow the creation of facts along such relational networks. Along the way, certain phenomenon are "black boxed" as facts, actants are enrolled, others resist, and certain entities succeed at becoming "facts" while others fail and disappear.

ANT has proved useful for encouraging an investigation into the role of nonhuman objects (including animals) in the creation of knowledge about nature, and for articulating the complex twists and turns that knowledge creation takes along interrelated networks. It has, however, also been critiqued in many ways, even from within STS (see Law 2004). Critics point to the simplistic portrayal of all objects as equal agents in knowledge construction, and to the lack of attention to larger social and political relations (i.e., class, gender) leading to seemingly flat networks where complex local and global power dynamics are not recognized. ANT has also been critiqued for discussing power in Machiavellian language, portraying simplistic and overly vertical relations of "power-over," ignoring the multiple and complex forms that power relations take (i.e., power with, power in spite of; Law 1991; Rocheleau and Roth 2007). Political ecology scholars have been at the forefront of this critique, finding much of the analytical insight of ANT useful, but the overly simplistic and structural nature of the networks problematic (Castree 2002; Birkenholtz 2009; Gareau 2005). A more thorough interplay of political ecology and STS has the potential for creating a more complex topography of power-laden networks imbued with social relations that are vertical and horizontal in nature (see discussion below).

Others, working within the "social worlds" theoretical framework, have followed the construction of scientific facts through shifting historical relations across different groups (social worlds) where commitment and action interact with compromise and conflict in affecting the success of certain scientific facts over others (Clarke and Star 2008). Here the focus is less on centralized power and "black-boxed" facts, and more on "gray boxes" of ambiguous and fuzzy "boundary objects" that stick together and travel with the help of a standardized package of tools and theories (Fujimura 1992).

Star and Griesemer (1989) proposed the concept of boundary objects to refer to "things that exist at junctures where varied social worlds meet in an arena of mutual concern." Boundary objects can be translated differently within different social worlds, but maintain enough common ground across groups to enable dialogue, thus facilitating cooperation without consensus (Clarke and Star 2008). The notion of boundary objects has been helpful in understanding how certain concepts or technologies travel across diverse settings, enabling negotiation. The boundary-object concept has expanded to include boundary institutions or organizations and boundary infrastructures—terms used to discuss objects and embedded institutions that cross larger scales than boundary objects might. Fujimura (1992, 1996) has expanded the work of boundary objects in another direction—as part of larger standardized packages, "gray boxes" that combine several boundary objects together with tools, theories, and methods. Standardized packages enable the creation of a more stable (social) fact than do flexible and ambiguous boundary objects. When a standardized package gains momentum, it can seem like the only game in town (e.g., the genetic model of cancer), traveling so fast that it can create a bandwagon effect (Fujimura 1988).

STS scholars have also contributed a "keenly honed" vocabulary to discuss relations among human/nonhuman, natural/social, and technological/natural/human domains (Whatmore 2002, 4). Such language includes Callon and Law's "hybrid collectif" (1995), Latour's naturesociety hybrids, quasi-objects, and "Middle kingdom" (1993), and Donna Haraway's "cyborgs" (1991). This new vocabulary rejects traditional boundaries and disrupts ontologies. Yet other work in STS has focused on boundaries—how they are policed on the one hand (i.e., what counts as science [Gieryn 1999]), or as discussed above, how they (as places, concepts, and organizations) can promote communication, negotiation, and translation across social worlds (Star 1995; Star and Griesemer 1989). And finally, STS scholars working primarily on the science-policy interface (Jasanoff 2004b) have introduced the language of co-production. According to Jasanoff (2004b, 2), co-production is "shorthand for the proposition that the ways in which we know and represent the world (both nature and society) are inseparable from the ways in which we choose to live in it." In a similar vein as Latour's "actor networks" and the language of hybrids, co-production provides both a metaphor and a frame of analysis for seeing science and society as being simultaneously co-produced. This language has become particularly useful in not giving primacy to either nature or society, either science or policy, as science studies has become more engaged with policy debates, as elaborated on below.

To summarize, STS scholarship has been invaluable in undressing "Science" (Nader 1996) and exposing its historical, social, cultural, and political context. In the process, several important arguments and areas of inquiry have arisen, all which intersect and overlap with many of the questions being addressed by political ecologists. These include:

- All knowledge is "local" and culturally/socially contextual, which means that scientific knowledge is situated (cultural) practice.
- Expertise is awarded, challenged, and contested differently in different contexts.
- Knowledge travels (circulates) through translations, packaging, and networks.
- Science and society are co-produced.
- Knowledge is inherently political.

Despite these contributions, the relevance of STS work to the myriad of contexts through which environmental knowledge circulates and is shaped would be enhanced through greater engagement with political ecology. There has been limited direct engagement by STS scholars with (1) messy "field-based" environmental scientific work such as ecology, conservation biology, wildlife ecology, forestry, agronomy, agricultural extension, animal sciences, and so forth, where knowledge is highly context-specific; and (2) the uneven power relations involved in the circulation and particularly the application of environmental knowledge on the ground. It is within such contexts that political ecologists often work, and where there is a growing interest in better understanding the knowledge politics linking university classrooms, funding agencies, and laboratories (mostly) in the developed world, with rural and urban field sites (mostly) in the "developing" world.

STS work has been focused primarily in the developed world, on the "big" sciences, and in laboratory settings. The need to direct attention to different social contexts where environmental knowledge is continuously produced through its application, packaging, and circulation has been recognized by a number of STS scholars. For example, Watson-Verran and Turnbull (1995) call on STS to move beyond a focus on *production*, to address the politics inherent in the *circulation* and *application* of Western science as a privileged knowledge system. They suggest that "the strength of social studies of science is its claim to show that what we accept as science and technology could be other than it is; its great weakness is the general failure to grasp the *political nature* of the enterprise and to work toward change" (138, emphasis added). They and others (e.g., Jasanoff 2004a, 28) argue for utilization of the strength of STS to better understand the production of other knowledges and to work toward devising ways in which different knowledge systems (including scientific) "can be made to interrogate each other."

STS scholars have subsequently recognized the need to more directly address power relations that crisscross the production-application-circulation nexus. Jasanoff, for instance, suggests that while Latour's networks provided a valuable approach to understanding the co-production of science and society, they are not political (Jasanoff 2004a). While power is said (by Latour) to be concentrated in "centers of circulation," there is no discussion about where these centers tend to be located (within the global political economy), and no discussion of norms and politics. Jasanoff also contends that STS work has not, in general "paid much attention to what happens when particular epistemic and material constructions of the world circulate through societies configured by very different historical and material constraints" (2004a, 28). Others have suggested that while STS scholarship has highlighted the relations that contribute to scientific knowledge production, it has been blind to (or at least silent on) what Law (1991) refers to as the "great distributions" of inequality, normally the purview of sociologists (race, gender, ethnicity, class). The exception to this is feminist STS scholarship, which has been invaluable in uncovering the various ways in which knowledge production, application, and circulation occurs in a world divided in various ways along lines of gender, class, ethnicity, and power.

WORKING AT THE INTERFACE OF KNOWLEDGE PRODUCTION, APPLICATION, AND CIRCULATION

In this edited volume, we seek to move beyond the starting points of the fields of political ecology and STS to where concepts and approaches are necessarily shared, combined, and created. This emphasis not only grows from intellectual developments in both fields (as described above) but also is a response to the changing ways in which environmental knowledge is implicated in environmental politics. Global interests in "local" environments, whether conservationist or developmentalist, have grown significantly over the past several decades (Zimmerer, Galt, and Buck 2004). As illustrated in the case described at the outset of this chapter, multiple knowledge claims are made about natural resources in different places around the world. Most of these knowledge claims are not produced wholly in place but elsewhere—traveling through a myriad of different institutions before being "adapted" to the realities of the place of concern. Understanding the mutual influences of production, application, and circulation of knowledges on environmental politics and practice is not solely an academic concern but one having significant policy and on-the-ground implications.

As discussed earlier, a major conceptual problem that has plagued conventional treatments of environmental politics has been the treatment of knowledge as a product—a product that retains its form as it circulates among the different contexts of production and application. In this volume, we seek to look beyond these boundaries to develop fuller understandings of a multi-sited and multi-actor environmental politics: a politics that is not constituted by isolated contestations among scientists over ideas, local stakeholders over resource access, and nations over control over markets or the media, but for which these realms are highly entangled. To effectively engage with this complexity, contributors to this volume seek to explore the connections among the development, application, and circulation of knowledge in environmental politics. While their vantage points differ, the themes lying at the intersections of production, application, and circulation are similar and include:

1. The myriad of influences and complex politics that shape *multiple environmental knowledges* as they circulate and are "applied" in conservation and development activities
2. The processes through which *society and nature are co-produced*: processes that are materially and discursively mediated through the activities of nonhuman and human members of the environment
3. The ways in which *knowledge is embodied* within and imperfectly translated across *power-laden social networks*

MULTIPLE KNOWLEDGE PRODUCTIONS

Environmental knowledge is produced not solely through scientific research but also by local people, resource managers, and government officials, who are influenced by circulating knowledge claims as they engage with local environments and with other humans over access to resources. There is a need to understand this broader arena of environmental politics—an arena that transcends the conventional foci of political ecology (resource politics) and STS (science policy and practice within Western institutional settings). Greater engagement across these two fields holds particular promise.

Such integration has already begun, taking on different shapes within political ecology and STS, and with promise for continued, more explicit exchange. As described above, political ecologists have contributed many critical analyses of the politics surrounding scientific knowledge claims in conservation and development but with only a subset of these engaging substantively with the methods and processes that underlie these claims. Therefore, the knowledge politics of conservation and development of most political ecology analyses remain contestations among fully formed knowledge claims. A deeper political analysis would also include the influences that shape how these claims arrive at the sites of contestation over resource access.

An emerging area of study within STS, "postcolonial technoscience," seeks to expand science studies beyond the (often assumed closed) boundaries of Western science and the nation-state, toward an engagement with the complex and multiple reconfigurations of knowledge, power, culture, capital, and science characteristic of an "emerging global order" (W. Anderson 2002, 643). In this way postcolonial technoscience echoes and expands on work within political ecology that has sought both to illustrate the value of local (nonscientific) knowledge production and to highlight the violent impacts of scientific practice on "third-world" people and environments. The violence occurs materially to nature (e.g., through the introduction of green-revolution technologies) and to people through loss of access to and control over resources. But the violence also occurs epistemologically by denying the legitimacy of other ways of knowing and managing nature.

A postcolonial STS (or an STS-informed political ecology) promises to take multiple ways of knowing seriously. It encourages an evaluation of the achievements, knowledge, beliefs, and methods of different and multiple knowledge productions in "symmetrical" ways. Harding (2003) discusses the value of such a turn within STS for exposing the *sameness* of all knowledge as essentially "local knowledge systems" (cf. Turnbull 2000) while exposing valuable differences across knowledge traditions. Work by Law (2004) and Mol (2002) highlights the importance of recognizing "multiple" knowledge constructions (and ontologies) as being more than one but less than many; as not one holistic whole, but neither a relativist disconnected multiplicity.[5] Work within political ecology has made similar calls for the recognition of multiple knowledges (Agrawal 1995; Goldman 2003; Murdoch and Clark 1994; Rocheleau 1991), but has also provided pointed critiques of the politics of processes claiming to bring scientific and "local" knowledges together (Brosius, Tsing, and Zerner 1998; Nadasdy 2003).

An increased interest in multiple, "nonscientific" knowledges within STS can also be seen in the calls made to include "citizens" in scientific discussions and decision making, particularly as related to risk (Irwin 1995; Irwin and Wynne 1996; Wynne 1991, 1992). More broadly, Jasanoff and Martello (2004) argue for the need to expand the range of knowledge sources contributing to global environmental governance. They highlight four specific reasons for why this should be: (1) the culturally, historically, and place-based specificity of much "local" knowledge "appears in the global context as more truly reflective of complex environmental realities"; (2) the incorporation of place-based "local" knowledge counters the simplifying, reductionist, and erasing tendencies of scientifically defined environmental standards; (3) local (more obviously "situated") environmental knowledges reflect a more holistic view of nature and culture; and (4) creating space for local knowledges in environmental governance could potentially create new "more environmentally aware social identities, thereby increasing support for robust management practices" (Jasanoff and Martello 2004, 339–40). This list reflects statements that have been made previously by political ecologists (Alcorn 1989;

Brookfield and Padoch 1994; Richards 1985; Dove 1996; Murdoch and Clark 1994; J. Scott 1998). Coming from an STS scholar long engaged in science-policy questions and working for more positive (and political) engagement of STS critique with policy production (Jasanoff 2004a; Jasanoff and Martello 2004), the points reflect a positive shift toward new (shared) ground involving the links between environmental knowledge and governance.

JOINT PRODUCTION OF NATURE AND SOCIETY

In this era of cloning, global climate change, payments for ecosystem services, and so forth, the blurring of the boundaries between "natural" and anthropogenic has become more obvious despite the fact that our understandings of what constitutes the "natural" has always been shaped by views of society (Castree 2005; Cronon 1996; Glacken 1967; Williams 1980). Political ecology, through its emphasis on the power dynamics and material transformations associated with conservation and development, has long treated nature and society as interrelated and coevolved. While much political ecology scholarship has arguably focused on interplay among well-defined material interests in conservation and development, there are many examples tracing how meanings attached to, and understandings of, nature have contributed to the politics of nature (Braun 2002; Carney 2001; Fairhead and Leach 1996; Ferguson 1990; Ingold 1987; D. Moore 2005; Neumann 1998; Zimmerer 1996). These treatments have emphasized that the relationship is constantly evolving through recursive historical processes. In this way, they share much with those of environmental historians (e.g., Cronon 1991; Merchant 1989; White 1983).

In studying the production of scientific knowledge, STS scholars have emphasized the blurring of the "animate" and "inanimate" and the "natural" and "human." Nature-society hybrids, cyborgs, naturesociety, and the collectif are all terms that have been used to describe the heterogeneous-combined, dialectically produced nature and society (Callon 1986; Haraway 1991; Latour 1993). In these and other ways, STS scholarship has expanded our vocabulary and ways of conceptualizing assemblages that are at the same time "natural" and anthropogenic.

The power of these concepts has been recognized by scholars within, and loosely connected to, political ecology for some time (e.g., Braun and Castree 1998; Demeritt 1996). Arguably, the theme of the "joint production of nature and society" has experienced the greatest level of exchange between STS and political ecology, although the direction of this exchange has, despite use of the term "political ecology" by STS scholars (Latour 2004), been largely from STS to political ecology. Scholars, not identified as either political ecologists or STS scholars, have mediated this exchange. These scholars, coming from political-economic (e.g., commodification of nature) or cultural landscape traditions, have contributed to a field of inquiry that can be identified as that concerned with "social nature" (Castree and Braun 2001). In brief, the social natures field analyzes the social construction of nature and landscape (Castree and Braun 2001; Braun and Castree 1998; Castree 2005; Ingold 2000; Whatmore, 2002). Western science is seen as a strong contributor to these constructions but one that is far from insulated from broader political and economic interests. Descriptions of these natural-social constructions have often relied on STS concepts such as hybridity (e.g., Swyngedouw 1999; Whatmore 2002; Zimmerer 2000a) and mutual construction (Bakker and Bridge 2006; Castree and Braun 2001; Ellen and Fukui 1996; Robertson 2006).

Despite the strength of STS concepts for analyzing material objects as assemblages of nature, technology, and human management, these concepts have been less helpful for analyzing the often politically fraught *processes* of assemblage, hybridization, and construction. These limitations have been duly noted by some STS scholars. For example, Sheila Jasanoff has identified the need within STS to direct greater attention to power and politics in processes of the "co-production" of natural and social orders (2004b, 2). Greater collaboration among STS and political ecology scholars at the interface between knowledge production and application holds great promise for improved understandings not only of the hybridized objects but also of the role of environmental knowledges in the heterogeneous and power-laden processes contributing to such co-productions.

THE PACKAGING, TRANSLATION, AND TRANSPORT OF KNOWLEDGE

Environmental knowledge is multiple, complex, and situated. Yet *certain* knowledge productions carry more weight than others. Certain formulations are recognized as "truth" and others as "falsehoods." Certain knowledge claims, theories, and statements travel quite far (from their place of origin), while others always remain only "local." In Latourian language, certain knowledge productions "win." But how, why, and for how long do certain ideas become or fail to become "common currency"[6] and assumed truths? Political ecologists have long asked this question, with an interest in the impacts (social and ecological) of the acceptance of certain knowledge constructs as "universal truth" and their subsequent application in faraway places. Often discussed as "conventional wisdom," such accepted truths have been studied and addressed as narratives—historical, cultural, political, and economic. While environmental narratives (e.g., desertification) might contain nuggets of scientific data, the argument was often successfully made by political ecologists that the science mattered little—it was the weight and power of the story (both metaphorically and structurally) in relationship to political interests and institutional prerogatives that kept the narrative alive—despite often competing scientific data to the contrary (e.g., Bassett and Zuéli 2000; Brockington 2002; Escobar 1998; Fairhead and Leach 1996; Forsyth 2003; Leach and Mearns 1996; Neumann 1998; Roe 1991; Stott and Sullivan 2000). While the narrative approach in political ecology has been useful in uncovering the history and political economy behind scientific truth claims in certain places, its portrayal of the processes that contribute to the success of certain knowledge claims over others has generally been limited—reduced to claims that stories are constructed to support powerful interests.

STS, on the other hand, has focused on the processes related to how, why, and where certain knowledge claims are accepted as "truth" and become mainstreamed while others are discarded. These questions have been asked in different ways by different STS scholars, and have provided invaluable tools for understanding not just the production of knowledge, but its subsequent circulation and then application. Some have focused on the importance of language, metaphor, and storytelling in the circulation of certain forms of scientific knowledge (Haraway 1989, 1997; Keller 1999; E. Martin 1991, 1994; J. Martin and Nakayama 1999). Here, as in political ecology, there is a focus on the power of the story, of the metaphor, not only in the circulation and application of particular scientific findings, but also in their very production. Others have focused attention on the successful creation of scientific facts through the building of alliances or associations along complex networks composed of nonhuman entities along with humans and their institutions (i.e., ANT). ANT shows how, through processes of negotiation and translation, the enrollment of actors is accomplished to gain support for a "fact."

Despite the discussion of "transgressions, displacements, dissidence," ANT work is primarily concerned with the stabilization of facts and less with the negotiations that take place. This approach has been useful for understanding the actions that go into the creation of "black-boxed" technologies and knowledge, but it is less helpful in understanding how messy collaborations are packaged; how less concrete "gray boxes," standardized packages, and boundary objects are formed across diverse groups and interests; and how these travel. Such processes are of more relevance to environmental science, where divisions between basic science, applied science, and scientific policy and applications are very diffuse.

All these different tools—narratives, metaphors, translation, boundary objects, standardized packages, and bandwagons—provide insight into the varied ways that knowledge becomes information, ideas become fact, and objects become artifacts. They provide different ways to understand "the contingencies and connections, multiplicities and messiness, involved in making knowledge into 'information'" (Waterton and Wynne 2004, 92). They also provide a promising avenue for more collaboration across STS and political ecology. For despite discussion of "social worlds" and various potential actors, much of the STS work remains focused largely on the internal world of science. Callon and Rabeharisoa (2003) suggest the need to move beyond this focus to look at relations between scientists and nonscientists in the production and dissemination of knowledge. They refer to such a move as "research in 'the wilds,'" yet the example they draw from is based within fairly organized public forums in the industrialized North interacting with medical science. Yet political ecologists and postcolonial STSers have already begun to conduct research "in the wilds"—at the interface of different social worlds where standards of epistemology, methodology, and expertise vary dramatically, as do language, culture, history, politics, education, economics, and livelihood. At such interfaces, the tools discussed in this section prove incredibly helpful for understanding how knowledge becomes information, how information is shared and transformed, why certain knowledge constructs travel within and across different social worlds and policy arenas, and how collaboration can happen without consensus.

ORGANIZATION OF THIS BOOK

This book is organized around the three major vantage points that not only dominate popular views linking environmental knowledge and politics but also have shaped prior approaches to the study of the politics of environmental knowledge. The three parts of the book approach environmental politics through the perspectives/lenses of the production (part 1), application (part 2), and circulation (part 3) of environmental knowledges. This structure is adopted not to reify these perspectives (in fact we seek to break them down) but as a conscious strategy to draw readers from these three spheres into productive intellectual spaces lying at their intersections (figure I.1). All contributions to this volume are located within these interstitial spaces, and as a result, there is significant overlap and exchange in contributions across these three parts. Three major themes cutting across the parts are those discussed above: the multiplicity of environmental knowledges; the joint production of nature and society; and the packaging, transport, and translation of knowledge.

A major goal of this volume is to elicit critical engagement with these themes by the reader. We hope to encourage this not only through the book's structure but through the qualities of both the contributors and their chapters. Contributors to this volume come from different disciplines and variously identify with the interdisciplinary fields of political ecology and STS. They are all scholars who are actively involved in research at one or more interfaces of the knowledge triad (see figure I.1). Their contributions not only explore important theoretical issues that are raised when working at these interfaces but illustrate these ideas with empirically grounded case material. The presentation of theory and case material by many of the most active scholars working at these interfaces provides a fertile landscape of inquiry, reflection, and debate that we hope will enrich the learning experiences of others working in this area at different levels of sophistication.

The chapters also provide useful and novel insights about method. The study of a more broadly defined environmental politics necessitates critical engagements not only with science policy and resource politics but with the discursive and material practices ("scientific" and otherwise) of the myriad of different actors using environmental knowledge. Contributors provide important examples of how to study an environmental politics that cuts across the production, application, and circulation trichotomy. In doing so, many of our contributors also point to different ways of producing and circulating environmental knowledge in conservation and development.

Each of this book's parts is composed of four to six chapters and is introduced by a short essay that relates the part to previous part(s) of the book, introduces the major theoretical questions explored by chapters in the part, and relates these questions back to the cross-cutting themes of the book as described above.

NOTES

This volume developed from presentations and discussions at the "Political Ecologies of Knowledge, Science, and Technology" workshop held in Madison, Wisconsin, on March 6–7, 2006. We thank the participants at the workshop and its sponsors, the Holtz Center for Science and Technology Studies and the Environment and Development Advanced Research Circle of the University of Wisconsin–Madison, for their contributions and support.

1. The direction, nature, and rate of environmental change are highly sensitive to geographic and historical context. Therefore it is difficult to easily translate findings from one site to another. While such translations are difficult in lab sciences, where experiments are highly controlled, they are especially problematic in the environmental sciences.

2. We use the terms here to refer loosely to a broad range of work situated in nature-society geography, cultural and ecological anthropology, environmental history, development studies, history of science, postcolonial science studies, feminist theories of science, sociology of scientific knowledge, and science studies. We do not assume bounded categories or set agendas with the use of these terms, but rather use them as guideposts for discussion across shared concerns and research interests.

3. For STS, three comprehensive handbooks (Hackett et al. 2008; Jasanoff et al. 1995; Spiegel-Rosing and de Solla Price 1977; see also Biagioli 1999) and additional edited volumes (Clarke and Fujimura 1992a; Jasanoff 2004b; Law and Hassard 1999; Law and Mol 2002; Pickering 1992; Reid and Traweek 2000; Star 1995) have outlined emerging trends in research foci and analytical styles within STS. More complete reviews of political ecology include those by

Forsyth (2003), Neumann (2005), Robbins (2004), Stott and Sullivan (2000), and Zimmerer and Bassett (2003).

4. Reflecting its social justice commitments, politics is generally treated by political ecologists as being shaped by the political-economic interests of major actors. The divergent interests held by the state and by local communities is a common theme (e.g., Peluso 1993). One critique of political ecology is that politics are overly simplified—seemingly driven by the clash of interests between social groups over the natural resource in question. Alternatively, politics can be seen as a much more open, contingent interplay of divergent motivations, powers, and strategies of different actors within and across major social groups (and over meaning as much as materials). This is particularly necessary for understanding the politics revolving around different types of environmental knowledge.

5. See also Strathern (1991) on "partial connections" between people within and across "cultural" worlds as always both multiple and singular, similar and different, and thus both local and global.

6. This term comes from Forsyth (2004), in a book review of Jasanoff (2004b).

Part 1

MATTHEW D. TURNER

PRODUCTION OF ENVIRONMENTAL KNOWLEDGE

SCIENTISTS, COMPLEX NATURES, AND THE QUESTION OF AGENCY

How scientists, conservationists, villagers, or developers utilize, transform, and protect the environment is shaped by how they understand it. How these understandings are produced is part of the politics of environmental knowledge. Scientific and lay understandings of the environment are shaped by a broader set of relationships beyond that of the relationship of the "scientist" and her object of study. Our knowledge of environmental processes is necessarily shaped by "outside" influences, given (1) the complexity and uncontrolled nature of most environmental processes compared to other objects of scientific study; (2) the importance of geographic and historical contexts in shaping the nature of environmental change in particular places; (3) the inclusion of human behavior as a factor (implicitly or explicitly) in most environmental analyses; and (4) the applied, policy-oriented nature of most environmental research. Clearly, the production of environmental knowledge is not immune from the clash of divergent interests in the realms of knowledge circulation and application. In fact, it is difficult to keep separate the conventional categories of knowledge production, application, and circulation in the realm of environmental politics.

Chapters in this part focus on the practices of scientists—the most privileged producers of environmental knowledge. All chapters treat "scientific practice" not as isolated, but as firmly situated with a broader social context within which environmental science operates. As such, scientists are influenced by the perceived needs of those who seek to "apply" environmental scientific knowledge (including themselves) as well as by the widely circulating knowledge claims made by scientists and others within and outside of their own fields. To study scientific practice isolated from processes of knowledge circulation and application violently abstracts it from the complex institutional realities of environmental scientific inquiry. All contributions explore areas of interface among these three realms.

Environmental knowledge formation is qualitatively different from classic laboratory fields due to its context specificity, multidisciplinary nature, uncontrollable research subjects, societal connections, and applied emphasis. These features increase the complexity of the research endeavor, the diversity of the knowledge claims, and the persistence of scientific uncertainty. Such complexity is associated with a *multiplicity of knowledge productions*. Knowledge claims by different types of actors, reflecting different spatial, temporal, disciplinary, and social framings, persist and compete for influence over how knowledge is applied and circulated.

Chapters in this part explore the multiple ways in which scientists respond to environmental complexity—reflecting the limits of disciplinary training, interdisciplinary frameworks, and common knowledges as well as presumptions about nature-society relations in the realm of "knowledge applications." Lisa Campbell's study illustrates how scientific uncertainty surrounding the size and natural history of sea turtle populations coupled with divergent assessments of the prospects for effective regulatory action have contributed to significant controversy over "sustainable harvesting" of sea turtles within the scientific community. These debates among ecologists are as much about the economics and politics of enforcement, regulation, and use as they are about the long-standing ecological parameters of "sustainable harvest" (or yield). Chris Duvall excavates the underlying logics of divergent claims made about ferricrete formation in West Africa from the colonial era to the present, showing that the disciplinary lenses (and intellectual traditions) brought to bear on the problem have strongly influenced variable characterizations of the problem. Tim Forsyth's treatment of soil-erosion research illustrates how abstract representations of the environment persist despite their ignorance of local context. Conventional scientific treatments of soil erosion in Thailand continue to rely on the universal soil loss equation (USLE)—a tool that abstracts from local context and, in so doing, poorly predicts or describes patterns of soil-erosion threat. Its continued popularity is in part due to the USLE's being seen as more scientific than alternative more "situated" scientific investigations just because of its abstraction from local realities (and seeming generalizability). Joan Fujimura uses the case of "systems biology" to show the strong role of analogy in cross-disciplinary science. Holism has proven recurrently attractive in the environmental and biological sciences. System approaches to holism in developmental biology have led to a reliance on machine analogies to model emergence from multiscalar biological and genetic processes. As with the USLE, such analogies may lead to misleading depictions when the dynamism of biogenetic processes diverge from physical, mechanic processes.

Fujimura's treatment of the reliance on machine analogy in "systems biology" also highlights another major theme in this volume: *the joint production of nature and society*. Not only is there a blurring of the nonhuman and human in how scientists think about complex biological systems, but these systems themselves are becoming increasingly cyborg- and hybridlike as a result of increases in technology and in our understandings of the role of living beings within humans. In such a context, even the ways in which the human body is understood change. Are humans individual organisms or communities of organisms? Mrill Ingram's chapter highlights the ways in which the answer to this question depends on our understanding of microbes. Her paper traces our changing relationship with microbes that inhabit our bodies, compost piles, and food. Rather than using chemicals to bluntly control or eliminate microbes, scientists increasingly see their goal as managing the relationship of human society with diverse microbial communities. Through this work, Ingram shows that the multiplicity of scientific engagements with the environment (as described above) is shaped by how human society is seen to be divided from "nature" or the "environment." For example, much work in conservation biology is based on the premise that there is a need to protect biodiversity from human society. Work within both STS and political ecology has challenged a clear nature-society divide. Scholars have increasingly interrogated the notions of "wilderness," "pristine," and "natural," showing how these and other invocations of isolation from human society are questionable. Still, nonhuman living and nonliving parts of our environment do not rely on human society for their existence, and they do change and respond separately from and against human designs. Therefore, the way in which the "agency of nature" can be understood (studied) and discussed has become a field of inquiry unto itself, about which there remains much debate.

A number of the chapters in this part explore the boundaries drawn between nature and society through the idea of nature's agency. The "agency" of nature has been used to refer to ecological objects and processes that simply exist independent of human thought, are not determined by human actions, react to external change, have causal powers, resist human epistemologies or actions, or display self-conscious intentionality. Nature's agency is inherently relational and always socially mediated (affected by politics, history, and culture). To make sense of the environment, scientists, resource managers, and land users divide and categorize that which they study, manage, and transform. While important, the categories that are created, named, and manipulated in the application of environmental knowledge do not determine ecological functioning. That is, "nature," or various aspects of it, acts in certain ways that often defy the categories, boxes, boundaries, and names we come up with.

STS scholarship has variously sought to diffuse and complicate the nature-society boundary. Language such as *hybridity*, *actants*, and *cyborgs* disrupts the boundary between nature and society and suggests new ways to discuss the agency of nature. Actor-network theory (ANT), in particular, introduces new ways of conceptualizing nature's "agency," thereby seeking a better balance of causal powers across the society divide. ANT-influenced political ecology and STS highlights the important roles that nonhuman "actants" play in the various social, political, and ecological networks we are engaged with. Such actants can include everything from black-boxed technologies (i.e., the telescope, microscope) to mammals, to microbes. In his chapter, Peter Taylor takes issue with prior ANT-inspired treatments of nonhuman actants, arguing that they work to reduce human agency, thereby ignoring the role of social power and politics in the reworking of nature-society relations. Agency is inherently a relational concept—nature's agency can only be thought of in the context of a society-environment relationship. In her exploration of microbe-human relations, Ingram further develops and illustrates a relational view of the nonhuman agency by showing how changing recognition of microbial agency over time has played a major role in the transformation of scientific understandings of human health, environment, and food science. Regardless of how it is defined, change is a critical part of the agency of nature's features. Duvall shows how limitations of human perception (temporal and spatial framing) have led scientists to view parts of nature as inert and constant (ferricrete), with important implications for interpretations of the impact of human activities on the landscape.

Despite the changing forms of agency asserted by various aspects of "nature," scientists are continually influenced by and working to *package and categorize nature* in such a way that knowledge about particular objects and processes travels (circulates). Given the local specificity of the society-environment relations, it is important to locate and recognize the influence of environmental knowledge claims packaged and circulated from elsewhere in the development of knowledge claims about particular "local" environmental processes. As described above, Forsyth shows how a universal model (USLE) developed in another part of the world (the United States) has strongly shaped how soil erosion is not only measured but conceptualized in Thailand. Duvall likewise shows how, in a region of the world where field research is scarce, narratives of ferricrete formation persist and recirculate over a period spanning the colonial and postcolonial periods. The "environmental narrative" literature has strongly influenced accounts of such circulations in political ecology. Both Forsyth and Duvall argue that "environmental narrative" accounts, while emphasizing how particular "scientific" explanations serve particular interests, have not critically engaged sufficiently with the methodology, logics, and data that underlie the production of scientific knowledge bound up within the narratives. In so doing, such accounts question the circulation and application of scientific arguments but not the production of these arguments themselves.

In similar ways, all chapters in this part present novel ways to understand the politics surrounding the production of environmental knowledge. They do so through shifting more attention to the areas of interface between the sphere of knowledge production and those of application and circulation. In so doing, they extend the political analysis of political ecology into the realm of the production of environmental knowledge (e.g., Duvall, Forsyth) and politicize prior STS accounts by not isolating the scientific inquiry from human society and the realm of knowledge application (e.g., Campbell, Fujimura, Ingram, Taylor). Taylor, through his concept of "intersecting processes," and Forsyth, through his call for "situated science," argue for changes in not only how we perform social science but how we perform environmental science so as to make it more relevant and meaningful to particular ecologies and people.

The contributions to this part do not resolve debates about topics such as nature's agency, complexity, or hybridity. The authors present different perspectives on these issues. But by providing a range of perspectives grounded in empirical examples, the chapters together provide the reader with a deeper, more nuanced understanding of the politics of environmental science through the lens of knowledge production.

1

TIM FORSYTH

POLITICIZING ENVIRONMENTAL EXPLANATIONS

WHAT CAN POLITICAL ECOLOGY LEARN FROM SOCIOLOGY AND PHILOSOPHY OF SCIENCE?

Political ecology has engaged with the politics of environmental science pretty much since it started. In the days of cultural ecology, researchers such as Harold Conklin (1954) showed that shifting cultivation was not necessarily as damaging as many governments or environmental scientists assumed, and that this kind of cultivation could both support livelihoods and have a beneficial impact on landscapes. Since the transition to "political" ecology, various researchers have emphasized how "science" always reflects politics. The British Political Ecology Research Group, for example, was formed in 1976 partly to assess how "science" was used to support state authority concerning decisions about energy or technology (PERG 1979). Blaikie and Brookfield (1987) urged more attention to how scientific explanations were constructed. And Thompson, Warburton, and Hatley (1986) famously showed how different "scientific" projections of environmental calamity or cornucopia in the Himalayas could be linked to the political worldviews of scientific institutions. Since then, a variety of political ecology texts have looked critically at how scientific explanations of environmental problems are affected by politics, and how we need to reassess these explanations in order to achieve more accurate, and more useful, environmental policies (e.g., Fairhead and Leach 1996; Bassett and Zuéli 2000; Forsyth 2003).

Despite this initial work on politicizing scientific explanations, the impact of these studies on many accepted concepts of environmental explanation remains weak. For some observers, the inability of political ecology to make environmental explanations more accurate and socially aware simply demonstrates the strength of the government institutions, discourses, and scientific networks that hold these explanations in place. For some other critics, however, there is a need to rethink how political ecologists have engaged with scientific knowledge. In particular, there is a need to move beyond analyzing the *social interests* of science (who benefits), and instead focus on the *explanatory potential* of science (how we explain problems). In other words, would political ecology have more impact on scientific explanations if it sought a more politicized engagement with how environmental cause-and-effect statements are made?

This chapter contributes to the interface of political ecology and science studies by exploring how political ecology may employ more politicized explanations for environmental problems. By so doing, this chapter also argues that political ecology should engage more constructively with debates within philosophy of science (concerning causality and the politics of explanation) as well as within the sociology of science (concerning which social interests are represented in science).

I use a case study from Thailand, and different approaches to explaining soil erosion, to illustrate this more politically situated environmental science. In particular, I look at the role of the so-called Universal Soil Loss Equation (USLE) as a scientific approach that both carries and suppresses politics.

WHAT'S WRONG WITH SCIENCE IN POLITICAL ECOLOGY?

Most environmental explanation invokes the use of "science" somewhere. Science in this context means a "hard" biophysical explanation of how environmental problems occur, or how they can be addressed. But the origin of this science, or its appropriateness for different contexts, is often not discussed.

"Science" is usually considered a source of technical knowledge constructed from trusted techniques, performed by qualified individuals, and subjected to rigorous testing and updating. This approach is sometimes called the scientific method or positivism because this method aims to create "positive" (or confident) predictions about the world based upon trends inferred from smaller samples of that world. Indeed, positivism is at its most powerful when it can make statements that are considered "laws." For example, the statement that "clean drinking water freezes at 0° centigrade at sea level" is a positive statement that is easily demonstrated and that most people are happy to accept as universally true.

Many environmental explanations have evolved using positivist science. Applications, such as climate change models and the USLE, were constructed using this method, and give positive predictions about environmental change, in terms of location, timing, and extent of change. Furthermore, one could say that most environmental explanation is usually conducted along the lines of generalized cause-and-effect statements that effectively become "laws" in practice. For example, it is commonly said that "deforestation causes floods" (China banned logging in 1998 in part to prevent floods), and that "overgrazing causes dryland degradation."

But these kinds of projections have also encountered problems in practice. For example, the USLE has been criticized by various parties for making "universal" predictions about soil erosion when it was developed mainly in the central plains of the United States, where rainfall intensity, slope length, and land-use practices are not always the same as in the tropics (Hallsworth 1987, 145). Climate change models have also been criticized as indicators of environmental risk because they have tended to refer to atmospheric changes (such as projected changes in temperature or storms) rather than how these important underlying physical changes may actually create risks as experienced by people living at the earth's surface (Demeritt 2001). In West Africa, many "universal" explanations of biodiversity or forest loss were unduly blamed on village-based agriculture without acknowledging the conservation practices of villagers, or the nonanthropogenic dynamics of forest cover on the borders between closed forest and savanna (Fairhead and Leach 1996).

Why is so much environmental explanation based on these kinds of inaccurate universalistic statements? Philosophers of science and sociologists of scientific knowledge have tried to answer this question in different ways.

For philosophers of science, part of the problem lies with the assumptions inherent within positivism. Furthermore, the practice of "positivism" has changed over the years, with implications for the confidence that scientists can make assertions about biophysical properties and the causes of environmental problems (see Harré 1993).

For example, some early positivist scientists—such as the German physicist Ernst Mach (1838–1916)—established "laws" by inferring that trends observed in existing data sets should also apply to similar data sets. Later on, however, the so-called Vienna School of the 1920s argued that positive laws existed only when trends in one data set were *verified* by comparison elsewhere. This approach was called "logical positivism." Yet even this approach was criticized by the philosopher Karl Popper (1902–94), who argued that positive generalizations should be created, first, by making *hypotheses* based on observed trends and, second, by trying to *falsify* them through experiments. Under this approach (also called "critical rationalism"), hypotheses should be considered true if it is not possible to falsify them. Today, Popper's approach is popularly still called "positivism." But clearly it is more ambitious than preceding forms of positivism, and some critics suggest it may encourage scientists to think too readily in terms of universal explanations (Yearley 1996).

For sociologists of science, any scientific explanations have to be seen alongside the social groupings and values that created them. Social groupings can include the expert bodies or scientific disciplines that define environmental problems or who can participate in the collection of information. Social values might include assumptions about particular activities that are considered problematic. For example, much research on forest conservation has been driven by the perspective of forests as wilderness (e.g., Perlman 1994; L. Brown 2001). This is clearly a well-supported and important outlook, but is relatively more powerful than alternative perspectives of open forests for local agriculture or livelihoods by forest-dependent peoples. These values and framings of environmental research affect which information is collected to make explanations of forest change and inevitably portray different actors in roles relevant to those values.

Moreover, social analysts of science highlight how environmental "problems" use words with diverse histories that can have multiple meanings in different contexts. For example, "wilderness" may be culturally specific because it classifies land as without users. "Deforestation" is commonly considered to be a problem. But various researchers, including groups as diverse as hydrologists and anthropologists, have pointed out that "deforestation"—as a term—is far too general and imprecise to describe all forms of logging or forest disturbance that occurs (indeed "forest" itself may also be too general; Hamilton and Pearce 1988). More attention is needed to define which impacts create which kinds of problems for various people. Together, these criticisms from philosophy and sociology of science indicate that it can be inaccurate to adopt universal explanations of environmental problems across diverse scales and meanings attributed to these changes. But perhaps a more pragmatic criticism is that these explanations can sometimes be very inaccurate! They frequently do not address underlying causes of environmental problems. And sometimes they support supposed "solutions" to environmental problems—such as land-use restrictions—that may have no long-term impact on problems, but which threaten poor people's livelihoods unnecessarily (Fairhead and Leach 1996; Forsyth 2003).

But can these problems be addressed within the frameworks of positivism? Would it be possible, for example, to give more consideration to diversifying the scale of environmental problems (e.g., making "universal" explanations more specific to different landscapes or land uses)? Similarly, might there be ways that researchers can diversify the collection and evaluation of scientific knowledge?

At the same time, the research agenda pursued by political ecology requires approaches that are inherently more political than simply diversifying the manner in which positivism is applied. For example, it is clear that the very experience of environmental problems, or the proposed solutions to them, is deeply imbued with meanings from either local people or from people living outside a region. Environmental change will always be framed in different ways, and these different perspectives influence the gathering of data and the creation of explanations. This situation does not necessarily mean that we have to "do positivism better" but instead that we have to look for a new way of explaining environmental problems that makes social and political framings a key part of scientific inquiry. What are the alternatives?

MAKING ENVIRONMENTAL SCIENCE MORE SITUATED

FROM CULTURAL ECOLOGY TO NARRATIVES

In order to politicize environmental explanations, we have to see how political ecology has attempted this in the past, and then see what lessons can be applied from sociology and philosophy of science.

Political ecology research has frequently pointed out the problems in positivist environmental explanation. As noted above, some early work in cultural ecology demonstrated that common assumptions about environmental fragility (or human behavior in fragile zones) simply did not hold true (e.g., Conklin 1954; Geertz 1963). In turn, this research gave rise to the study of "environmental adaptation" and "institutions" as examples of human behavior that can reduce degradation in locations where degradation was expected. Indeed, much community-based natural resource management has looked at how the perception and management of resources are shaped by local norms and practices, rather than universal assumptions.

But other research within political ecology has also theorized how—and with whose inputs—environmental explanations have emerged. Two strands have been prominent. The Cultural Theory work of Thompson, Warburton, and Hatley (1986) famously showed how predictions of deforestation in the Himalayas varied by a factor of sixty-seven (even ignoring some apparent typing errors in some predictions). Accordingly, they argued that organizations will select or represent data to support worldviews. Cultural Theory, however, proposed that worldviews could be identified in predetermined ways based on the principles of how far individuals want to follow rules ("grid") or act communally ("group"). Cultural Theorists then proposed that different worldviews could be mapped relatively easily onto individualist actors (such as corporations), egalitarians (NGOs), hierarchists (governments), and fatalists (powerless groups such as hill farmers). For many analysts of environmental knowledge, these categories were too reductionist (Forsyth and Walker 2008, 22).

In contrast, a "narratives" approach to environmental knowledge emerged within political ecology. An environmental narrative is a well-known and convenient explanation of environmental processes that is widely accepted as truth, but that contains important simplifications and errors. Narratives are based upon a more poststructuralist and historical analysis to indicate how environmental explanations carry many hidden normative values, which in turn reflect the selective participation of different actors in the past and present (Roe 1991, 288). Maarten Hajer (1995) referred to narratives as "storylines," in which diverse physical events and processes are ordered into convenient explanations that also include concepts of blame and urgency according to dominant social groups. Fairhead and Leach's study of deforestation in Guinea, West Africa (1996), argued that government agencies and some international NGOs were using a narrative to blame deforestation on local farming practices and, accordingly, to control deforestation by restricting historic agricultural practices including fire. Consulting alternative sources of information such as historic photographs and oral histories revealed that local villagers had actually helped create islands of closed forest. Politicizing scientific explanations, therefore, depends on showing who shaped them and demonstrating the political implications of using them. Similar findings for explaining dryland degradation have been found by Turner (1993) and Bassett and Zuéli (2000).

PHILOSOPHY OF SCIENCE

Research within these themes has demonstrated the problems of separating environmental science from the contexts in which it is made or implemented. But how do these studies help day-to-day environmental management?

For example, the usual purpose of analysis such as Cultural Theory is to show that there is little point in following one dominant viewpoint when there are other worldviews that might challenge it. Narrative analysis also provides more historical and context-specific analysis of how certain explanations have become reified. But students and many environmental policy makers become frustrated when they realize that these approaches do not necessarily answer questions such as whether climate change *is* happening or whether deforestation *does* cause water shortages.

Some other work within STS and political ecology has addressed this dilemma by seeing how social solidarities influence scientific explanations themselves. In effect, this uses insights from philosophy of science, as well as the sociology of who participates in scientific processes and how.

Bruno Latour's *Politics of Nature* (2004), for example, suggested political ecologists need to ask two questions about scientific "laws": How many are we? Can we live together? These questions demonstrate that building scientific conventions (i.e., explanations) results from how many people are included and willing to live with the emerging rules. These ideas strongly echo the work of the seminal philosopher of science Willard Quine (1908–2000), who argued that the social frames (or "truth conditions") for scientific statements are crucial for understanding how scientific "truths" are assessed.

In a similar vein, Funtowicz and Ravetz (1993) argued that scientific "certainty" reflects the fixity of scientific disciplines or professional networks. They argued that "certainty" (as opposed to "uncertainty") arises when outside opinions or experiences are excluded from discussion. These ideas have been applied in political ecology to examine how powerful bodies such as government agencies might create scientific explanations and models of citizenship simultaneously in order to control both landscapes and human activities. One example is "scientific forestry," which maximizes timber production through concentrated plantations and silviculture. The scientific knowledge linked to this type of forestry has been alleged to exclude alternative uses of forest land in countries such as India and Thailand (Agrawal 2005; Forsyth and Walker 2008).

There are other ways of linking social participation to causal explanations. Theorists known as semantic and institutional realists, such as John Searle (1995), have argued that truth statements (including scientific explanations) can arise only when they are organized like a sentence, with a subject, object, beginning, middle, and end. For example, people living on floodplains can explain floods in terms of the changes they see—such as the coexistence of upland agriculture and deforestation. But a more accurate explanation of floods might involve factors outside of this framing.

But again, one concern about this approach is that it is still too relativist because it still links explanations to social viewpoints. Instead, critics have asked if it is possible to acknowledge social influences on explanations, but also to develop explanations that can travel between contexts (e.g., Blaikie 1999; Forsyth 2003). Is this possible?

The following example illustrates one way of more transparently linking social solidarities and truth claims. It compares a positivist approach to political ecology with approaches that combine an awareness of social solidarities with physical measurements to achieve more situated ways of explaining environmental problems that are more accurate and socially representative.

EXPLAINING SOIL EROSION IN NORTHERN THAILAND

THE UNIVERSAL SOIL LOSS EQUATION

The highlands of northern Thailand is one of the most researched regions in Asia because of its proximity to Communist countries and its diverse range of problems, such as opium production, deforestation and land degradation, and poverty.

Classically, people of northern Thailand are classified into lowland Thai (*khon-muang*), who traditionally inhabited irrigated intermontane basins, and upland minorities (so-called hill tribes), who historically practiced shifting cultivation in the uplands. Some uplanders lived in Thailand for as long as the lowland Thai (such as the Karen). Others migrated to Thailand from neighboring Burma, Laos, and China during the twentieth century (such as the Hmong and Mien). Some studies of shifting cultivation identified two styles of shifting cultivation. "Rotational" cultivators (such as the Karen) used agricultural plots around semipermanent settlements with fallow periods of some seven to ten years. "Pioneer" cultivators (such as the Hmong) typically did not live in semipermanent villages, but instead grew crops around villages for some ten to twenty years until the exhaustion of land forced them to find a new site for a village (Grandstaff 1980). These distinctions, however, are now blurred. The uplands are inhabited by both Thais and minorities, and shifting cultivation has been replaced by more permanent and commercialized agriculture.

Soil erosion has been described as a problem in this region for a long time. Deforestation is blamed for erosion, for it reduces the canopy cover of soil and disturbs soil. Cultivation takes place on steep slopes, which encourages erosion. Sediment from the hills is blamed for silting up lowland dams and rivers. Erosion of soil is also considered to reduce its water-holding properties, which is seen to exacerbate lowland water shortages. Some observers fear that historic "pioneer" shifting cultivators may not understand the long-term impacts of permanent agriculture. In 1987 the Bangkok think tank Thailand Development Research Institute (1987, 296) wrote: "Whereas slash and burn agriculture was once more closely attuned with the ecosystems exploited, it now causes untold ecological damages. . . . In the process, major watersheds are being denuded, with increasing silt loads washed down into the nation's rivers [and] silting up dams."

In positivist terms, most research focused on soil erosion in northern Thailand has used the universal soil loss equation (USLE). This equation was formulated in the United States after the dust bowl of the 1930s and was intended to predict levels of erosion and hence to allow farmers to keep soil loss within acceptable levels (USDA 1961).

$$A = R \times K \times LS \times C \times P,$$

where A is potential soil loss; R is rainfall and runoff; K is soil erodibility; LS is the slope length-gradient factor; C is the crop/vegetation and management factor; and P is the support practice factor (such as soil-conservation measures).

The Thai government has published maps based on the USLE that show where annual rates of erosion are estimated to be many times above sustainable limits. Jantawat (1987, 13) estimated soil erosion in northern Thailand to be "higher than the world and Asian averages" at the remarkably precise figure of 933.67 tons per square kilometer per year. Manu Srikhajon and Ard Somrang (1980) estimated less precise categories of "severe" and "very severe" erosion on 16.9 percent and 5.8 percent of northern Thailand, respectively.

Experts using the USLE have also framed research in terms of whether farmers perceive erosion to be a threat. Some farmers are only now getting used to living in locations for more than twenty years. In the late 1980s, for example, Harper and El-Swaify (1988) found that 37 percent of farmers studied explained that the reason they did not use conservation was because they did not believe erosion to be a serious problem. Similarly, Pahlmann (1990, 99) noted that 43 percent of the upland (ethnic Thai) farmers she interviewed in Nan Province "did not perceive they had a problem of soil erosion at all" while only 23 percent considered it to be a serious problem. To illustrate the argument, Pahlmann (1990, 82) showed how farmers were keen to identify problems such as weeds, insects/pests, and lack of water or land, whereas "soil erosion" came last in problems mentioned by farmers. As a result, the government and some development agencies have urged planting grass (vetiver) strips to reduce erosion on steep slopes. (Grass strips effectively reduce the length of slope, one of the causes of erosion cited in the USLE.) Sometimes cultivation on steep slopes is also banned.

But is the USLE really "universal"? It was developed to address problems in the central plains of the United States. In the tropics, smallholder agriculture often divides slopes into different plots; bamboo and grass create barriers to water flow; rainfall tends to fall intensely during rainy seasons; farmers might use crops with different cropping cycles that do not overlap with rain; and farmers might use diverse crops and soil-conservation practices.

Moreover, the USLE tends to indicate only sheet erosion on agricultural land. Sheet erosion occurs evenly across a soil surface, such as agricultural fields, and the USLE does not consider other, often geomorphological, processes of erosion such as gullying. Statistics generated by the USLE unsurprisingly tend to show that erosion occurs mostly under agriculture, and less so under forest cover. As a result, policy makers tend to assume agriculture is the primary cause of erosion, and that reforestation is a useful solution to erosion.

Against this, however, research has shown that erosion and sedimentation are frequently controlled by nonanthropogenic activities such as geomorphological processes, and that gullies can occur on both forested and agricultural land. Sometimes gullies can also be caused by nonagricultural activities such as road construction and plantation forestry (Calder 1999). Consequently, the USLE is not an accurate tool for predicting erosion in many areas, and indeed may give rise to management solutions that are inappropriate. Perhaps unsurprisingly, one soil-erosion assessment using the USLE in northern Thailand predicted rates of soil loss 104 times greater than those actually observed in runoff plots! (Thitirojanawat and Chareonsuk 2000).

SITUATED SCIENCE: COMBINING SOCIAL INCLUSION AND BIOPHYSICAL EXPLANATION

The positivist approach to soil erosion using the USLE had the attraction of giving highly quantified predictions of the amount of erosion and its likely location. These predictions have been useful to the Thai government in describing "problems" in remote locations, which have justified some kind of state presence and intervention. The downside to this approach, however, is that the USLE predictions are neither very accurate nor very relevant to local people's definitions of problems.

An alternative approach has been ethnographic studies of how upland farmers perceive and respond to erosion (e.g., Kunstadter, Chapman, and Sabhasri 1978). But these studies rarely speak across the boundary between positivists and anthropologists. One might ask, for example, whether erosion still *is* a problem or not? And if so, what to do about it? There are ways, however, to combine the different insights arising from asking local people with the more generalizing predictions of positivism.

First Case Study: The Akha

Francis Turkelboom conducted a study of the Akha people in the far north of Thailand that combined physical analysis of erosion and social perceptions (Turkelboom 1999; Turkelboom, Poesen, and Trébuil 2008). This study therefore used farmers' own observations to answer questions that are usually posed under positivist science. The methods used for this research included field surveys and participatory discussions with farmers.

The study showed that sheet erosion was increasing due to agriculture, largely because of tilling (or digging) on steep slopes. Gully erosion also occurred on agricultural land and under forest cover. These problems, however, were divided differently among households. Farmers owning higher slopes tended to lose soil fertility. But those on lower slopes often gained soil fertility because of sedimentation from above. Perhaps most important, the study also showed that upland agriculture was unlikely to add significantly to river sedimentation in the lowlands because only about one-third of agricultural slopes fed stream networks (the remaining two-thirds were deposited on slopes where soil would not be moved again).

The study also suggested that erosion from new commercial crops was not as bad as expected. The study measured erosion using different techniques. Local farmers were asked to identify slopes most at risk, and then tracers indicated how much soil was moved through tilling. The study also estimated the impacts of storms on erosion, and made physical measurements of gully erosion. These measurements indicated that rain-fed rice (the most "traditional" Akha crop) had the highest rates of erosion: sixty tons per hectare per crop cycle. Maize and beans were least erosive with median soil losses of, respectively, nineteen and ten tons per hectare per crop cycle. These rates are certainly high and are likely to cause declining soil fertility. But they provide more information about the causes of erosion than many assumptions.

Newspapers in Thailand often blame the commercial production of cabbage for erosion. Turkelboom's research, however, showed that erosion under cabbage lay between the two extremes above. The research showed that the cropping systems used by the Akha meant that most cabbage was planted before the start of the rainy season. Consequently, the months when cabbage had only a small surface area were generally during the dry season, when rainfall erosion was least. Moreover, if the price of cabbage fell, farmers would abandon fields, allowing them to be invaded by grass and hence protected against erosion.

Finally, the study also showed it is important not to generalize about farmers. Rather than treating all farmers equally, Turkelboom (1999, 208–11) identified five levels of entrepreneurialism or concern for soil conservation within the village. These varied among so-called secure investors (who owned paddy fields and fruit plantations), profit maximizers (adopting high-risk crops such as cabbage), diversifiers (farmers who mix rice cultivation with limited cash crops), survivors (those who cultivated only rice on a short-term basis), and dropouts (who relied solely on wage labor and petty business). Typically, secure investors maximize agriculture on flat, terraced land, and have less risk of erosion. Diversifiers and survivors, however, are less secure and have to diversify cultivation between terraced land (when they can access it) and steep slopes. These groups are most at risk from erosion because they have to use sloping land most intensively. So-called dropouts do not themselves claim landownership, but often carry out the tilling for other, richer, farmers. The survivors accounted for some 30 percent of the village. The point of this classification is to acknowledge that soil and crop management does not take place uniformly across single ethnic groups, and that there is great diversity among and within households.

Second Case Study: The Mien

A second study also shows ways to combine local values with scientific measurements. I conducted this study among the Mien in the far north of Thailand. The Mien cultivate dry and irrigated rice, corn, soya, ginger, peanuts, and limited coffee, oranges, and lychee at an altitude of 700–800 meters (Forsyth 1996; updated by fieldwork, 2004–5). This study asked how farmers responded to soil degradation, and used historic aerial photographs, geographic information system (GIS), soil-erosion measurements, and interviews with farmers.

As with Turkelboom's study, erosion was undoubtedly occurring, and this happened most where agriculture had been most concentrated on steep slopes. Erosion was measured to be between twenty-four and sixty-four tons per hectare per year, which at its highest level is generally considered very high. But, as with other studies, farmers said they were aware of the risk of erosion on slopes, and so tried to avoid using the steepest slopes.

This research also indicated that erosion was composed of both sheet erosion and gullies, which form on both forested and agricultural land. Evidence also suggested that these gullies were mainly the result of long-term weathering of underlying granite. According to the GIS, there was a relative absence of slopes between 10 and 20 percent—giving a hummocky appearance—which is characteristic of an "all-slopes topography" associated with granite (Twidale 1982). These gullies may therefore be important contributors to lowland sedimentation, yet be unrelated to upland agriculture. Indeed, villagers preferred to keep gullies vegetated in order to harvest plants (and in particular banana trees, which were fed to pigs), and so human activities might even reduce the erosiveness of gullies.

Most importantly, the research also indicated that farmers were willing to avoid steep slopes because they were known to be locations of high soil erosion. Instead, farmers tended to farm the flatter slopes more intensively in order to maximize food production. The implication of this finding is that overcultivation, rather than erosion per se, was the main hazard facing upland farmers. Consequently, the emphasis on erosion as the problem in the uplands is not necessarily accurate, because farmers are more worried about declining soil fertility from various means. Moreover, the common assumption that upland agriculture is the primary cause of erosion is flawed, because farmers take steps to avoid it and there are diverse nonagricultural causes of erosion. Seeking local opinions about environmental problems, and using diverse, nonuniversalist methods of generating knowledge, has therefore provided less simplistic environmental explanations. Moreover, these new explanations offer means to assist upland farmers more effectively.

Making Situated Environmental Science

So how do these two case studies indicate a different, more situated, form of environmental science?

First, they are different from the universalistic environmental explanations because they employ a different epistemology than positivism. They do not seek to infer generalizations about environmental causality across different physical and social contexts. Instead, these studies used local framings of environmental problems to shape the explanation of physical processes. In terms of philosophy of science, they adopted insights from institutional or semantic realists to create explanations that reflect the viewpoints and experiences of selected social groups. As a result, the universalistic predictions of the USLE are not useful in these contexts. Moreover, the general assumption that upland agriculture must cause, and be damaged by, erosion has also been challenged.

Second, the studies are different from purely ethnographic accounts of soil erosion. The studies certainly drew upon a sensitive analysis of how different people experience soil degradation. But rather than leaving the analysis there, these studies tried to explain the origins of and potential solutions to soil degradation in terms that allow practical interventions. Moreover, the studies also produced quantified measurements of erosion and the effectiveness of the USLE that might—in principle—be used by government agencies to revise approaches to soil erosion in general.

Making environmental science more situated does not therefore mean looking for local verification of universalistic assumptions (as with the USLE), or understanding how different social solidarities (upland farmers or government soil scientists) explain environment in different ways. Instead, a situated environmental science means understanding how social solidarities and values shape the generation of knowledge, and using this to create a more diverse form of explanation that can be directed to deliberatively selected development objectives. This approach does not imply that social solidarities should be studied alongside unreconstructed environmental science. Rather, it means understanding the mutual connections between solidarities and truth claims, and therefore increasing the amount of deliberation about the underlying values, assumptions, and objectives of who makes these truth claims and for what purposes. In turn, this deliberation might change the identification of spatial scales and affected people within environmental research.

CONCLUSION

This chapter argues that political ecology has achieved much success in illustrating social influences on science, and especially concerning how environmental truth claims are linked to social networks and solidarities. But to have more impact on general environmental practice, political ecologists need to move beyond analyzing the *social interests* of science (focusing on who benefits), and instead focus on the *explanatory potential* of science (how we explain problems). To achieve this, political ecologists need to engage more formally with debates within the philosophy as well as sociology of science. Some of the implications of this chapter are summarized in table 1.1.

Looking at the explanatory potential of science has important implications for political ecology. At present, too much environmental explanation is based on universalistic science derived from a positivist epistemology. Positivistic statements are often inappropriate when applied across diverse ecological and social contexts because they frame environmental problems in selective ways, are frequently inaccurate, and encourage land-use policies that restrict local livelihoods in unnecessary ways.

Yet political ecology loses some of its political impact if it conducts research only on the relationships of social solidarities to knowledge claims. To be clear, this work is necessary and valuable to indicate how knowledge is generated and legitimated. But if political ecologists are to engage more critically with dominant scientific explanations, they need to engage with debates about environmental causality as well as social interests. As Blaikie (1999, 144) noted, "A counterweight to the deconstruction of science must also be provided. . . . By adopting an epistemology which avoids relativism and unreconstructed pluralism, it may be possible to address specific audiences in languages they recognize to identify real and feasible choices."

A key way to achieve this approach is to consider the role of social interests in shaping the norms and framings implicit within any environmental explanation. How far are these shared by different people? For example, most people would agree with the positivist statement that "clean drinking water freezes at 0° centigrade at sea level." Yet most people would also agree that "the sun rises in the morning." This second statement is an apparent truth, but is clearly inaccurate when we think of the solar system. Yet this second statement works for most people (who do not leave the earth's atmosphere) because there is no need to challenge it. So many environmental cause-and-effect statements are like this second statement. People assume explanations such as "deforestation causes floods" to be true because they seem to work for their purposes. But adopting different purposes, or different frames—often from previously ignored social groups—means we can challenge these assumptions *and* understand environmental change in more effective ways.

An important implication of this approach is that environmental science must necessarily become more deliberative than commonly practiced. For example, the objectives and basic framings used to underpin scientific research will need to be opened to greater scrutiny. The social inclusion and normative values guiding research will have implications for what kinds of causal statements will emerge and for what objectives. But how to do this?

TABLE 1.1 *Comparison of universalist and situated environmental science*

	Universalistic science	Studies of social interests in science	Situated science
Summary	Application of causal "laws" or assumed cause and effect	Studies of how different scientific statements reflect different social solidarities	Connecting social solidarities with causal statements to create more diverse and relevant explanations of problems
Example	Universal Soil Loss Equation (USDA 1961)	Ethnographic studies of perceptions and practices within localities or networks (Conklin 1954)	Hybrid science, blending studies of social framings of problems with diverse scientific assessments of environment (Forsyth and Walker 2008)
Epistemology	Positivistic inference	Knowledge in context	Institutional analysis of truth claims
Typical methods	Creating an equation from samples, and applying this to new locations	Ethnography, historical and contemporary text and discourse analysis, physical assessments of environmental problems	Ethnography, historical and contemporary text and discourse analysis, physical assessments of environmental problems
Focus	Controlling biophysical changes	Understanding different knowledge networks	Addressing deliberatively defined problems
Expert bodies	Formal bodies such as applied soil scientists, government land development agencies	Formal and informal expert bodies including villagers, citizens	Formal and informal expert bodies including villagers, citizens
Governance structure	Scientific authority located in expert bodies, communicated to policy makers	Diversifying truth claims according to social solidarities that uphold them	Diversifying truth claims, increasing deliberation about how, for whom, and for what objectives explanation is desired.

(*continued*)

TABLE 1.1 *(continued)*

	Universalistic science	Studies of social interests in science	Situated science
Political implications	Can link scientific knowledge and expertise to statemaking and control over landscapes and people	Makes the experiences and impacts of different solidarities and dominant discourse more transparent	Challenges dominant explanations; proposes alternative explanations

The philosopher of science Steve Fuller has argued that this approach to science should be called a "politically oriented social epistemology," which holds that analysts should have a priori normative grounds for directing the generation of knowledge (Fuller 2000; Remedios 2003, 11, 21). This approach is different from so-called interest-oriented social epistemology, which focuses instead on the different interests represented within different knowledge claims. Political ecologists can discuss what values should direct the generation of knowledge in different contexts (Forsyth 2008). Poverty alleviation and social justice? Environmental conservation for its own sake? Knowledge will always reflect social values about what kind of world is acceptable. Increasing the political debate about how values shape knowledge—and vice versa—will expose the hidden politics within scientific statements. Considering how values influence explanations might also lead to a more flexible, and more useful, environmental management based on positivism alone.

LISA M. CAMPBELL

2 DEBATING THE SCIENCE OF USING MARINE TURTLES

BOUNDARY WORK AMONG SPECIES EXPERTS

Sustainable use is central to conservation policy as espoused by the International Union for Conservation of Nature (IUCN). Defined as using renewable natural resources "in such a way that does not threaten a species by overuse, yet it will optimize benefits to both the environment and human needs" (http://www.iucn.org/themes/ssc/susg/), sustainable use is one indicator of a shift in conservation policy, away from exclusionary practices of restricting access or "fortress conservation" (L. Campbell 2000; Adams and Hulme 2001). The rationale for sustainable use is primarily economic, and emphasizes the need to make wildlife conservation valuable enough to compete with alternative habitat uses and to offset the direct and opportunity costs of protecting wildlife. Sustainable use can also generate funds to fuel conservation projects, and proponents argue that refusing to put a "price tag" on wildlife may result in wildlife being treated as valueless. The rise of sustainable use can be interpreted as both a response to the challenges of wildlife conservation in practice (L. Campbell 2000) and the turn to neoliberal environmental governance (McAfee 1999).

In spite of its proliferation in policy, sustainable use has proven controversial for the wildlife conservation community.[1] In this paper, I consider how conservation experts interested in marine turtles view sustainable use. Specifically, I examine how science is invoked in evaluating possibilities for using marine turtles as resources, and the outcomes of a "discursive struggle" (Bryant 1998) among scientists interested in marine turtle conservation (hereafter experts). I draw on theory from both political ecology and STS to explore science as contested terrain among scientists themselves. At stake is the overall policy direction for marine turtle conservation.[2]

POLITICAL ECOLOGY MEETS SCIENCE STUDIES: BOUNDING KNOWLEDGE, ASSERTING POWER

The control of discourse, knowledge, and ideas has been a central concern of political ecologists (Blaikie 1999; Watts 2000), who trace how knowledge disperses through complex networks to produce political-ecological outcomes. Scientists have been identified as powerful actors in controlling ideas (Fischer 2000), and some of the most influential works in political ecology have challenged scientific claims about the environment, casting such claims as narratives or orthodoxies (Leach and Mearns 1996). In this light, science and environmental policy are mutually constitutive, and political ecology addresses the "scientific legitimatization of environmental policy" (Forsyth 2003, 4).

Environmental narratives are perpetuated due to the nature of science and the institutions that act on it; both are resilient. Narratives are not simply discursive devices, however, as they inform policies that are (often) implemented. In failing to acknowledge alternative, often local, understandings of environmental change, they have sometimes been implicated in environmental degradation and social injustices (Forsyth 2003). Thus, political ecologists have attended to the plurality of environmental knowledge (Watts 2000), and local, indigenous, or traditional knowledge is often cast in contrast to scientific knowledge (Leach and Mearns 1996; Nadasdy 1999; Goldman 2007). Although this work is revealing, it often assumes a model of "monolithic" science, one that masks the many debates that occur among scientists themselves. This is one area where STS can inform and enhance political ecology.

STS attends to science as a social process and questions claims of objectivity. For conservation biology the issue of objectivity, and specifically the line between science and values, has been debated. Among conservation biologists the debate generally relates to whether crossing this line is a problem (Shrader-Frechette 1996; Blockstein 2002) and, if it is, how conservation biologists can refrain from doing so (Lackey 2007). In contrast, STS scholars explore the relationship between science and values with the assumption that these are inseparable. Takacs (1996), for example, examines how biologists began strategically promoting the concept of "biodiversity" in the late 1980s to reduce the focus on specific species; the importance of values in this strategy was often explicit. In their study of the Ecological Society of America (ESA), Kinchy and Kleinman (2003) argue that the ESA draws on "historically resonant discourses" that emphasize both the value-free nature and utility of science, in order to position ecologists as authorities on the environment. Although "the organization routinely reconstructs a fact-values boundary that keeps advocacy apart from science" (Kinchy and Kleinman 2003, 881), many of its members recognize that boundaries between science and values are blurred; Gray and Campbell (2009) found similar attitudes among scientists interested in marine protected areas.

Attempts to distinguish between science and values or between "good" and "bad" science are examples of boundary work, or how scientists and institutions patrol and defend the realm of what counts as science (Gieryn 1995). Resulting boundaries are products of a process: "what science 'is' at a given time and place results from complex negotiations among scientists and those allies whose allegiances they would enroll, or who would enroll them" (Takacs 1996, 114). Gieryn (1995, 393) describes the bounding of science and nonscience as "a contextually contingent and interests-driven pragmatic accomplishment." As will be shown here, marine turtle experts engage in boundary work among themselves, promoting their own views on sustainable use as based on science and those of their opponents as driven by values. At stake in this "discursive struggle" is the categorization of marine turtles as worthy of strict preservation or, alternatively, as resources to be used and managed. While the struggle is fought primarily in the language of science, other factors, including attitudes toward markets, uncertainty, and intrinsic (and other noneconomic) values attached to marine turtles, underlie the debate.

MARINE TURTLES, THE MARINE TURTLE SPECIALIST GROUP, AND SUSTAINABLE USE

Marine turtles are globally distributed, long-lived animals, often categorized as charismatic megafauna (Mast 2005; figure 2.1). All seven species are on the IUCN's Red List of Threatened Species[3] and the Convention on International Trade in Endangered Species of Wild Fauna and Flora (CITES) lists all species in appendix I, thus prohibiting international trade in marine turtle products by treaty signatories. Marine turtles are protected under a variety of other international agreements (Richardson et al. 2006).

The Marine Turtle Specialist Group (MTSG, established 1966) of the IUCN's Species Survival Commission (SSC) is one of the key organizations involved in marine turtle conservation.[4] It influences policy making through its role in compiling data and making recommendations on species status for the IUCN Red List, and through publications such as its *Global Strategy for the Conservation of Marine Turtles* (MTSG 1995) and its manual *Research and Management Techniques for the Conservation of Sea Turtles* (Eckert et al. 1999).

The MTSG's current position on sustainable use is a cautious one. The *Strategy* (MTSG 1995, 3) is most supportive of nonconsumptive use, like ecotourism, while consumptive use is portrayed as problematic: "Too frequently . . . wide use by a growing human population, coupled with the migratory nature and slow rates of natural increase of these animals, has resulted in most utilization being non-sustainable. . . . Although this Strategy recognizes that utilization of marine turtles occurs in many areas and does not oppose all use, it does not support non-sustainable use." Although the *Strategy* does suggest that small-scale, localized marine turtle egg harvesting might be done sustainably, and calls for the provision of guidelines in a techniques manual, such guidelines have not

FIGURE 2.1 Hawksbill sea turtle. Photograph courtesy of Caroline Rogers.

been developed. Recent consultations with select MTSG members have identified five "burning issues" for marine turtle survival, and direct take of both eggs and adults is listed second (http://www.iucn-mtsg.org/hazards/).

The MTSG's position on use has evolved over time. In the late 1960s, the MTSG deemed science-based management the solution through which international trade could be perpetuated, and called for research into commercial rearing of marine turtles (IUCN/SSC MTSG 1969). The MTSG's interest in mariculture dissolved in the 1970s, a shift Donnelly (1994, 5) attributes to biological uncertainty and problems with trade controls, rather than "any philosophical opposition to using the species." Although mariculture operations have been few, they often have been at the center of the sustainable-use debate (L. Campbell 2002b). For the past decade, debate has focused on various proposals by Cuba to CITES to trade hawksbill turtle shell with Japan, with acrimonious exchanges between the proposals' proponents and opponents (figure 2.2). Controversy over the MTSG's role in the related CITES discussion led the chair of the SSC to dissolve and reconstitute the MTSG under new leadership in 2003. It is with this sense of discord in mind that I explore the views of marine turtle experts on sustainable use.

FIGURE 2.2 Cuban fishers landing marine turtles, including hawksbills. Photograph courtesy of Brendan Godley.

VIEWS OF MARINE TURTLE EXPERTS ON SUSTAINABLE USE

The analysis in this paper is based on thirty-eight interviews with experts in marine turtle biology or conservation, conducted in 1995 (see L. Campbell 2000, 2002b for further details). Interviewees were male ($n = 21$) and female (17), from North (31) and Central (7) America, working as academics (14), in government (10), and in NGOs (11) (3 classified as "other"). Most had degrees in biological sciences (33). I update interview results based on my fifteen years as a participant observer in the marine turtle conservation community, where I am an appointed member of the MTSG and an elected member of the board of directors of the International Sea Turtle Society. I first consider the scientific arguments for and against sustainable use, and then the role of other issues in informing such views. With Gieryn's contention (1995, 393) that "what science 'is' at a given time and place results from complex negotiations among scientists" in mind, I am particularly interested in how the sustainable-use debate, and boundary work within that debate, has shifted over time.

THE SCIENCE OF SUSTAINABLE USE IN 1995

The Challenges of Using Marine Turtles

Interviewed experts agreed unanimously on two things: conservation programs must be based on good science, and marine turtles pose challenges for their sustainable use.

The challenges with using marine turtles relate to life-history characteristics, data availability, and the status of populations (table 2.1). Experts identified long generation times and migrations as the most problematic life-history characteristics. Long generation times result in low recruitment to the reproductive age class, and in long lags between taking action and seeing results, making it difficult to evaluate the impacts of conservation measures. Marine turtle migrations were seen as equally problematic, because (1) all threats along the migratory route may not be known; (2) migration itself makes it difficult to determine the magnitude of individual and cumulative threats; and (3) national user groups, types of use, and management techniques along the route may be conflicting. Direct harvesting throughout the migratory range was the most frequently identified threat: "they pass through incredible expanses of water and so the possibility for exploitation in each area that they cross is incredible. And the sum of those threats are [*sic*] really daunting" (E3). For some experts, cumulative threats made use unsupportable, while for others, they posed challenges for management: "You only have access to one portion of the life history. Somebody may be doing something completely different and they may or may not be compatible" (E7).

Experts were concerned with lack of data for marine turtle populations (with

TABLE 2.1 *Expert views on consumptive use of marine turtles*

Limitations on using marine turtles	No.	Respondents
Long generations/delayed sexual maturity	16	E1, E3, E4, E6, E7, E8, E9, E11, E12, E15, E16, E17, E24, E29, E35, E37
Marine turtle migrations	12	E3, E6, E14, E15, E16, E17, E19, E24, E29, E31, E35, E36
Data availability	12	E3, E8, E9, E14, E16, E18, E19, E21, E24, E31, E32, E35
General biology/life-history characteristics	7	E7, E8, E15, E19, E26, E32, E36
Declining populations	5	E3, E8, E10, E13, E37
Views on use: adults versus eggs	**No.**	**Respondents**
Adult harvest is difficult	9	E7, E10, E11, E12, E13, E15, E20, E31, E32
Adult marine turtles can be harvested	6	E1, E6, E16, E23, E26, E30
Adult marine turtles cannot be harvested	5	E9, E12, E13, E24, E37
Egg harvesting is possible	20	E1, E2, E3, E6, E7, E8, E9, E10, E11, E12, E16, E20, E23, E24, E26, E30, E31, E32, E36, E37
Egg harvesting is questionable/difficult	5	E3, E13, E14, E24, E25
Wildlife in the market	**No.**	**Respondents**
Economic value is convincing argument for conservation, wildlife "pays for itself"	15	E4, E8, E9, E11, E12, E14, E15, E17, E18, E19, E20, E21, E23, E35, E37
Current economic system, specifically demand and competition, will not lead to conservation, even if wildlife is valued under it	12	E3, E5, E14, E17, E18, E19, E23, E26, E27, E28, E29, E31
Not all wildlife values can be measured economically	6	E11, E14, E15, E17, E27, E35
Positions on uncertainty	**No.**	**Respondents**
Uncertainty dictates caution, err on side of nonuse	19	E3, E4, E6, E7, E8, E9, E10, E11, E12, E13, E14, E19, E20, E21, E25, E27, E33, E35, E36
Make do with information that exists, experiment with use	10	E1, E5, E14, E16, E18, E23, E26, E28, E30, E34
Would rather not have use, but uncertainty should not be used for nonaction	4	E14, E31, E32, E37

(continued)

TABLE 2.1 (*continued*)

Uncertainty not an issue, marine turtles cannot be used	3	E15, E17, E29

Views of "others" on use	No.	Respondents
Both pro- and anti-use factions exist in MTSG	12	E9, E10, E12, E14, E15, E19, E21, E23, E25, E31, E35, E36
MTSG is opposed to use	9	E1, E4, E5, E9, E16, E18, E26, E30, E32
MTSG is not opposed to use	4	E3, E6, E8, E24
Views of MTSG based on emotion	8	E1, E4, E5, E16, E26, E30, E31, E32
Views of MTSG based on science	4	E3, E6, E8, E24

Notes: Text of transcribed interviews was coded, with emerging codes shown in column 1. The number of respondents expressing a particular view is in column 2. Specific interviewees (coded by number: e.g., E1 refers to expert interview 1) expressing a particular view are in column 3.

data gaps often arising from life-history characteristics), especially "in many of these places where there is an interest in harvesting, no, the information isn't available" (E6). For six experts, the most important data gap was that regarding population regulation: "we still don't have a very satisfactory model for sea turtle life history. . . . Fundamentally we do not know what affects recruitment because it happens at the end of this ghastly dispersal. First pelagic dispersal which is impossible to deal with, and then there must be some sort of recruitment from that pelagic dispersal stage into the smallest class of settled juvenile stage. We just don't know what affects that" (E16). With such information, "we could begin to really assign reproductive value of each class" (E24) and know which classes are more "harvestable." The importance of lack of data as an argument against use is reflected in the following quote: "I think that most people who have given it [sustainable use] serious thought know that it's something that we can't really evaluate because of our lack of data . . . clearly in the context of modern conservation science . . . one has to assess the area of sustainable use. . . . We're all just kind of saying 'yeah, boy, it would be great. Gee, I wish we knew more so that we could really talk about it knowledgeably'" (E26).

The status of marine turtle populations, both absolute and relative, was also a concern for some experts: "The problem now for most sea turtle populations is that they've undergone such an extreme decline . . . the populations probably have to be protected for a long while, until they rebound, and then a sustained yield would be possible" (E10). For other experts, concerns were linked to our ability to "know" status. For example, E4 believed that further research is required to "find out how many turtles you actually have" and that "we're overestimating our population sizes in most cases." Four experts saw signs of both decline and increase in different locations, but some were cautious about interpreting such data: "Yes, it looks like for the number of nests the trend is upwards, both for green turtles and for loggerheads. But it's a very short time period. You never know what's going to happen in the next decade" (E19).

Using Eggs versus Using Adults

Experts were asked specifically about the feasibility of harvesting adult marine turtles versus their eggs. While only five experts rejected adult harvesting under any circumstances, seven agreed that harvesting adults is problematic and the least acceptable form of marine turtle use (table 2.1). Experts saw adult harvest as difficult due to their value to the overall population: "When you get something that reaches that stage of development, it's already made it through so many factors, you know, mortality factors, and survived all of those threats that I have a problem when you start taking mature adults and reproducing animals" (E13). Lack of data was seen as particularly problematic, making it "pretty hard to come up with a harvest scheme that makes sense as an adult." Given the many unknowns, "the harvest of adults on a nesting beach is difficult to support anywhere, and certainly not as a conservation tool" (E9).

Six experts believed adult harvesting is possible. E6 pointed to "some populations in the world which can sustain harvests because they've been protected for long periods of time, and never really heavily exploited." E26 identified overall good health of a population as a criterion for harvesting, which can be determined by looking at the ratio of neophyte to non-neophyte nesters, an indicator of adult recruitment. Harvesting scenarios suggested by this group included male-targeted and temporally and regionally delineated harvesting to reduce overall impacts and facilitate enforcement. These experts saw adult harvesting as problematic, but not impossible.

Marine turtle experts were generally more tolerant of egg harvesting, as eggs are "reproductively designed to take a hit" (E12). In some places eggs are naturally "lost," as exemplified by highly eroding beaches in Suriname: "You have a certain percentage of eggs that are doomed by geological cycles . . . this is a takeable surplus. Either let it wash out to sea or you can take it. And if by taking them, and selling them in town, and using that money to make sure that eggs above the high tide line are marked off and are saved, you're better off than if you kept everyone away and didn't do anything. So you can apply a straight logic in cases like that" (E23).

A second "straight logic" case is the mass nesting by ridley turtles (called *arribadas* in Spanish) where the density of nesters leads to the destruction of previously laid eggs (L. Campbell 2007b), creating "an ideal situation to compromise with use" (E11).

When there is no such loss or destruction of eggs, harvest levels are difficult to calculate, but experts were more willing to compromise "if one does have to make choices" (E7). Modelers were identified as the scientists to inform such choices, and one modeler supported the viability of egg harvesting: "Usually you can sustain a fairly large harvest of eggs because of the high mortality of the hatchlings anyway—they're not worth very much to the population—so you can eat some" (E10).

While most experts saw egg harvests as more feasible, five experts disagreed. For example, E24 questioned current assumptions about early life-stage mortality and suggested hatchlings have a higher survival rate than most people believe. Thus, egg harvesting might not be substituting for presumed mortality. Some experts resisted the idea of "overproduction": "The more produced the better . . . [this is] the reproductive strategy that these animals have put in place" (E13). Finally, some experts were concerned with harvesting the "right" eggs in overproduction scenarios: "You can't guarantee that this clutch that you're allowing to be harvested would have been destroyed. This may have been the clutch that would have survived . . . maybe the egg clutch you allowed to be taken was the fittest, . . . that had the better genes" (E14).

While most of the experts opposed to egg harvesting also opposed adult harvesting, there were some exceptions. For example, E26 believed that monitoring the impacts of adult harvesting is easier, and thus decisions more easily reversed. E23 cited specific examples of adult harvesting that hadn't resulted in extinction and of egg harvesting that had severe impacts.

Scientific Uncertainty

While experts agreed that conservation policy should be based on "good" science, the variation among their views on use reflect different attitudes toward scientific uncertainty. Four distinct groupings emerge (table 2.1).

First, the largest group of nineteen experts could not support consumptive use of marine turtles because of scientific uncertainty. E3 is a good example of experts in this group: "And no one really understands the population regulation. . . . And until you really understand that, I wouldn't, I wouldn't advocate any kind of use" (E3). This is an important trade-off; E3 also stated that she considered sustainable use a valid wildlife conservation tool. For experts in this group, scientific certainty takes precedence, and most suggested that if data were available, they would have "no problem" with consumptive use. Some of them expressed optimism regarding the potential to collect needed data with improved technology, while others were more skeptical.

Second, a group of ten experts recognized knowledge gaps, but advocated making the best of existing information. E16, for example, argued against getting "bogged down" in the notion of sustainability: "It's not something you can prove. It's something that you . . . accept on a contingent basis until you disprove it. . . . But in point of fact you're never going to be 100 percent sure that it's sustainable." E28 worried about defending a no-use position based on uncertainty and thus "reinforcing this position that we don't know enough to manage." E16 saw appeals to uncertainty in a more sinister light: "The proposition that the only way to proceed is to, first, study and know everything, and [then] choose the correct solution and implement it, I think is just not a real world solution. And I believe it's used now to in fact obstruct progress." Overall, experts in this group believed that some data gaps will remain, that use schemes fail or succeed on socioeconomic issues, and that conservation in the real world requires risk-taking and adaptive management.

Third, a group of four experts were concerned about the biological limitations on use and were less accepting of uncertainty than experts in the second group. Their preferences would be to limit consumptive use, and all were actively involved in doing so at specific sites. However, socioeconomic need sometimes overruled their concerns when alternatives to use were unfeasible. More than all others, these experts took the approach of making use "tolerable."

A fourth group of three experts represented the extreme of the no-use side. For them, information was not a limitation on sustainable use; information confirms that turtles cannot be used, not as adults and even questionably as eggs. For this group, biology was everything: "I'm considered an uncompromising kind of a person and rigid. . . . I don't really care about the political aspects of it. I don't even care about the cultural or social aspects of it. I just know what's feasible and what isn't feasible" (E29).

THE SCIENCE OF SUSTAINABLE USE IN 2009

Over the past fifteen years, more information about sea turtles has become available. Migrations are better understood (Godley et al. 2008), population models are more sophisticated (Chaloupka 2002; Heppell et al. 2005), and we are better able to "age" turtles (Snover and Hohn 2004). In addition, some nesting populations have shown signs of recovery (e.g., green and leatherback turtles in the Caribbean and Atlantic [TEWG 2007; Troëng and Rankin 2005; Broderick et al. 2006]; olive ridleys in the Pacific [Abreu-Grobois and Plotkin 2007]). Here, I consider how improved understanding of migrations and increasing trends in some populations have affected the sustainable-use debate.

Migrations

Satellite tagging illuminates where turtles go from the point of tag application to points between, until the tag or turtle expires (Godley et al. 2008), and genetic analyses of mitochondrial DNA allow turtles found in foraging areas to be "matched" to their nesting beach of origin (Bowen and Karl 2007). Thus, the problem of "not knowing where sea turtles go" has been partially resolved; technology has made turtles at sea more legible (J. Scott 1998).

Improved understanding of migrations has bolstered claims to turtles as a "shared" resource, to be managed as such (L. Campbell 2007a). While sharing could conceivably involve use (e.g., allocating harvesting quotas based on relative contributions of nesting beaches to shared foraging populations), the dominant argument is that countries wanting to use marine turtles are impinging on the rights of those who want to see them protected. In the case of Caribbean hawksbills, for example, Mortimer, Meylan, and Donnelly (2007, 18) conclude that genetic data show "harvests in any part of the Caribbean impact the species throughout the region. . . . Thus, any nation that opens or promotes harvest could be undermining badly needed efforts to conserve the species at other sites." Godfrey et al. (2007) challenge the veracity of Mortimer et al.'s claim, specifically those related to harvesting and sustainability, and the challenge illustrates that scientific debate among experts continues. Knowing where turtles go has not resolved debates about sustainable use, though it has changed the nature of the argument.

Increasing Populations

Over the same fourteen-year period that some nesting populations of marine turtles have been increasing, several adult and egg harvests have been eliminated (L. Campbell 2007a), including the "straight logic" case of egg harvesting in Suriname, and an adult green turtle harvest in Caribbean Costa Rica, adjacent to a nesting beach with a twenty-six-year upward trend (Bjorndal et al. 1999). In January 2008 Cuba announced the end of its marine turtle fishery. Increasing nesting populations, thus, have not translated into new opportunities for sustainable use. Rather, the meaning of such increases has been questioned. For example, Bjorndal and Bolten (2003) argue that green turtles in the Caribbean need to be as numerous as they were in (at latest) pre-Columbus times before populations can be considered "recovered" and harvesting an option. Another response has been to de-emphasize the importance of nesting populations as the indicator of population status (see L. Campbell 2007a). Finally, genetic information has led to the definition of "management units," or subsets of populations that share genetic haplotypes. Use programs thus need to be assessed in terms of their impacts on (by definition smaller) genetically determined subpopulations.

UNDERSTANDING EVOLVING DEBATES ABOUT SUSTAINABLE USE: SCIENCE, VALUES, AND CONSERVATION POLICY

During interviews, experts stressed the importance of science in informing their views on sustainable use, and listed specific biological constraints. Although some of those constraints remain (i.e., long generations), some unknowns have become "known," and some nesting populations have increased. Yet sustainable-use programs are fewer than they were in 1995, and there has been no progress in the MTSG regarding guidelines for use. Here I reflect on the reasons for this, again beginning with 1995 interview data.

While interviewed experts agreed that science should inform conservation policy, and discounted the role of "other" issues, they discussed a number of "other" issues at length. These can be broadly categorized as beliefs about markets and values associated with marine turtles.

Beliefs about Markets

The need to "get people on side" with wildlife conservation was identified by fifteen experts as the key impetus for sustainable use (table 2.1). "I don't think that very many people see the preservation of the natural world for its own sake . . . they don't see it as a sellable idea" (E4). Even when they disagreed in principle, experts saw a need for "incentives to protect these places and animals" (E17). "Selling" wildlife conservation was seen variously as desirable, necessary, and lamentable: "You are . . . confronted implicitly, if not explicitly, with the fact that [if] you can't prove that it's economically beneficial in the narrow sense, then it's not important. And there's a time when the proper response to that is 'oh yes I can' and you go ahead and prove it. There's another time when the response to that is 'no, it's important for another reason.' And I think that we have rolled over a little, we—the whole big conservation group—have rolled over a little too easy on this" (E27). Experts had various concerns regarding monetary valuation of wildlife. Some were concerned about the ability of the market to accommodate diverse values, for example, "the old idea of just knowing that they're there" (E35). Other experts saw the extended value of species to ecosystems as difficult to estimate because the full function of species in complex ecosystems is unknown. Some raised the problem of human preference: "I don't think people are smart enough to choose what animals they want to keep around" (E17).

A second group of experts was concerned with the economic system itself, rather than its ability to accommodate wildlife. E15 reflected on the nature of the capitalist system, and suggested that arguments for sustainable use rest on a false assumption: "that you can put an economic value on an animal and thereby you are going to conserve it . . . that you won't just exploit it to extinction and then just go on to something else" (E15). Twelve experts expressed concern about the roles of demand and competition in capitalist systems; use is seldom sustainable, because the act of supplying a product increases demand. Experts were also wary of the "slippery slope" of relative value. For example, E19 pointed to land values in overpopulated coastal areas; as property prices increase, government incentives to sell off publicly protected land will increase, as protecting wildlife will no longer be the rational economic choice.

Science versus Values in the Use Debate

Interviewees were not asked directly about their "feelings" for turtles, but some volunteered them, unprompted. Many experts talked about the aesthetic appeal of turtles—a "cuteness" or "huggability" factor—and a few admitted that aesthetic, emotional, or other values affected their own views. As E15 put it: "I don't want to see sea turtles butchered." A second group of five experts explicitly denied any emotional basis of their views. They claimed they would have "no problem" with consumptive use of marine turtles if the data could support it: "If it were really a sustainable use . . . I don't have a strong doctrine against 'no sea turtle shall be killed'" (E24).

While most experts described their own views as based on science, they saw alternative views as based on values. For example, five experts felt that "the conservationists without the biological training" (E6) were emotional about turtles, while biologists were not. Similarly, those who opposed consumptive use characterized their opponents as greedy and exploitative. Although some experts did engage with scientific arguments related to use (e.g., E16 rejected the "too many unknowns" argument; E23 questioned preferences for egg harvesting), most experts distinguished themselves as scientifically objective, versus others who are values-driven.

The Shifting Terrain of Science, Values, and Conservation Policy

With the elimination of many use projects (and more specifically the withdrawal of Cuba's hawksbill harvesting proposal to CITES) since 1995, the sustainable-use debate has lost some of its focus. It has not, however, disappeared. Current debates about Red List assessments provide some insights as to the status of the use debate.

As some nesting populations have shown signs of increase, the status of marine turtle populations and how status is assessed have become increasingly visible points of contention for the MTSG (Mrosovsky 1997, 2003; Meylan 1998; Webb and Carrillo 2000; Broderick et al. 2006; Seminoff and Shanker 2008). For example, although the IUCN listed the hawksbill turtle as critically endangered in April 2008, the draft Red List assessment generated much discussion on the MTSG e-mail list when it was circulated in June 2007. MTSG members raised a number of concerns, including the utility of global assessments for widely distributed species showing different regional trends, the appropriateness of the IUCN Red List criteria for marine turtles, and whether the designation of critically endangered was accurate. Those who believed the designation was inaccurate saw this outcome as a function of the criteria used in assessments, rather than their misapplication, and argued that the critically endangered listing undermined scientific credibility. These critiques elicited some boundary work by the assessment's defenders. For example, one defender dismissed the assessment's critics as "funded by the Japanese."[5] Another labeled most contributions as "polemic and melodramatic" and motivated by ideology. Several contributors rallied to the defense of the assessors, interpreting any debate as disparaging of their work, even though very few criticisms addressed the specifics of the assessment.

As Red Listing is a primary function of the MTSG, controversy around it is likely to continue and potentially increase as marine turtle populations show signs of recovery; part of this controversy will arise from the implications of such increases for sustainable use, as illustrated in the following contribution to the e-mail-list discussion: "A very powerful message is sent when a species is listed as threatened or critically endangered. I know, we are scientists and not politicians. . . . But I can tell you right now that if a species were to be 'upgraded,' then some governments would push for a resumption of trade in that species, and with no particular regard for the geographic range of the turtles caught and traded." In an era when there are few sustainable-use projects in practice, debates about the Red List can be read as a proxy for debates about sustainable use as a management tool.

CONCLUSION

When I originally undertook this research in 1995, I found the diversity of views on sustainable use among interviewed experts surprising, given the MTSG's unenthusiastic position. Several interviewees believed consumptive use has potential as a conservation tool, but few had published their views. Two explanations may account for the differences in individual views and MTSG policy. The first is that some experts may have been unwilling to oppose utilization because of its prominence in policy statements at that time; recall the expert (E26) who suggested that "in the context of modern conservation science . . . one has to assess the area of sustainable use," but who opposed marine turtle use because of lack of data. Alternatively, those who stated that they were in favor of experimenting with use may not have felt free to express their views openly, and some implied this was the case. The evidence of boundary work within the group suggests there are consequences to expressing such views, for example, being dismissed as "greedy" or "exploitative."

That there are gaps in our understanding of marine turtles is an important feature of the boundary work in this case, for a number of reasons. First, all experts can claim that their views are supported by science, or at least that there is no scientific certainty that their views are incorrect. As long as the science is uncertain, there are few incentives to address "other" issues that might inform one's own view, and very few interviewed experts did so. Second, scientific uncertainty means that there is little to be gained, definitively, in debating the science of use. Thus, the boundary work done by marine turtle experts is most often focused on distinguishing between science and values (usually emotional attachment or capitalist greed), or science and advocacy. This is not to say that arguments about science (and accusations of "bad" science) do not occur, but they are difficult to resolve; when experts debate the science, their motives for doing so, rather than their arguments, are usually questioned. Third, as knowledge gaps about marine turtles are filled, expert values may be exposed. For example, with genetic data available to describe the origins of foraging hawksbills in the Caribbean (Bowen et al. 2007), the debate about the Cuban harvest becomes one of rights. While such debates may be equally contentious, the views of scientists will not necessarily be privileged, and there may be more clarity regarding the "real" issues that are being contested.

STS helps in understanding the boundary work among marine turtle experts, and reveals a fractured rather than monolithic science, or scientific community. How then does this enhance a political ecology perspective on marine turtle conservation? Elsewhere I have described a marine turtle conservation narrative, one that adopts the language of sustainable use and community-based conservation, while continuing to rely on the "tools" of traditional conservation, that is, parks and protected areas. This is accomplished via the promotion of nonconsumptive use via ecotourism (see L. Campbell 2002a). While I have discussed some of the tensions in the narrative and labeled it a "narrative in transition," it might also be considered a "narrative of least contention," one that emerges not because of the strength of or unanimity in the scientific community, but because it masks disunity. Similar to Jeanrenaud's work (2002), this offers a very different vision of the monolithic scientific community that political ecologists often confront (e.g., Leach and Mearns 1996). The foundations of "dominant" narratives may be shakier than we imagine.

Where political ecology enhances the STS perspective is in revealing what is at stake in the discursive struggle outlined here. The boundary work among marine turtle experts is not so much about claiming scientific ground as it is about determining how marine turtles are managed. Since its inception, the MTSG trajectory on marine turtle use suggests that those on the nonuse side of the struggle have been the victors, particularly in the battle against large-scale commercial operations. Although the most recent version of this battle—Cuba's proposal to trade hawksbill shell with Japan—appears to be over and mariculture is mostly a nonissue in 2010, debates over use are likely to continue. Turtles and their eggs are used throughout the world, both legally and illegally (L. Campbell 2003) and, while large-scale commercial operations have been a lightning rod in the use debate, conservation policies also target these other uses. For example, the Inter-American Convention for the Protection and Conservation of Sea Turtles lists eliminating all consumptive use of marine turtles as a primary goal, with no consideration of whether use might be sustainable (L. Campbell, Godfrey, and Drif 2002). However, in some of the places where marine turtles are used, populations appear to be increasing, and if these trends continue, pressure to use marine turtles (or to use them more intensively) may also increase (L. Campbell et al. 2009). The nonuse stance in the MTSG to date means that there is little accumulated knowledge that might guide any new use programs, should they arise.[6] Even when existing harvests are tolerated, they have seldom been monitored or studied for what might be learned about using marine turtles as resources (but see L. Campbell 1998; L. Campbell, Haalboom, and Trow 2007). Perhaps ironically, the Cuban monitoring program was considered commendable (Fleming 2001). How and whether marine turtle conservation policy will evolve in the face of population increases remains to be seen, although recent debates about the hawksbill listing suggests any evolution will be contentious. In the post-Cuban-hawksbill-harvesting era, the Red List process itself may replace large-scale commercial harvests as the primary site of debate, one that will be cast as scientific, but that has already been revealed as involving much more.

NOTES

1. I focus on the direct consumption of wildlife through harvesting or hunting. Wildlife viewing via tourism, often (and problematically) labeled nonconsumptive use (Meletis and Campbell 2007), is another way of using wildlife and is addressed elsewhere (L. Campbell 2002a).

2. The data presented in this paper are reproduced, with permission, from an article published in *Ecological Applications* (L. Campbell 2002b).

3. The Red List has leatherbacks (*Dermochelys coriacea*), hawksbills (*Eretmochelys imbricata*), and Kemp's ridleys (*Lepidochelys kempii*) as critically endangered; greens (*Chelonia mydas*) and loggerheads (*Caretta caretta*) as endangered; olive ridleys (*Lepidochelys olivacea*) as vulnerable; and flatbacks (*Natator depressus*) as data deficient (http://www.redlist.org/).

4. Membership in the MTSG is restricted to those invited to join based on their expertise.

5. As in other areas of ocean conservation, many marine turtle scientists and enthusiasts vilify "the Japanese." In this case, Japan's desire to import hawksbill shell to serve their traditional *bekko* industry fuels anti-Japanese sentiment. This is problematic for the MTSG, the leadership of which is sensitive to its Japanese members and to the need to collaborate on regional conservation. Many MTSG members work against anti-Japanese sentiment, but it continues to be expressed in formal and informal settings.

6. There are several regions where such possibilities exist. For example, both Broderick et al. (2006) and Chaloupka and Balazs (2007) have suggested some level of green turtle harvesting might be feasible at Ascension Island and Hawaii, respectively.

3

JOAN H. FUJIMURA

TECHNOBIOLOGICAL IMAGINARIES

HOW DO SYSTEMS BIOLOGISTS KNOW NATURE?

The field of systems biology is capturing the attention of scientists, funding agencies, and private industry. This postgenomic approach appears to diverge from earlier genetic approaches in its attempt to grapple with complex systems in biology.[1] Systems biology aims to analyze the vast amount of information produced by the genome sequencing projects and other collections of data to better represent biological complexity that could not be understood through a focus on genes. This chapter analyzes systems biology approaches that develop and use computational tools to map, navigate, and attempt to materially replicate this complexity.[2] It explores how scientists' engagement with complex systems can often lead to blurring, discursively and materially, of distinctions between mechanical and biological systems. It ends with a discussion of the implications and potential consequences of this approach.

By now, systems biology incorporates a range of different approaches. This chapter focuses on the role of mechanistic models and principles in modeling living organisms.[3] It examines the use of metaphors and languages taken from engineering models of the complex systems of automobiles, airplanes, and robots to study complex living systems. The chapter concludes that systems biology represents the outcome of a series of *movements back and forth across the machine–living organism border*. Analogous movements can be seen in various "systems approaches" across the environmental sciences, including climatology, soil science, and ecology. I call these "technobiological imaginaries,"[4] and this chapter argues for a careful analysis of their historical production, specifically around the question of *what is lost in translation* at these border crossings.

Each version of machine and living organism contributes to the final vision of what constitutes "mind," "body," and "nature" in systems biology. As social and historical studies of science have shown, particular versions of social arrangements often become embedded in conceptions and technologies of nature and machine. Individuals or groups author each concept and technology in particular contexts. This chapter contributes to an examination of authorship and the ways such authorship becomes embedded in technologies as regulatory practices that in turn produce particular kinds of minds, bodies, and nature. In order to do so, we have to understand which versions of machines and which versions of nature move back and forth, and when, across the machine-nature border in the production of systems biological knowledge. By examining these multiple border crossings, we may learn about how what we know to be nature and machine is constituted.[5] I refer here not just to representations, but also to material natures and machines. The promises of systems biology—that is, fabricated organs, drug treatment regimes, and the "healthy" body—are material productions and interventions.[6] Or as my sociological forebears noted, representations are real in their consequences.[7]

This chapter also discusses to what extent the machine–living organism border is itself a production of biological investigations throughout the twentieth century. In the past, biologists defined what is living in opposition to their understandings of what is human-made. What consequences will research and applications in synthetic biology and stem-cell biology have for this opposition?

This chapter is based on ethnographic research in scientific laboratories and at scientific meetings, interviews with researchers, and documentary analyses of published literature over a period of nine years.

SYSTEMS BIOLOGY: ADDRESSING THE PROBLEM OF BIOCOMPLEXITY

In May 2000 the National Science Foundation issued its call for research proposals on the theme of "biocomplexity" that emphasized "the interplay between life and its environment, i.e., from the behavioral, biological, social, chemical and physical interactions that affect, sustain, or are modified by living organisms, including humans." Like ecological and environmental sciences, systems biology is a field of research that attempts to address biocomplexity in biological systems.

Systems biology research began in the late 1990s and has taken different forms. Some focus on multiple causality and interdependencies. Others use the language of flows, inputs, and outputs. Systems biologists generally want to connect networks, pathways, parts, and environments into descriptions of functional processes and systems. Although many systems biologists have focused on functioning organisms more than on environments, some systems biologists have incorporated environments into their models.

Systems biology approaches also differ by their theories and methods, including those borrowed from systems theory, mathematics, statistics, computer science, artificial intelligence, physics, engineering, robotics (which includes computer science and engineering), and social science (network theory). Understandably, there are disciplinary contests about both epistemic aims and methodological approaches among those researchers who call their work "systems biology" (Fujimura 2003; Calvert and Fujimura 2009). Despite these disciplinary contests, systems biology appears to be gaining influence in the production of our present and future biological and medical research. It has attracted attention from scientific journals, academic institutions, and private industry in the United States, Japan, and the United Kingdom. At my home institution, the new Wisconsin Institute for Discovery has selected systems biology research from a host of applicants as one of its five key research themes.

Systems biology models biological complexities as organized systems in order to understand them. It seeks to explain how organisms function by using information on DNA, RNA, proteins, and other agents to develop systematic models of biological activities. A major methodological emphasis is the modeling and simulation of biological systems according to particular principles or rules. There are differences in terms of which rules are dominant in any system, which system is in focus, and which elements are included in a particular system. Computational experimentation is a major emphasis of systems biology. Experiments include introducing perturbations into simulated systems to see what happens under different conditions. The information used in the simulations is often taken from databases.

Systems biologists aim to help make sense of the bits and bytes of information produced by the transnational genome projects, but they also claim that their field is not reductionist and is instead a reaction to the failure of molecular biological approaches to provide a satisfactory understanding of the operation of biological systems. Some systems biologists argue that systems biology attempts a more holistic approach toward explanation. Like *systems*, *reductionism* and *holism* are "fighting words" in the history of biology. Philosophers of science and biologists have been engaged in debates about reductionism and holism for at least the last hundred years. I acknowledge these debates and the multiple meanings of these terms, and I use the terms here as the actors I interviewed used them. Generally, the debate is between parts and wholes. Some systems biologists use the term *reductionism* to refer molecular biologists' attention to the parts of the organism, including DNA, RNA, and proteins, and their functions, as contrasted with their interest in understanding how parts are organized into systems because systems explain more than individual parts.[8] Others argue that systems biology approaches can now experimentally as well as conceptually explore phenomena such as emergent properties in developmental biology, in contrast to the late twentieth century when wet-lab molecular biological reductionist methodologies were required because one could not vary more than one component at a time and keep track of the results. New computational and biological technologies have produced a multivariable developmental biology, which may allow developmental biologists to experimentally explore the holistic theories of Bertalanffy and Weiss (Gilbert and Sarkar 2000, 7–8).

One question I examine in this chapter is whether some systems biology approaches deploy another form of reductionism.

SYSTEMS BIOLOGY MODELING: ANALOGIES TO MACHINES, CYBERNETIC SYSTEMS, AND ARTIFICIAL INTELLIGENCE

A systems biology approach considered here attempts to meld ideas and methods from control engineering, mathematics, artificial intelligence, and robotics to model biological systems. This approach uses detailed molecular pathways as well as systems rules to frame an organism's functions and organism-environment interactions. The new systems biology speaks of a *holistic* or *organismic materialism* that researchers consider to be holist as opposed to reductionist.

In this fashion, systems biology models of the functioning and development of organisms engages older discourses of "wholism" (e.g., Bertalanffy 1933, 1952; Weiss 1955) and organicism, as demonstrated by the journal *Science*'s March 1, 2002, special section on systems biology. The introduction to the section was entitled "Whole-istic Biology" and referenced, as the source of this word, Ludwig von Bertalanffy's 1967 introduction to his book *General System Theory*, which included his writings that dated back to the 1930s. Like Bertalanffy, these approaches focus on regulation, on interactions among parts, on properties as emerging from these interactions, and on laws that govern organization of parts and processes.[9]

However, this citation process may be more rhetorically strategic than an actual description of the research approach (Fujimura 1996, 2003). My investigation of systems biology reveals a host of approaches that bring different traditions of research, different epistemologies, and different ontologies to the exploration of organic systems. Some systems biology approaches argue that physics has the right tools; others frame their studies in terms of the mechanistic ontologies of cybernetic systems, others in terms of biological "wholisms." Some still argue that reductionist methodology is the correct approach, while many argue that reductionist genetics cannot answer questions of organismal functioning.

Despite these differences, mechanical analogies and cybernetic systems dominated the principles guiding systems biology modeling and simulation, especially in the period 2000–2007. For example, the language of circuitry is prominent in systems. Gene regulatory networks, metabolic networks, and signal transduction networks are also part of systems biology discourse and work with mechanical systems analogies.

This mechanistic analogy is further combined with control theory in this form of systems biology. Control theory is "the mathematical study of how to manipulate the parameters affecting the behavior of a system to produce the desired or optimal outcome" (see http://mathworld.wolfram.com/ControlTheory.html). In engineering, control theorists work on problems such as traffic flow and traffic control. For example, Hiroaki Kitano (2002b, 1662), director of the Institution of Systems Biology in Tokyo, Japan, describes the knowledge objects of systems biology as "seek[ing] to know [. . .] traffic patterns, why such traffic patterns emerge, and how we can control them." Leroy Hood, director of the Systems Biology Institute in Seattle, Washington, similarly likens his systems biology to solving problems in mechanical systems where "the behaviors of the different kinds of elements involved in automotive mechanics—mechanical, electrical, and control—would be integrated and compared to the model prediction" (Hood 2002, 24–26, see also Ideker, Gaitski, and Hood 2001). Kitano explains control and robustness using a Boeing 747 as his model. The 747 has an automatic flight control system that maintains a robust flight path (direction, altitude, and velocity of flight) against perturbations in atmospheric conditions. Kitano (2004a, 829) goes on to say that "although there are differences between man-made systems and biological systems, the similarities are overwhelming. Fundamentally, robustness is the basic organizational principle of evolving dynamic systems, be it through evolution, competition, a market niche or society."

CYBERNETIC THEORY AND BIOLOGICAL SYSTEMS

Control theory is a close relative of cybernetic theory.[10] Their view of the human-machine relationship is more dominant in some versions of systems biology than is Bertalanffy's version of systems theory. Developed by scientists (e.g., Norbert Wiener, John von Neumann, Claude Shannon) working during World War II in operations research to develop war-related technologies, cybernetics was framed as a command-and-control communication system. It incorporates information theory and systems analysis. In information theory, "communication and control were two faces of the same coin," and "control is nothing but the sending of messages which effectively change [control] the behavior of the recipient" (Mindell, Segal, and Gerovitch 2003). Biologists who also worked in operations research used some of this language to rethink problems in biology. Indeed, some of them used cybernetic theoretical and technological framework to rework biological representations of life itself.

According to Donna Haraway (1981–82), Lily Kay (2000), Evelyn Fox Keller (2000, 2002), and N. Katherine Hayles (1999), the application of cybernetic theory to biological problems led to the formulation of molecular biology's view of the bodies of humans and other animals as information systems, as networks of communication and control. They further argue that cybernetic technologies alone were not capable of transforming biology; that language, metaphors, and analogies were critical for this reshaping of biology.

Most significant in this reshaping was cybernetic science's framing of problems of complexity as a way to think about how to control complex systems. For example, Keller (2002) argues that molecular biology began to seriously engage cybernetic theory because researchers believed that traditional embryology could not adequately explain the development of whole complex organisms. However, the molecular genetics of that era also was thought incapable of explaining the complexities of development in part because of the limitations of Crick's dogma (DNA makes RNA makes protein). Keller argues that the problem of complexity led to the use of cybernetic theory to produce a view of the organism as a machine or a set of regulatory networks.

CONTROL AND DESIGN

In 2000 cybernetic theory reappeared as control and design in some versions of systems biology. Kitano's institute's approach aims at what they call the control-and-design method, a version of cybernetic theory's command-and-control. Control-and-design methods are especially aimed at producing designed biotechnologies, including promises of more effective cancer treatments and reverse-engineered organs. For example, by modeling biological systems as mechanical systems, Kitano (2002b, 1662) hopes to delineate how the "mechanisms that systematically control the state of the cell can be modulated to minimize malfunctions and provide potential therapeutic targets for treatment of disease."

Kitano also argues that biological systems models will be used to simulate potential effects and inefficiencies of drug therapies.

> [Biological systems] models may help to identify feedback mechanisms that offset the effects of drugs and predict systemic side effects. It may even be possible to use a multiple drug system to guide the state of malfunctioning cells to the desired state with minimal side effects. Such a systemic response cannot be rationally predicted without a model of intracellular biochemical and genetic interactions. It is not inconceivable that the U.S. Food and Drug Administration may one day mandate simulation-based screening of therapeutic agents, just as plans for all high-rise building are required to undergo structural dynamics analysis to confirm earthquake resistance. (Kitano 2002b, 1664)

Working with California Institute of Technology control engineer John Doyle, Kitano uses control-and-design methods to represent and then biologically reverse-engineer architectures of an organism. The idea is to use reverse engineering to eventually produce organs for transplantation and other applications (Kitano 2002b; Csete and Doyle 2002; see also Noble 2002). Doyle begins his reverse engineering by analogizing complex technologies, like the Boeing 747, to biological organisms, based on the assumption that they are alike in systems-level organization.

ROBUSTNESS

Complex mechanical systems such as those in 747s are engineered for robustness to allow them to adapt to and cope with environmental changes for optimal functioning. In order to promote robustness, control engineers try to build systems using four key parameters: feedback systems; redundancy for multiple backup components and functions; structural stability, where intrinsic mechanisms are built to promote stability; and modularity, where subsystems are physically or functionally insulated so that failure in one module does not spread to other parts and lead to systemwide catastrophe. Csete and Doyle (2002, 1664) applied this concept of robustness to biological complexity: "Convergent evolution in both domains produces modular architectures that are composed of elaborate hierarchies of robustness to uncertain environments, and use often imprecise components. . . . These puzzling and paradoxical features are neither accidental nor artificial, but derive from a deep and necessary interplay between complexity and robustness, modularity, feedback, and fragility. This review describes insights from engineering theory and practice that can shed some light on biological complexity." Kitano (2003, 125; 2004b) has similarly proposed a model of cancer as a robust system that resists traditional drug therapy. "At the cellular level, feedback control enhances robustness against possible therapeutic efforts." This control protects normal cells, but cancer cells—once they have been transformed from normal cells—may also have a similar robustness. In control-engineering language, "Computer simulations have shown that a cell cycle that is robust against certain perturbations can be made extremely fragile when specific feedback loops are removed or attenuated, meaning that the cell cycle can be arrested with minimum perturbation" (Kitano 2003, 125). Kitano argues for a "'systems drug-discovery' approach that aims to control the cell's dynamics, rather than its components" (125). He calls for a "unified theory of biological robustness that might serve as a basic organizational principle of biological system," a unified theory that would be "a bridge between the fundamental principles of life, medical practice, engineering, physics and chemistry" (Kitano 2004a, 834).

LIMITATIONS ON ANALOGIES BETWEEN CONTROL ENGINEERING AND BIOLOGICAL SYSTEMS

The control-engineering approach appears to be a top-down, engineered systems approach to biological organisms that begins with particular design requirements and principles. In contrast, biologists argue that biological organisms are ostensibly the results of evolution and that organisms, species, and evolving environments are historically contingent products. Molecular biologists have preferred bottom-up approaches because they argue that they better capture the complexities and idiosyncrasies that are the results of locally contingent evolution (Fujimura 2005; Calvert and Fujimura 2009).

Systems biologists acknowledge the historical contingency of biology systems by incorporating evolutionary theory in models of complex, robust biological systems. For example, Kitano (2004a, 829) connects robustness to changes in the environment and in genes. "My theory is that robustness is an inherent property of evolving, complex dynamic systems—various mechanisms incurring robustness of organisms actually facilitate evolution, and evolution favours robust traits." Systems biologists further recognize that biological regulatory systems cannot be fully analogized to mechanical systems. "For example, current control theory assumes that target values or statuses are provided initially for the systems designer, whereas in biology such targets are created and revised continuously by the system itself" (Kitano 2002a, 208). When discussing the development of anticancer drugs, Kitano theorizes the robustness of cancer disease states where cancer cells can alter themselves and their surroundings to promote their survival. The system he theorizes is a very complex one, far from car mechanics.

REPRESENTATIONS OF THE COMPLEX HUMAN BRAIN: FROM MACHINES TO BIOLOGICAL SYSTEMS TO MACHINES

Kitano was originally trained in physics and computer science. At one point in his career, he helped develop Sony's AIBO, a robotic pet dog with humanlike neuronal control systems that would enable the robot to learn and develop. *Aibo* is a Japanese word that means pal or friend and an English acronym for artificially intelligent robot. Kitano's Symbiotic Systems Laboratory (1999–2004) also developed PINO, a pint-sized walking humanoid robot, and an artificial voice-recognition system. For Kitano, robots were laboratories for his efforts to improve artificial intelligence software. They were "symbiotic systems" to aid the development of artificial intelligence. "Current research is aimed at the development of novel methods for building intelligent robotics systems, inspired by the results of molecular developmental biology and molecular neuroscience research." For Kitano, "symbiotic intelligence" was a complex biological system analogized as a cybernetic system.

> The underlying idea is that the richness of inputs and outputs to the system, along with co-evolving complexity of the environment, is the key to the emergence of intelligence. As many sensory inputs as possible as well as many actuators are being combined to allow smooth motion, and then integrated into a functional system. The brain is an immense system with heterogeneous elements that interact specifically with other elements. It is surprising how such a system can create coherent and simple behaviors which can be building blocks for complex behavior sequences, and actually assemble such behaviors to exhibit complex but consistent behavior. (Kitano, http://www.symbio.jst.go.jp, 2001)

Rodney Brooks, MIT computer scientist and the director of MIT's Artificial Intelligence Laboratory, is credited with convincing the computer science world of the benefits of studying biological systems such as the human brain for developing robots and other intelligent machines. Brooks said that an understanding of the complex organization of biological systems could be used as the groundwork for establishing complex robotic systems.

> How is it that biological systems are able to self organize and self adapt at all levels of their organization—from the molecular, through the genomic, through the proteomic, through the metabolic, through the neural, through the developmental, through the physiological, through the behavioral level. What are the keys to such robustness and adaptability at each of these levels, and is it the same self-similar set of principles at all levels? If we could understand these systems in this way it would no doubt shed fantastic new light on better ways to organize computational and post-computational systems across almost all subdisciplines of computer science and computer engineering. Thus our grand challenge is to find a new "calculus" for computational systems that let us begin to control the complexities of these large systems that we are today building on an ad hoc basis, and holding together with string and baling wire, instead of with genuine understanding. (Brooks 2004)

In contrast to Brooks, Kitano became steadily less interested in robots and more interested in using computational tools to do systems biology. Kitano is now using control engineering and robotics to model living systems. He has moved from producing robotics that emulate biological systems to simulating biological systems using robotic systems.

BEYOND THE SEAMLESS BORDER BETWEEN HUMANS AND MACHINES

The fascinating conclusion to the human-machine analogy is that systems biology appears to be that *representations of biological systems and of engineered systems are converging in some kind of symbiotic interaction*. This convergence makes sense when we remind ourselves that human scientists have been building what we know of both biological systems and engineered systems, and the analogies between the two, since at least the seventeenth century (e.g., Hammond 1997; Otis 2002; Morus 2002; Westfall 1978). John von Neumann used a formulation of how the human brain worked as his model for the first digital computer. It appears that systems biology, the most recent biological approach to understanding biocomplexity, is the outcome of *movements of representations back and forth across a machine–living organism border*.[11]

Why should we care about this human-machine interaction? I argue that we need to examine closely what happens at this human-machine border for *what is lost in translation* in the production of particular systems biology models. Although many scientists argue that science produces only approximations of nature anyway, some philosophers of science argue that knowledges are not only partial, but also particular (e.g., Barad, Haraway). All approximations are not equal. Thus, representation may produce different consequences depending on what is gained and what is lost in this border-crossing production process.

This idea is perhaps easier to grasp when we consider the production of technologies, as contrasted with biological models. For example, the idea of AIBO as friend and robot epitomizes the emotions and thoughts that preceded the production of this robot. AIBO designers were optimistic about the potential for AIBO to help them learn how to design intelligent systems and its potential for producing robots that could be friends to humans—the ultimate in human-machine interaction. One idea guiding this development was that robots could provide friendship and care to both children and older people in Japan, where the birthrate has fallen to produce mostly one-child families and little family support for older adults. Although the view of AIBO as a tool for designing intelligent systems or as a design for human companionship seems innocuous or even positive, some technology researchers have argued that AI and robotics researchers endow their designs with specific characteristics and definitions that they define as "human" (Suchman 2001). That is, AI designers design systems to embody their own particular ideas about emotionality, embodiment, sociability, the body, subjectivity, and personhood. Suchman (2001, 7) is especially vigilant about "the ways in which autonomous machine agency might be consistent with regulatory practices aimed at producing certain kinds of humans and excluding others." That is, we have to consider what kinds of humanities technologies provide and what kinds they exclude.

In the case of systems biology, particular forms of cybernetic and information theoretic models are incorporated into some systems biology models. Historians of cybernetics have argued that cybernetics was designed as a system meant to control and dominate, to achieve and affect power. Donna Haraway, for example, views cybernetics as "command-control systems" "ordered by the probabilistic rules of efficient language, work, information and energy" (Haraway 1981–82, 246). She reads capitalist mentalities and theories of male dominance into cybernetic systems' exchange, and use of information for a particular end is an example of capitalist mentalities and theories of male dominance. Cybernetics contributed to this capitalist view of animal behavior the idea that information was the key commodity of exchange. Thus, human and animal behaviors are read in terms of engineering, labor sociology (the organization of labor of Frederick Winslow Taylor), linguistics (semiotics, to understand how systems of signs affect behavior patterns), philosophy, and operations research.

My point is that biological theories and representations of nature are human productions and often contain particular ideas taken from human understandings and goals. Many social studies of science have pointed to other examples where particular forms of social organization have been used to construct biological representations (see, e.g., Strathern's *After Nature* [1992]). The mappings of social onto biological are never one-to-one but instead heterogeneous in their outcomes. Cybernetic theories used to develop systems biology models are one set of technobiological imaginaries used to produce renderings of nature.

So what is lost—and what is retained—in translation when cybernetic and systems theories are used to produce systems biology models? What kinds of ideas and representations do not become part of the production of systems biology models? For example, what representations of nature would be crafted if modelers used ideas of symbiosis or coexistence in place of ideas of dominance and hierarchy?

Symbiosis is the coexistence of organisms of different species in interdependent relationships where each benefits the other. For example, human bodies have multiple species of microorganisms living within them in relatively harmonious and mutually beneficial relationships. Indeed, microbiologists tell us that human functioning requires many of the microbes in our intestinal systems. Thus, the human body may be viewed as a composite of multiple organisms living together in an ecological community. While some biologists and ecologists argued for this view of biological functioning in early and mid-1900s, symbiosis was ignored by the waves of enthusiastic efforts to understand how genes work, using molecular-biological research, which flourished from the late twentieth century until very recently. Molecular biologists used cybernetics ideas to study the human body as a conglomeration of molecules hierarchically controlled by what Keller called the "master molecule," DNA, whose directions are translated by RNA to produce particular proteins, which constitutes a cell's body and parts, and onward to ultimately produce a human body. That human bodies begin from cells seems to have disappeared from this master-molecule origin story. This is, in part, because the discovery of the structure of DNA led to a massive effort to study how it worked, leading to the technocratic big science project to map and sequence "the" entire human genome, which has in turn yielded more complexities.[12] As noted above, the first official systems biologists were physicists who designed their ideas of systems biology to work from and with the massive collection of information produced by the various genome projects. This information was primarily in the form of DNA, RNA, and some proteins, so their systems biology focused on designing networks of molecules, beginning with genes.

More recently, however, systems biology models that view bodies as ecologies of multiple organisms are gaining attention and credibility as researchers have begun to say that genes cannot answer the complex questions of human biologies. These ecological models view human systems as constituted of many different organisms and environments interacting within and beyond the human skin—environments within and beyond. They add to early top-down approaches (discussed above) a diversity of biological entities and a specificity of biological processes. Thus, for example, complex systems models are now considering the roles of diet, nutrition, and microbial factors in the development of complex diseases and in the efficacy, metabolism, and toxicity of drugs in human populations (e.g., Nicholson and Wilson 2003, 669). Nutrition has been considered to be a significant factor in disease susceptibility, progression, and recovery, but often was not taken seriously because direct causal links were difficult to investigate in humans. Some systems biology models now propose to accomplish such modeling in mammalian—for example human—physiological systems.[13]

Much of what I described as "systems biology" at the beginning of this chapter appears to be a genetic-reductionist approach. It wants to be the "middle ground," connecting genetics and biochemistry with development and organismal physiology by using computer simulations and engineering expertise. Nevertheless, it maintains the epistemological primacy of the gene, in part because it applies systemic approaches to genomic information in databases. In this section, I have described attempts by other systems biologists to challenge earlier versions of systems biology.

CONCLUSION

This chapter examines some technobiological imaginaries used by systems biologists as they grapple with complexities in biology. Systems biology aims to represent gene networks, cells, organs, and organisms as systems interacting with each other and with their environments. It employs representations of complex biological networks to abstractly model these interactions. Molecular networks include protein-protein interactions, enzymatic pathways, signaling pathways, and gene regulatory pathways. Cells are envisioned as elements in a cell-signaling pathway and in a cellular computing system. Epigenetic explanations are framed as elements in the organism's environment that turn on or off networks of genetic signals. Development is often discussed in terms of reverse engineering. Biological systems then are viewed as *engineered systems*, which have traditionally been described by networks such as flow charts and blueprints. According to Alon (2003, 1866), "remarkably, when such a comparison is made, biological networks are seen to share structural principles with engineered networks." For example, three of the most important principles shared between biological networks and engineered networks are "modularity, robustness to component tolerances, and use of recurring circuit elements."

In place of Alon's amazement, I argue that it is not at all remarkable that biological networks and engineered systems share these principles. Mechanical systems have been analogized to model biological systems and vice versa since at least the seventeenth century in Euro-American biology. Most recently, molecular biology—which itself owes much to the cybernetic model—has produced the information in genomic databases that systems biology uses as the material it molds and simulates. More significantly, the border between representations of organism and machine has been crossed multiple times in both directions, and systems biology is only the latest new field or "discipline" that authorizes and promotes such border crossing. In place of amazement, we need to examine the production of these similarities.

Like nature, biology is a historical object and subject. It has been constructed of bits of things piling up, the accumulation of information, the sedimentation of ideas and objects. The human genome projects of the end of the twentieth century enabled the collection of masses of information through mechanization, lots of labor and love, private and public funds, and competition. To make knowledge of the collected information, however, researchers are searching for other tools, a change in methodology and epistemology. Some systems biologists are attempting to provide rules and principles to organize these bits of information into systems that help to explain the function and dysfunction of organisms. Some of these rules, concepts, and principles are borrowed from artificial intelligence, robotics, computer science, mathematics, control theory, and chemical engineering. The borrowed principles and rules provide the means to organize and explore information, to create models that can be used to test different values of parameters. However, these imported rules, concepts, and principles are not simply mechanical terms. They also are epistemologies that create new ontologies, that is, new realities.[14] They can be used to manipulate systems to produce different natures, new biologies. These new biologies are and will be produced through human and material agencies.

Systems biology increases both the quantity and kinds of exchanges among expertises that move across the border between human bodies and machines. For example, robotic-engineered systems have taken much of their form from representations of biological systems, and now robotic engineering concepts are being used to model living organisms. This further supports the conclusion that the historical production of knowledge and technologies of living organisms and engineered systems, and the rules used to analyze and manipulate them, are co-produced forms.

The multiple border crossings make it difficult to tease out the translations between forms of representation. Yet examining these crossings is necessary to the study of systems biology. This is where the social study of biology can contribute to the analysis of the process as well as to the production of biological knowledge.

Disentangling the materialities of engineered and biological systems helps delineate the various border crossings in order to *understand which representations and realities are lost in the translation*. By "lost," I mean more than "loss." *Translations can distort, transform, delete, and add*. I do not refer to a loss of an "original" form of reality and a gain of a "false" form. I refer instead to the multiple realities, the multiple biological forms, which could possibly be created through the work of systems biologists. We know from recent bioengineering feats that biotechnologies are able to create different potential forms of reality. Bioengineers cannot always control their engineering as well as they would like; nevertheless they can produce new forms of life.[15] Thus, understanding what happens at the border crossing between engineered and biological systems is crucial for understanding technobiological imaginaries and their potential consequences. Although Kitano, Doyle, and Hood understand that the Boeing 747 and the automobile are too simple to emulate existing biological systems, they nevertheless use principles from mechanical systems to model biological systems. In the process, their models may excise whatever cannot be translated into the instrumental and technical terms of control engineering as they calibrate between existing biological organisms and virtual, artificially created advanced technologies. This excision does not make their productions any less material or real. But it may make them different.

The principles used by systems biologists frame which biological realities are created. If engineering and command-control principles continue to dominate systems biology models of living organisms, what will they produce? What will their sociomaterial consequences be (Fujimura 2005)? The control-engineering approach appears to be a top-down, engineered systems approach to biological organisms that begins with particular design requirements and principles. In contrast, biologists argue that organisms are ostensibly the results of evolution, which means that the organism and the species as well as the evolving environments are historically contingent products. Such historical contingencies mean that multiple biological forms are possible. Despite the dominant hierarchical master molecular theory behind much of molecular biology, molecular biologists have used bottom-up approaches because they believe that they better capture the complexities that are the results of historical evolution. Can scientists use mechanistic models to think about nature and simultaneously keep in mind that these mechanistic models are only one slice into understanding complexity? Or does mechanism limit our abilities to see and account for complexities? The very real practices of producing new biomaterialities through stem-cell engineering and synthetic biology often use systems biology models as their guides.

NOTES

1. Alternative approaches to those discussed in this chapter include developmental systems theory (e.g., Oyama 2000a, 2000b) and Kauffman's complexity theory.

2. Synthetic biology is an example of current efforts to materially replicate biological complexity (Calvert and Fujimura 2009).

3. New developments include two other versions of systems biology, one that is steeped in specific biological systems or topics (e.g., the biology of aging) and another that is framed along the lines of the ecological sciences.

4. This term is taken from my earlier article on scientists' "future imaginaries" (Fujimura 2003) and has synergy with Shelia Jasanoff's concept "sociotechnical imaginaries," which was the basis for a conference on "Sociotechnical Imaginaries" held at Harvard University, November 14–15, 2008.

5. See Traweek (1992) for border crossings in physics.

6. To acknowledge the agency of the material or biophysical does not mean that one accepts the readings of biologists, for example, as perfect understandings of those materialities. As the literature in science and technology studies has shown, the practices that produce biological knowledge are formulated and performed by humans acting within cultures, social institutions, professions, career strategies, technical styles of practice, and novel protocols. Beyond the poststructuralist assumption that biology frames nature through particular lenses,

science studies has demonstrated empirically how those frames and the particular readings have been produced in many different cases.

7. W. I. Thomas's words were, "If men define situations as real, they are real in their consequences" (Thomas and Thomas 1929, 572).

8. *Systems*, *holism*, and *reductionism* have also been politically charged terms in biology in terms of politics writ large. For instance, Richard Lewontin, Richard Levins (Lewontin and Levins 1985), and Susan Oyama (2000a, 2000b; Oyama, Griffiths, and Gray 2001) brought a Marxist perspective to their articulation of developmental systems theory in which genetic reductionism was morally suspect (i.e., it could lead to racism and sexism) and false because it was undialectical. They argued that it is impossible to "bridge" genetics and development because they are complementary. The general systems movement—led by Boulding (1956) and Rapoport (1986)—was fundamentally a New Left critique of 1950s science, which was perceived to be blinded by overspecialization and therefore not socially conscious. The antidote was interdisciplinarity and a more holistic perspective (Hammond 1997). For yet another example, ecology was considered "a subversive science" in the 1960s not because of the environmental movement per se (which had relatively tenuous links to academic ecologists), but because it worked outside of established disciplinary niches and insisted on surveying a broad swath of disciplinary territory.

9. See Hammond (1997) for a history of general systems theory.

10. See Mindell, Segal, and Gerovitch (2003) for a discussion of the relationship between cybernetics and control theory.

11. Similarly, social theories and biological theories have been analogized in previous centuries. Symbolic interactionism learned much from animal and plant ecology. Functionalism took as its model the biological functioning of organisms.

12. There is no "the" human genome, but many different genomes, both among humans and among other animals and plants.

13. See Denis Noble (2006) on the biology beyond the genome.

14. See Mol (1999) on multiple ontologies, and Fujimura (1996) and Thomas and Thomas (1929) on consequential realities.

15. E.g., the beltline pig was an early bioengineering experiment that produced a "monster." Monsters are always a possibility with such experiments. It is also important to remember that researchers at the Roslin Institute researchers made 277 attempts before successfully producing to Dolly, the cloned lamb, who then lived only six years before being euthanized for severe health problems.

4

PETER J. TAYLOR

AGENCY, STRUCTUREDNESS, AND THE PRODUCTION OF KNOWLEDGE WITHIN INTERSECTING PROCESSES

The label "political ecology" has been used in many ways, but this volume is primarily aligned with inquiries into environmental degradation that first became active during the 1990s, in which the complexity of social and environmental dynamics are analyzed in terms of intersecting and conflicting economic, social, and ecological processes operating at different scales. These *intersecting processes* range from the local institutions of production and their associated agroecologies, the social differentiation in a given community and its social psychology of reciprocal expectations, through to national and international political economic changes. Common features of political ecology's rich descriptions are that they connect local struggles and changes around land, labor, and other resources to disputes over roles and responsibilities; draw upon the historical background of the current processes; highlight the dynamics related to inequality; and attend to critical developments in the larger political economies (Watts and Peet 1993; Taylor and García-Barrios 1995; Taylor 2005, 159–65).

During the 1970s the sociology of scientific knowledge (SSK), a field now subsumed in science and technology studies (STS), introduced an angle of inquiry that remains common in STS: what does it mean in actual practice for people to establish and modify scientific knowledge? (This *practice-oriented epistemology* was SSK's antidote to the emphasis in philosophy of science on ideals about how scientists ought to proceed in justifying or refuting knowledge.) Since the late 1980s the STS literature includes many rich descriptions of the diversity of things scientists do and the resources they use in the production of scientific knowledge: scientists employ or "mobilize" equipment, experimental protocols, citations, the support of colleagues, the reputations of laboratories, metaphors, rhetorical devices, publicity, funding, and so on (Latour 1987; Law 1987; Clarke and Fujimura 1992b, 4–5). Such *heterogeneous construction*, as I call it (Taylor 2005, 102ff.; alternatively *heterogeneous engineering*, Law 1987; or *assembling the collective*, Latour 2004, 16) lies at the center of *actor-network theory* (Science Studies Centre 2004).

It has become popular in actor-network theory to describe human, other living beings, and nonliving things alike as actors or *actants*. The move to ascribe agency to nonhumans may seem attractive to researchers wanting to give more acknowledgment to the ecological dimensions of political-ecological dynamics. After all, these dimensions tend to have been eclipsed by the political-economic emphasis in political ecology. In this chapter, however, I resist this move and pull in a different direction. More insight and more interesting questions follow, I suggest, from exploring parallels between the frameworks of intersecting processes and heterogeneous construction without ascribing agency to nonhumans.

The critical interpretation of early actor-network theory in the first section and the parallels explored in the other two sections reflect the direction I have pursued since the late 1980s as a teacher and writer interested in both STS and political ecology (Taylor 2005). A review of developments in actor-network theory or of discussions of nonhuman agency in STS and political ecology lies beyond the scope of this chapter. For readers wanting a wider view on these topics, some recommended entry points include Downey and Dumit 1997; Forsyth 2003, especially 77–102; Latour 2004, 2005, especially 10–11; Law and Hassard 1999; Perkins 2007; Pickering and Guzik 2008; and other essays in this volume.[1]

LOWEST-COMMON-DENOMINATOR AGENCY: WHAT'S (NOT) IN THE MIND OF SCIENTIFIC AGENTS?

Through an STS-style interpretation of the early work of Latour and Callon, this section shows that, although the playfulness of accounts of actants might seem to animate the discussion of the nonhuman contributions, such accounts can reduce everything to a lowest common denominator and dull the analysis of human purposes, motivations, imagination, and action. The interpretation also introduces a theme that informs the other two sections: the psychology of agents is an arena in which researchers are implicitly arguing about the production of knowledge in relation to social causality as well as to social actions that are conceivable or favored. Expressed in another way: STS researchers are arguing about knowledge production in relation to the *structuredness* of society as well as to the actions of human agents in the production and reproduction of structuredness (figure 4.1).

> On Christmas Eve of 1976 in the Bay of St. Brieuc in Brittany, deep down in the water thousands of scallops were brutally dredged by fishermen who could not resist the temptation of sacking the reserve oceanographers had put aside. French gastronomes are fond of scallops, especially at Christmas. Fishermen like scallops too, especially corralled ones, that allow them to earn a living similar to that of a university professor (six months' work and good pay). Starfish like scallops with equal greed, which is not to the liking of the others.

These were the words of Bruno Latour, in *Science in Action* (1987, 202). He continued:

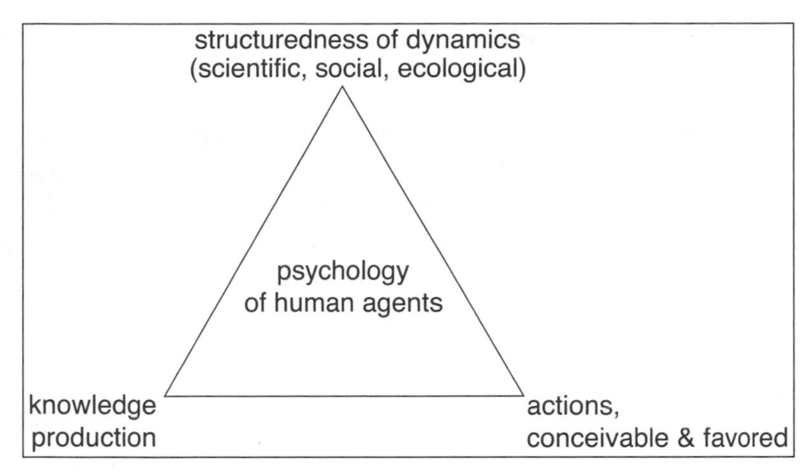

FIGURE 4.1 Schema to illustrate the theme that discussions about the psychology of human agents can be interpreted in terms of the three-way relationship among knowledge, action, and the structuredness of dynamics within which the agents are embedded.

> Three little scientists sent to the St. Brieuc Bay to create some knowledge about scallops love scallops, do not like starfish and have mixed feelings about fishermen. Threatened by their institution, their oceanographer colleagues who think they are silly and the fishermen who see them as a threat, the three little scientists are slowly pushed out of the Bay and sent back to their offices in Brest. Whom they should ally themselves with to resist being rendered useless? Ridiculed by scientists, in competition with starfish, standing between greedy consumers and new fishermen arriving constantly for dwindling stocks, knowing nothing of the animal they started to catch only recently, the fishermen are slowly put out of business. To whom should they turn to resist? Threatened by starfish and fishermen, ignored for years by oceanographers who do not even know if they are able to move or not, the animal is slowly disappearing from the Bay. Whom should the scallops' larvae tie themselves to so as to resist their enemies? (202–3)

The situation Latour described was first presented by his colleague, Michel Callon (1986, 219–20): "The researchers place their nets but the collectors remain hopelessly empty. In principle the larvae anchor, in practice they refuse to enter the collectors. The difficult negotiations which were successful the first time fail in the following years. . . . The larvae detach themselves from the researchers' project and a crowd of other actors carry them away. The scallops become dissidents. The larvae which complied are betrayed by those they were thought to represent." In Latour's and Callon's descriptions the same language was being employed for fishermen, scientists, scallops. They were all agents, actors, or, the term Latour and Callon favored, actants. This equivalence was not just playful language; it was a matter of method. Callon again (1986, 200–201): "The observer must abandon all a priori distinctions between natural and social events. He must reject the hypothesis of a definite boundary which separates the two. These divisions are . . . the result of analysis rather than its point of departure. . . . Instead of imposing a pre-established grid . . . the observer follows the actors in order to identify the manner in which these define and associate the different elements by which they build and explain their world, whether it be social or natural."

In these quotations Latour and Callon (hereafter L&C) were clearly anthropomorphizing—animals, here the scallops, act just as much as scientists in the Bay of St. Brieuc do. In classical anthropomorphism, however, animals' behavior is explained as if they had goals like humans (or, more generally, as if they feel, imagine, and represent like humans) and behaved accordingly. L&C departed from this in two ways.

1. The image of human cognition is reduced to humans' having simple goals, specifically, to resist and to overcome resistance—a form of simple agonistic behavior. For L&C, scientists use laboratories, technical artifacts, allies, and other resources to shift the world, working against its inertia and against others trying to shift the world in different directions (Latour 1987).
2. Equivalence in the terms describing the actions of humans and animals: they all "resist." To act, to be an agent, is to resist.

The terminological equivalence allowed L&C to oppose other commentators on science who would have scientists (or other humans) be the only source of resistance. It also ensured consistency in a larger scheme, evident in subsequent texts of L&C, that extends beyond human and other living agents to include technological objects. Objects resist, so, if "act" is equated with "resist," objects, such as a scalloping dredge, can, like humans and scallops, be actors, agents, actants.

An obvious objection to L&C's anthropomorphism is that STS accounts depend on the scientists to reveal the animals' cognition, perhaps even to reveal their behavior (Collins and Yearley 1992, 312ff.). This conceptual flaw seems to create a big problem for the actant program: if the program cannot be operationalized, its empirical adequacy can hardly be established. Significantly, the equivalency of actants has not been put on an empirical footing—we know nothing about scallops' cognition and little about their behavior two decades after

L&C's 1980s work. Nevertheless, the actant terminology has become very popular in STS. The plausibility of the actant-anthropomorphism must be drawing from another source.

My broad-brush interpretation of actant-anthropomorphism's plausibility begins from the assumptions about human goals and cognition (item 1) underwriting this approach. These assumptions are similar to those of behaviorist psychology, which has always attempted to minimize the role of internal mental representations in explaining an organism's behavior. Provided only that the organism is internally motivated to satisfy its appetite, provision of food in experimental situations can reinforce the desired behavioral responses. (Equivalently, when electric shocks are used for negative reinforcement, the organism only has to be a pain avoider.) Similarly, L&C's image of scientists, building networks in response to the stimulus of others building competing networks, reduces the psychology of cognitive agents to a bare minimum. All that L&C need to assume is that scientists seek to accumulate resources, resulting, if successful, in "centers of calculation," "obligatory passage points" (Callon 1986) and their becoming "macro-actors" (Callon and Latour 1981). Governed only by this egocentric metric of resource accumulation, these agents do not have any practical imagination about constraints and facilitations influencing their possible action, let alone about the possible structuredness of those constraints and facilitations.

L&C's scientists are, admittedly, more scheming than the pigeons or rats described by behaviorists. Nevertheless, on the explanatory (not descriptive) level, the psychology of these scientists is minimal. It is as if a coach of an American football team commanded the players to move the ball up the field against the resistance of the opposing team and asked them to refer only to that objective. No anticipation of the coordinated responses of other players, either on their own team or the opposing one, could be used by the team's players to decide on their moves. Such a team would, against most opponents, fail to score.

Behaviorism is a dirty word; few social scientists or humanists admit to this disposition. Latour's counter to this accusation is that we should assume a minimal psychology to allow the agents to show us how they think about the world, what they see as resources, and whom they see as allies (pers. comm., April 1993). Minimal psychology is a methodological assumption, not Latour's belief about actual agents. This distinction and rationalization, however, obscure what I see at stake about how L&C's approach invites STS researchers to think about agents.

Consider this question: What guidance does L&C's psychologically minimalist method give us about the forms that agents' action can take? As a negative answer, the "no mental representations" dictum ensures two things.

1. Agents are not internally bound—inborn dispositions, cognitive constraints, individual creativity, and so forth cannot determine action and belief.
2. More important, agents are not Socially determined—with nothing in the mind of scientists, there is no place for interests, determined by the agents' class (or other) position in the Social Structure, or for other external influences to reside. (*Social* and *Structure* are capitalized to denote a gross and relatively static view, something given while the science in question develops, e.g., "In Capitalist Societies . . .")

Given this absence of both internal and external constraint, it might seem then that anything goes; that every action is spontaneous and contingent. Latour (1994), however, pulls us back from such an extreme position. *Technical* mediations—"interruptions," "translation," "black boxing," "delegation"—commonly modify an agent's course of action. Not anything can be done in science, technology, and society. (In fact, humans are not humans without technical mediations; Latour 1994, 64.) Notice, however, that the resistances are technical; there is no mention of sociological mediations, involving, say, ideology, socialization, or dominant metaphors.

Now what was at stake in the origins of L&C's actant program is clearer: From every angle possible the idea of agents' actions being socially determined had to be opposed (or made more difficult to conceive). Technical mediations are stressed precisely because they are not social mediations, and the minimal psychology of L&C's agents helps them resist Social determination. The key issue here is not whether L&C are behaviorist insurgents in the STS ranks, but that we can view them as social theorists supporting a particular argument about relations between agents and society. They are telling us how agents' sociality influences their actions, and how society, in turn, is influenced by those actions. Let me tease out that interpretation.

L&C's method called for us to describe the heterogeneous networks of resources and allies that scientists in action mobilize as they resist other scientists in action (Latour 1987; Taylor 2005, 93ff.). The psychology of these agonistic resource-accumulators is minimal; their actions cannot be determined by Social Structures. The *sociality* of these agents, however, is not minimal; in L&C's descriptions agents are embroiled in contingent and ongoing mobilizing of networks of resources and allies. This descriptive focus tends to keep causality distributed across networks, not concentrated inside socially autonomous agents. ("Tends" because L&C's individuals remain at the center of the networks. If the networks become strong, L&C want us to see the responsible agents as macro-actors, who were once microactors and are always vulnerable to becoming so again [Callon and Latour 1981].)

As a program of social theory, L&C's method cannot be sustained consistently. The resilience of at least some, if not most, of the strong networks will ensure their persistence for some period of time. Persistent networks can be viewed as social structure (of a small *s* kind). More subtly, any regularities in the opportunities and constraints that agents experience as part of their sociality invite interpretation as social structuredness. Pursuing this interpretation, we could ask how agents' actions generate, maintain, and undermine that structuredness. Indeed, the agents themselves might consciously identify at least some of these regularities or structuredness. The issue of social determination of the production of knowledge that L&C had hoped to banish is thus resurrected, albeit in a *distributed* rather than direct form.

To some readers, distributed social determination of the production of knowledge will sound like recent actor-network theory (as articulated, e.g., by Latour 2005). My critical interpretation of L&C's program may nevertheless stimulate some readers to take a fresh look at their own ascriptions of agency to nonhumans. If so (or even if not), the critique invites readers to ask of any given account of knowledge production what it implies about the psychology of the human agents, the structuredness (if any) of the scientific, social, and/or ecological dynamics, and the actions conceived or favored by the knowledge producer (figure 4.1). In a nutshell, I advocate asking what we are supposed to be able to do with any knowledge claims.

AGENCY IN HETEROGENEOUS CONSTRUCTION

Neither a thin view of human agency nor steering attention away from social structuredness is a necessary correlate of a heterogeneous constructionist perspective in STS. It can be construed in other ways pertinent to the project of STS and political ecology guiding each other's steps ahead. Drawing from analysis of my experience in socioenvironmental research (Taylor 2005), this section raises three points, which complement the critique of the previous section and provide groundwork for the final section.

NATURE NEEDS NO AGENCY ON ITS OWN

Actant accounts, we have seen, risk reducing everything to a lowest common denominator, dulling the analysis of human purposes, motivations, imagination, and action. Political ecologists wanting to give more attention to the nonhuman contributions to ecological dynamics might, therefore, look for alternatives. In that spirit, notice that, under an STS framework in which knowledge is an outcome of heterogeneous construction, nonhuman things have causal weight even if the concept of agency is reserved for humans. For example, in research on a salt-affected irrigation region (Taylor 2005, 102): "'Technical' considerations, such as the assumption of income optimization [in the model of farms], and 'social' considerations, such as the [geographic and institutional] separation of the modeler from the farmers, had implications for each other in practice. 'Local' interactions were connected with activities at a distance. For example, the modeler and the principal investigator decided not to pursue sociological inquiry into how farmers change, which meant that the content of and conduct of the survey of farms and farmers could remain unchanged."

The interconnections were made especially visible when the modeler in the study team attempted to modify the knowledge production by considering objectives other than income and the processes through which farmers actually change. His efforts were limited by diverse considerations: the Principal Investigator's specialized training and status relative to the modeler; the Research Institute's specialization; the availability of software; transportation for visits to the region (which lay 300 kilometers from the institute); the length of time the study was funded; and so on. In general, the outcomes of scientific work, which include theories, technologies, readings from instruments, collaborations, a person's reputation, and so on, are accepted because the networks of linked components or resources make the outcomes *difficult to modify in practice*. Any person—any human agent—attempting to modify the network-supported knowledge will encounter the causal weight of nonhuman things, but this weight will depend on the *linkage* of those things into the network.

Causal weight through linkage is even more evident if we emphasize not only the heterogeneous components making up networks, but also their construction over time (Taylor 2005, 129–31). Such temporality or temporal emergence, now emphasized by many STS authors (Latour 1994; Pickering and Guzik 2008), is central to the political-ecological picture of intersecting processes, a picture to be elaborated in this chapter's last section.

HUMAN AGENCY IN HETEROGENEOUS CONSTRUCTION IS DISTRIBUTED

The first section introduced the theme that the psychology of agents is an arena in which STS researchers are implicitly arguing about the production of knowledge in relation to social causality and social actions that are conceivable or favored (figure 4.1). This interpretive theme should be applicable to any STS account of knowledge production, so how does it apply to accounts of heterogeneous construction? As a start, notice that the psychology of scientists (or knowledge-producing agents) can be viewed as distributed, just as social causality is distributed: "Focusing on agents' contingent and ongoing mobilizing of webs [networks] of materials, tools, and other people [means] that the character of their agency can be interpreted as *distributed* over those webs, not *concentrated* mentally inside socially autonomous units whose ideas or beliefs are key to the order they impart on the world. That is, although agents work with mental representations of their worlds and can speak about motivations, the malleability of those representations and motivations is not a matter simply of changing beliefs or rationality" (Taylor 2005, 103).

Distributed psychology is by no means passive. "Achieving some result in the material world requires human agents to be engaged with materials, tools, and other people" (103). To build the model of the irrigation region, the modeler "had to engage with pasture growth, government sponsorship, an agricultural extension system, and so on. Moreover, materials, tools, and other people confront scientists with their recalcitrance. So scientists project themselves into possible engagements out in the world in order to imagine what will work easily for them and what will not. . . . Through their imagined [and actual] engagements, people build up knowledge about their changing capabilities for acting in relation to the [diverse, practical] conditions in which they operate" (103).

This picture clearly follows the emphasis in sociology of scientific knowledge on what it means in actual practice for scientists to establish and modify scientific knowledge and activity. The social action favored by accounts of heterogeneous construction is a politics from below, in which scientists try to make changes at specific points in the networks by mobilizing concrete alternative resources (i.e., materials, tools, and other people). Opportunities and constraints persist over time—society is structured—when scientists (and other agents) are unable to mobilize alternative resources (or are unable even to see that alternatives exist). Shifts or restructuring of this social structuredness occur, in part, through the actions of scientists combining with contributions of diverse other agents to produce new knowledge and other scientific outcomes. Such actions, in turn, modify the materials, tools, and other people that agents can imagine mobilizing.

The relationships between psychology of agents, production of knowledge, social causality, and social actions are rendered more concrete in the account of the research on the salt-affected Kerang irrigation region (Taylor 2005, 106).

> The modeler wanted to consider sociologically realistic processes of how farmers change. But his ability to produce results that paid attention to such processes was constrained by his distance—geographically, organizationally, and conceptually—from the farmers' domain of social action. This geographic

and organizational distance was, in turn, related to the centralized character of government and intellectual activities in the one major city of each Australian state, something constrained by the previous 200 years of development. Toward the end of the project, the modeler contemplated a move counter to that centralization; namely, to live and work in the Kerang region as an agricultural consultant. He was aware that this move would raise practical issues such as purchase and maintenance of a car, long-distance access to computer facilities and libraries, ways to keep abreast of discussions about the wider state of the rural economy, and other [personal] considerations. . . . The modeler decided not to move, which meant that the representation of the Kerang region he was able to produce facilitated the making of policy based on simple economic grounds. This outcome did not flow from a political or intellectual commitment to economically based technocratic rationality; many practical, not just intellectual or ideological, considerations would have been entailed in producing a different result.

HUMAN AGENCY IN HETEROGENEOUS CONSTRUCTION IS NOT SIMPLY DISTRIBUTED

The preceding picture of knowledge production raises a serious question about agency—in this case the agency of STS *researchers*. What difference can be made by STS interpreters of science who employ the framework of heterogeneous construction? The prospects seem limited. The framework (as I construe it) does not support big, attention-shaping themes like L&C's "human, other living beings, and non-living things alike are all actants," or Latour's "we have never been modern" (1993). Moreover, although accounts of heterogeneous networks may not discount the structuredness of society (see above), they tend not to spell that out in ways that guide scientists' imagining about their capabilities for acting (Taylor 2005, 154–56; but see Clarke 2005). It is possible that showing scientists commonalities among networks from similar situations might help them negotiate the complexity of their own networks. Yet each case of science-in-process seems to have its own idiosyncratic complexity. In any case, the combined contributions of multiple agents produce outcomes (and thereby modified networks) that may differ from what the scientists intended. Finally, according to this framework, scientists are always already mobilizing diverse resources and, in advance of acting, are imagining what will work easily for them and what will not—there seems little room for STS descriptions of their heterogeneous construction to add something they do not already know that informs their subsequent work.

These limitations acknowledged, let me identify three possible roles for STS interpreters who employ the framework of heterogeneous construction. First, they can make scientists more self-conscious about their distributed agency and the particularities of their heterogeneous networks. Most scientists, when encouraged or prodded by interaction with others, are able to tease open the distributed aspects of their agency (Taylor 2005, 148–56). We should note, however, that the need for such prodding arises because scientists routinely employ representations or ways of speaking that, at least temporarily, push the distributed complexity of their networks into the background. In a similar complexity-backgrounding manner, they also employ simple themes that help crystallize how they will try to mobilize different resources with specific ends in mind. Political ecologists, for example, emphasize the idea that reversing environmental degradation requires finding ways to counteract the effects of inequality. Guiding oneself using simple themes might sound like concentrated, not distributed, agency, but from the perspective of heterogeneous construction, we can view such self-representations and discourses of concentrated agency as *resources* for scientists' knowledge production and other scientific activity. Not the *only* resources, of course, which points to a second role for STS interpreters of science: to nudge scientists to move more readily between discourses of concentrated agency and examination of the wider range of resources they have to mobilize to act. The possibility of holding simpler ideas about concentrated agency in dynamic tension with the distributed complexity of actual practice opens up a third possible role for STS interpreters: to inject themes about social structuredness into accounts of the heterogeneous complexity of knowledge production (Taylor 2005, 154–56). On this last count, exchanges between political ecology and STS might be fruitful; this is the subject of the chapter's last section.

PARALLEL PROCESSES IN POLITICAL ECOLOGY AND STS

Political ecology departs from both actor-network theory and heterogeneous construction in STS in that it gives explicit attention to social structure (in the form of the political economic processes in which any place-based environmental degradation is embedded). Combining heterogeneous construction in STS with political ecology might yield a framework that, while reducing the full complexity of intersecting processes and heterogeneous construction, resists tendencies to suppress that complexity, highlighting, instead, its structuredness.

SOCIAL STRUCTURE

The features of intersecting processes accounts in political ecology, sketched in the chapter's opening paragraph, can be illustrated through a synopsis (from Taylor 2005, 160–61) of research undertaken by García-Barrios and García-Barrios

(1990) on severe soil erosion in a mountainous agricultural region of Oaxaca, Mexico.

The soil erosion appears to be associated with the undermining of traditional political authority after the Mexican revolution, yet the twentieth century was not the first time soil erosion had occurred. After the Spanish conquest, the indigenous population collapsed from disease, and the communities moved down from the highlands and abandoned terraced lands, which then eroded. The Indians adopted laborsaving practices from the Spanish, such as cultivating wheat and using plows. As the population recovered during the eighteenth and nineteenth centuries, collective institutions evolved that reestablished and maintained terraces and stabilized the soil dynamics. Because such landscape transformation introduced the potential for severe slope instability, it needed continuous and proper maintenance. The collective institutions revolved around the mobilization of peasant labor for key activities, first by the church and then, after independence from Spain, by the rich Indians, *caciques*. These activities, in addition to maintaining terraces, included sowing corn in work teams, and maintaining a diversity of maize varieties and cultivation techniques. The caciques benefited from what was produced, but were expected to look after the peasants in hard times. Given that the peasants felt security in proportion to the wealth and prestige of their cacique and given the prestige attached directly to each person's role in the collective labor, the labor tended to be very efficient. In addition, peasants were kept indebted to caciques, and could not readily break their unequal relationship. The caciques, moreover, insulated this relationship from change by resisting potential laborsaving technologies and ties to outside markets.

The Mexican revolution ruptured the reciprocal and exploitative relationships by taking away the power of the caciques. Many peasants migrated to industrial areas, returning periodically with cash or sending it back; rural transactions and prestige became monetarized. Monetarization and loss of labor meant that the collective institutions collapsed and terraces began to erode. National food-pricing policies favored urban consumers, so corn was grown only for subsistence needs. New laborsaving activities, such as goat herding, which contributes in its own way to erosion, were taken up without new local institutions to regulate them.

My schematic of the García-Barrioses' account (figure 4.2) captures the idea that no one kind of thing, no single strand on its own, is sufficient to explain the currently eroded hillsides. The character of intersecting processes can be summarized in the following terms: There are processes of different kinds and scales, involving heterogeneous elements, which are interlinked in the production of any outcome and its ongoing transformation. This means that such situations lack definite boundaries; what goes on "outside" continually restructures what is "in-

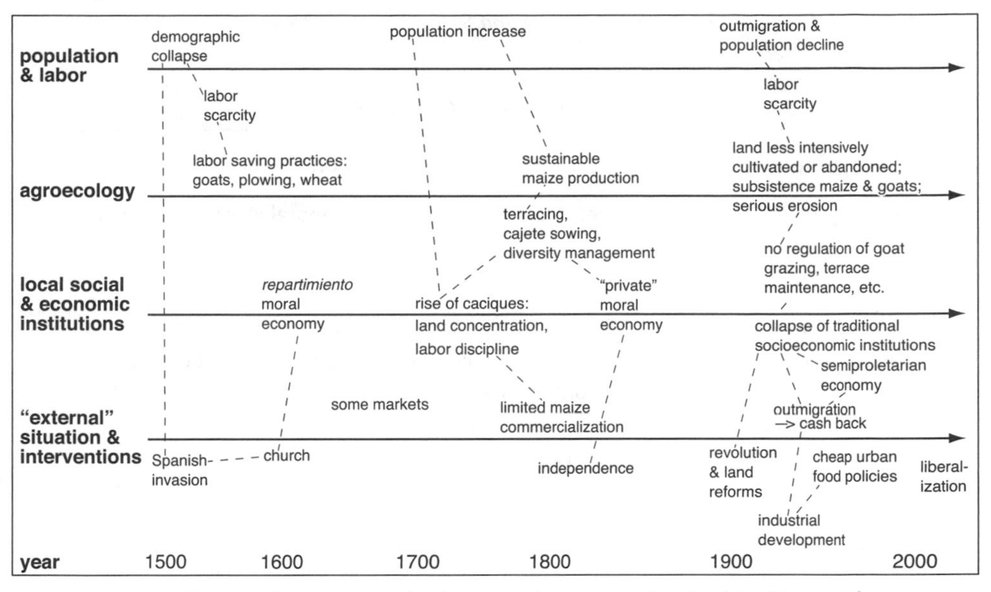

FIGURE 4.2 Intersecting processes leading to soil erosion in San Andrés, Oaxaca. The *dashed lines* indicate connections across the different strands of the schema. See text and Taylor (2005, 165), for discussion.

side"; and diverse processes come together to produce change (Taylor 2005, xiii). Notice also that heterogeneity within intersecting processes includes differentiation among unequal agents; scale crossing and ongoing restructuring means that historical contingency is significant; and the structuredness of such processes is not reducible to micro- or macrodeterminations (Taylor 2005, 161–62; see also Mitchell 2002).

If STS highlights the temporal dimension of network building, a picture can be drawn akin to political ecology accounts of intersecting processes. The interconnected strands that develop over time as knowledge gets established could be underlying unmediated reality; agents' creative process; language, tools, and work organization; and wider sources and audiences.[2] These are processes of different kinds and scales, involving heterogeneous elements, interlinked in the production of any outcome and its ongoing transformation. No one kind of thing, no single strand on its own, is sufficient to explain the established knowledge.

Such parallels suggest that STS could well look to the well-developed cases in political ecology for models of ways to address the complexity of intersecting processes in the production of knowledge (table 4.1). (STS researchers might also learn from the kind of social theorizing implied by intersecting processes, but that topic lies beyond this chapter.[3])

TABLE 4.1. *Comparison of political ecology and STS with respect to three features of intersecting processes*

	Political ecology	STS
Heterogeneous components	Yes	Yes (in networks)
Ongoing (re)structuring	Yes	Temporal emergence widely discussed, but structuring not central to most accounts[a]
Embedded in wider dynamics	Yes (especially political-economic dynamics)	Accepted, but not well developed analytically or central to most accounts[a]

[a]But see Akera (2007) on "ecologies of knowledge."

DYNAMIC TENSIONS

Political ecology is *political* not only in the sense of paying attention to the political-economic processes in which environmental degradation is embedded. It is also political in the sense that many political ecology researchers are interested in changing the situation studied, typically to counteract the effects of inequality at the same time as reversing the environmental degradation. These two senses can be linked if we ask how scientists produce the knowledge needed to help them engage in change (recalling figure 4.1's three-way relationship). Self-reflexive change agency constitutes a third sense of political. Of course, debates about the methods and theoretical frameworks used to gain knowledge are often asked within the field of research itself (e.g., Peet and Watts 1993 for political ecology). However, in the context of this volume, let us incorporate STS perspectives, extending here the discussions of agency from this chapter's first two sections.

In an intersecting-processes approach to political ecology, agency is distributed across different kinds of agents and scale, not something centered in one class or one place. Just as in accounts of heterogeneous construction, this kind of political ecology favors the idea of multiple, smaller engagements linked together within the intersecting processes. This observation invites us to think not only about the agency of the people studied but also, reflexively, about the agency of political-ecological researchers producing knowledge. On the level of research organization, intersecting-processes accounts highlight the need for transdisciplinary work grounded in particular locations. At the same time, to the extent that accounts like the one of Oaxacan soil erosion show an *intermediate complex-*

ity—neither highly reduced nor overwhelmingly detailed—they preserve a role for some kind of social scientific generalization and thus influence beyond the specific situation (Taylor 2005, 163–65).

In this telegraphic summary of agency in political ecology a further parallel with STS emerges. Under heterogeneous construction, scientists move between discourses of concentrated agency and acknowledgment of the complexities of their distributed agency; political ecologists might similarly hold intermediate complexity accounts of intersecting processes in dynamic tension with recognition of the need for research and policy to be grounded in the particularities of the place. Let me put this parallel together with the earlier ones around structure to build up a framework for viewing accounts that scientists use to communicate to others as they seek to establish knowledge and pursue their inquiries. The framework makes room for moves that reduce the full complexity of intersecting processes and heterogeneous construction, but resists tendencies to suppress that complexity. It also reflects the dynamic tension between distributed and concentrated psychology of human agents.

Starting at the top left in figure 4.3, we can view research as a "dialogue," involving concepts and evidence, between researchers and the situations they study. (By using the term *researcher* in place of *scientist* we can accommodate the work of STS researchers as well.) This dialogue may involve simple, broadly applied formulations, for example, "population growth leads to environmental degradation" (or population *decline* in the Oaxacan case), as well as accounts that have a level of complexity characteristic of political ecology's locally centered but translocal accounts of environmental degradation. Such complexity can, however, be difficult to convey so that members of an audience digest and know how to take it up in their own thinking and inquiry. Either for this reason, or because simple formulations seem to provide effective rhetoric when mobilizing campaigns for social change, political ecology researchers may look for ways to move upward across the *bridge* from intersecting processes to simpler formulations (middle of top row). In the opposite direction, teachers of political ecology may introduce scenarios or themes that, while simple enough to convey readily, open up issues and point to greater complexity and to further work needed in particular cases (Taylor 2005, 174ff.). The same bridging or tension between simple and complex applies to STS; STS also involves a dialogue with a situation studied, namely, the social interactions involved in establishing knowledge (top right).

The simple-bridge-complex arrangement can be projected horizontally and applied to the practice of researchers (middle row). A simple formulation would refer to the way that researchers highlight their dialogue with the situation studied (and the veracity of the resulting knowledge) when speaking, writing, and

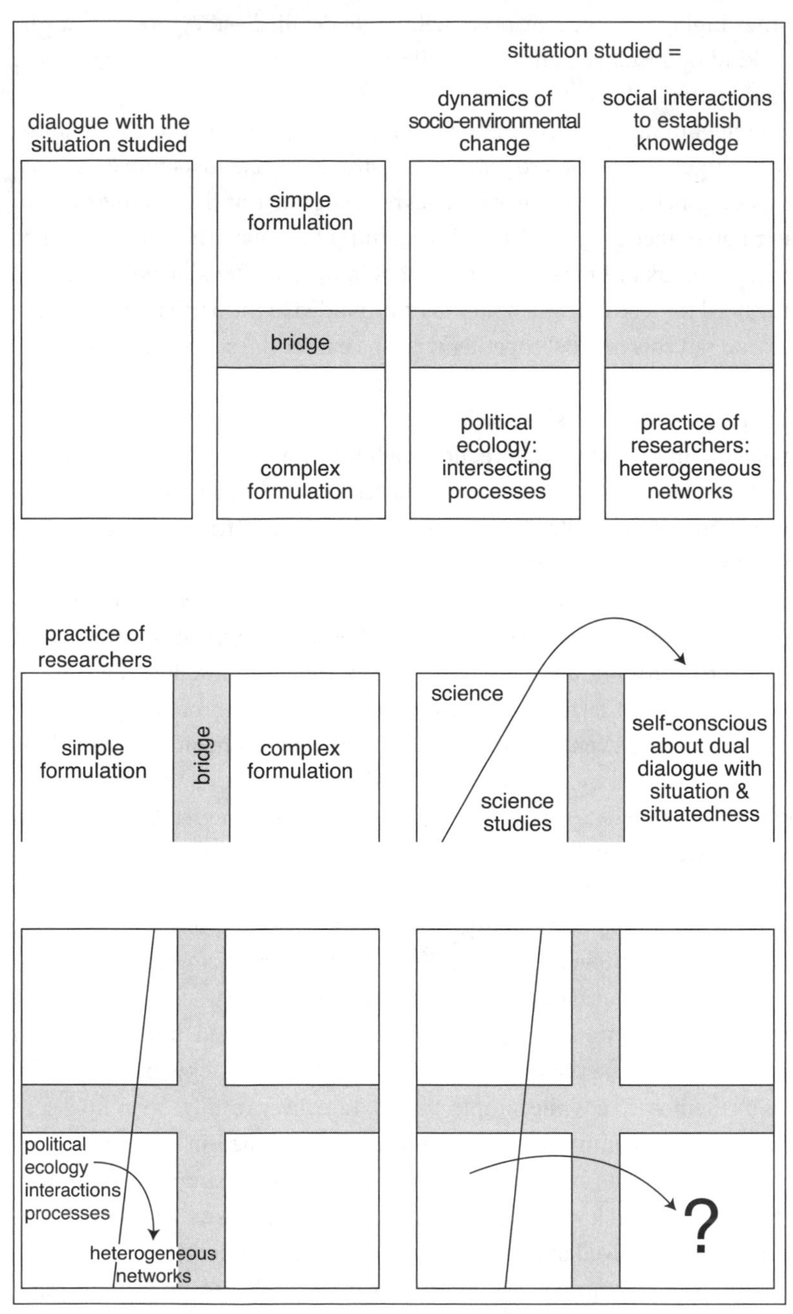

FIGURE 4.3 Framework that articulates and negotiates dynamic tensions between (apparently) simple and complex (i.e., intersecting processes) accounts of the situations studied in political ecology and the social situatedness of political-ecological researchers. See text for discussion.

attempting to influence policy and politics. The more complex formulation would involve attention to the ways that particular researchers are always already negotiating diverse practical considerations as they try to establish knowledge and continue their work and lives (i.e., are always heterogeneous constructors of knowledge). The arrow (middle row, right) denotes the question of how to bridge from the simple formulation to a self-conscious attention to the dual dialogue, that is, with the situation studied as well as with the situation in which the researcher is enabled to act.

The bottom row combines the top two into one framework that articulates and negotiates tensions between (apparently) simple and complex accounts of the situations studied in political ecology and the social situatedness of political-ecological researchers. The arrow in the bottom left refers to the use of political ecology to guide STS, as sketched in this chapter. The arrow to a ? in the bottom right refers to the challenge faced by researchers who want to bring the practice-oriented epistemological concerns of SSK to bear on combining the intersecting-processes approach of political ecology with the heterogeneous network-building of STS. I characterize this challenge as one "of using your knowledge, themes, and other awareness of complex situations and situatedness to contribute to 'a culture of participatory restructuring of the distributed conditions of knowledge-making and social change'" (Taylor 2005, 203). In very broad terms, this summarizes the kind of social action I see as favored by this chapter's picture of agency in the production of knowledge within—and about—intersecting processes. It is consistent with this picture (and the chapter's word limit) that I do not prescribe the means of meeting this challenge, but end with an open question (from Taylor 2005, 201; see also 203ff.): "Clearly, more work is needed on what I and other agents can do—but not alone, nor solely through our accounts of the world—to contribute self-consciously to the ongoing restructuring of the dynamics among particular, unequal knowledge-making agents whose actions implicate or span a range of social domains."

NOTES

1. Space prevents me from discussing ways that the frameworks promoted or reviewed by these and other authors are challenged by or complementary to the perspectives in this chapter, but I welcome being drawn out through e-mail or other exchanges.

2. See http://www.faculty.umb.edu/peter_taylor/IPHetCon.pdf.

3. In brief, an intersecting-processes approach to political ecology offers perspectives on some STS approaches that have begun to have currency in political ecology (approaches over and above the idea of nonhuman agency critiqued in the first section of the chapter). Political ecology researchers have begun to borrow many other concepts from STS: "boundary object," "networks," "co-production," "hybridity," "standardized packages," "black boxes," and so

on. There is a tendency in STS, however—as in sociology more generally—to seek explanatory mileage from labeling. One way to disrupt that tendency is to recast concepts in terms of dynamics, and here intersecting processes are helpful. For example, to refer to "hybrids" of nature and culture presupposes things that were separate before being hybridized. Instead, the "quasi-objects" that are hybridized (Latour 1993) can be reframed as stabilizations—perhaps transient—within *ongoing* intersecting processes in which heterogeneous components are linked together. This reframing means that, whether something is referred to as natural or social, or whether such labels are eschewed (Latour 2004, 2005), its significance as a component (or *resource*) lies in *how it is linked with* many other diverse elements (see the second section of this chapter; also Taylor 2005, 101ff.). Similarly, for any co-production of, say, Science and the State (Jasanoff 2004b). Indeed, any specific researcher who wants to contribute to changing a process of co-production might find it helpful to tease out the specific heterogeneous components and interlinkages. This would reveal multiple sites of potential engagement for themselves in relation to others (Taylor 2005, 131–32). An intersecting-processes approach also means that any entities or structures posited in social theory will not be well-bounded, but can be seen as aspects of "ongoing change in the structure of situations that have built up over time from heterogeneous components and are *embedded or situated within wider dynamics*" (Taylor 2005, xiii). The combination of all three features—ongoing structuring, heterogeneity, and embeddedness—gives the intersecting-processes approach to political ecology the potential to be a fertile site for social theorizing (in ways that go beyond the emphasis of Latour [2004, 2005] and others on disturbing the social/natural divide). See Taylor (2005, 248ff.), for a programmatic sketch of agency, psychology, and social theorizing within intersecting processes.

5

MRILL INGRAM

FERMENTATION, ROT, AND OTHER HUMAN-MICROBIAL PERFORMANCES

Where they were once silent and poorly represented, microbes are now noisily and prolifically performing their way into the scientific imagination. Challenges scientists have faced in identifying microbes in their natural habitats, culturing them, and interpreting their behavior have been overcome in the last several years through new genetic and information technologies (Perkel 2003). Metagenomics, for example, which involves techniques for analyzing genetic material collected directly from the environment, allows scientists to move beyond isolated explorations of single, cultured microbes (Handelsman and Smalla 2003). Less than 1 percent of soil microbes are easily cultivated in a lab; metagenomics, along with microbial community modeling and information technologies, have allowed scientists to analyze genetic information from groups of microbes, revealing impressive microbial genetic diversity in the environment (figure 5.1). Microbial communities within the human body are also being explored via these new technologies, allowing scientists, as one MIT researcher put it, "to delve into the fact that we're not alone, we're symbionts" (Powledge 2006). These discoveries about microbial diversity have been accompanied by research revealing microbes as social organisms, interacting with each other and their environments in complex ways (Kim et al. 2008).

In this chapter, I describe research leading to new characterizations of microbes and microbial behavior. The emerging scientific discourse provides an opportunity to observe how new technology can drive scientific discovery, and how scientists employ metaphor as they work to process new data and information on environmental phenomena (E. Martin 1987, 1994). These new microbial metaphors, moreover, conflict with deeply embedded "antimicrobial attitudes" and technological practices long prevalent especially in medicine and agriculture.

Might this new science drive a shift in prevailing attitudes toward the invisible organisms with which we share our world? To explore this question I detail challenges to change in agricultural politics presented by an increased climate of fear about food safety and threats of pathogenic microbes. In a well-traveled and interconnected world, we must continue to work out desirable human-microbial relationships, and an examination of the "multiplicity of knowledge productions" (as Matt Turner describes in his introduction) about microbes reveals conflicted conversations about consumer rights, food safety, agricultural policy, and land use.

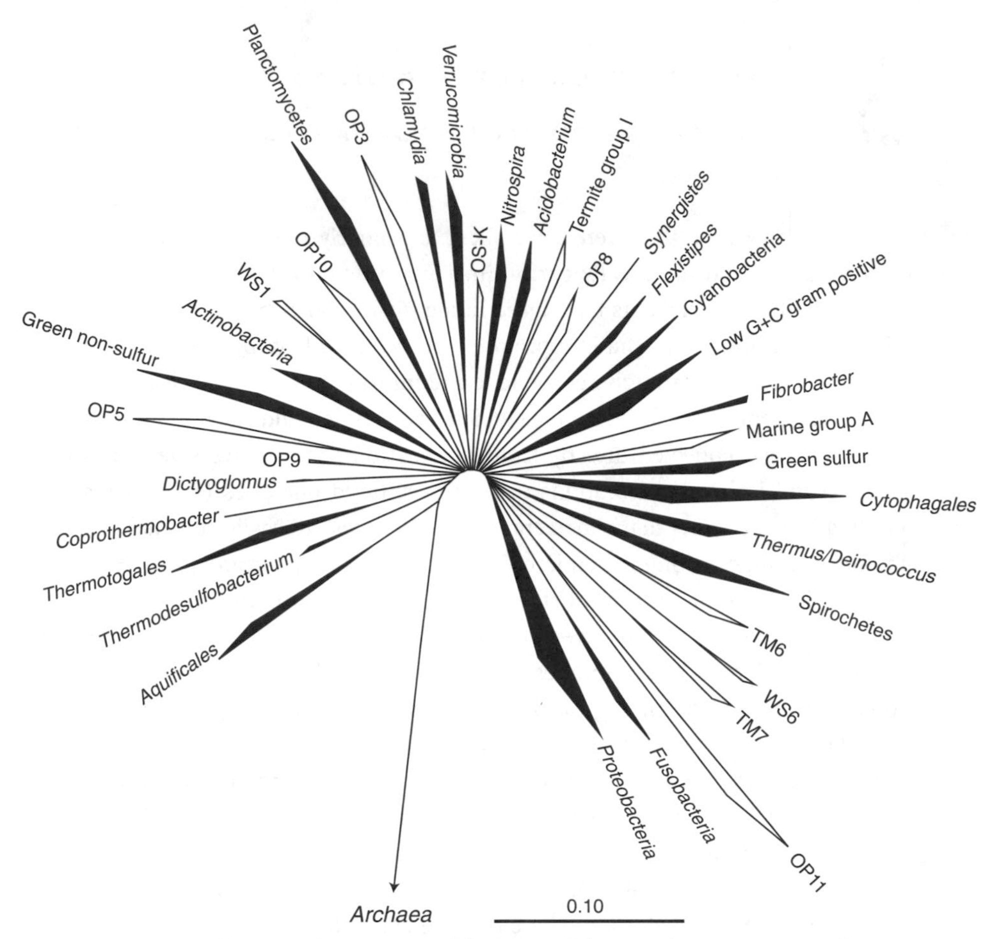

FIGURE 5.1 Evolutionary distance tree of the bacterial domain showing recognized divisions and putative (candidate) divisions as of 2003. A similar tree created in 1987 had only twelve branches, indicating the tremendous growth of recognized types of bacteria. Division-level groupings of two or more genetic sequences are depicted as wedges. The depth of the wedge reflects the branching depth of the representatives selected for a particular division. Divisions that have cultivated representatives are shown in black; divisions represented only by environmental sequences are shown in outline. The scale bar indicates 0.1 change per nucleotide. Reproduced by permission from Norman Pace.

My goal here is not only to illuminate different ways of knowing microbes and their associated technologies, but also to lay out the societal implications of those ways of knowing. This is a story of human-microbe hybridity, with a moral to tell of who stands to win and to lose according to choices we make about the technologies that facilitate those human-microbial relationships. At the end of the chapter I discuss some conditions for positive and productive human-microbial relationships, or performances, and why they are so necessary.

NONHUMAN AGENTS AND HUMAN-MICROBIAL PERFORMANCES

I begin with a discussion of nonhuman agency in networks, and how an understanding of emergent, relational agency can help in assessing the equity of different human-nonhuman networks. Microbes offer a particularly valuable actor to consider in discussions of nonhuman agency because they so beautifully illustrate the concept of human-nonhuman hybridity. Microbes are not just on us and in us, but also actively "make" us. Microbes participate in our breathing, eating, drinking, and digesting, and the more one thinks about the links between food, health, culture, and agriculture, the more impossible it becomes to draw clean lines between human and microbe.

The importance of recognizing agency on the part of the nonhuman has been a central tenet of science studies literature and fundamental to the work of many scholars, many of them geographers, who argue for the need to bridge the nature-culture divide and to acknowledge the ability of the nonhuman to "kick back" (Ivakhiv 2001; FitzSimmons and Goodman 1998; Latour 1993; Lulka 2006; Philo and Wilbert 2000; Roberts 1995; Whatmore 2002; Whatmore and Thorne 1997). However, this focus on nonhuman agency may lead us astray if it leads us to emphasize human and nonhuman as preconstituted and separate subjects. As Sarah Whatmore (2002, 157) has warned, the separation of nature and human risks "reiterating an *a priori* distinction between separate worlds in need of some kind of remedial re-connection." For her, this separation leads to "a residual humanism that restricts the reconfiguration of ethical practice to terms in which the best the nonhuman can get is to be better represented by us (as lesser humans)." Instead, she argues, we need to emphasize the active producing of relationships, "the movements and rhythms of heterogeneous association." This is a valuable concept in comprehending the infinite fuzziness of any boundary between microbe and human. Microbes require us to rethink agency as something not individual, but as performative and emergent, as a creation resulting from interspecies interaction.

Karen Barad offers "posthumanist performativity" as a way to think about nonhuman agency. She is working to contend with how discursive accounts of the world continue to evoke what she sees as very passive understandings of matter. Her solution, like Whatmore's, is to move away from a focus on the nonhuman as a separate, ethically and agentially equivalent subject, and to focus on the activity that results through relationships and interaction. "Bodies come to matter through the world's . . . performativity," she writes (Barad 2003, 10). "Intra-activity is neither a matter of strict determinism nor unconstrained freedom. The future is radically open at every turn." She continues: "Agency is not aligned with human intentionality. . . . Agency is a matter of intra-acting; it is an enactment, not something that someone or something has . . . [it] is not an attribute whatsoever" (15).

Ideas like these suggest that agency need not be thought of something that is divided up among a number of actors. It does not diminish with an increase in the number of agents, but is emergent, and results from interaction. This understanding means that our task in exploring the agency of the nonhuman is not simply to track connections and "prove" associations that connect human and nonhumans in the world, but to analyze and evaluate different emergent properties of networks, and to consider the ethical and political implications of different associations and the technologies that enable those associations. This is an important element in the work of tracing networks, an endeavor that has been thoroughly critiqued for failing to contend with the power fields that shape networks and enable, and limit, their potential.

We can get a better sense of performative agency by contrasting it with agency as conceived by Steve Fuller. Fuller (1994, 749) has challenged work like Bruno Latour's "high agency narratives" as running the risk of "dissipating the emancipatory power of newly constructed agents—the formerly suppressed voices." Fuller argues for a concept of agency that involves scarcity, that the more agents there are, the less agency each can have such that "agency loses its decisiveness in bringing about outcomes" (749). This conception of agency might be understood to inform the work of Michael Pollan (2001), who has developed several intricate and compelling stories about the coevolution of humans and plants. Pollan's stories illustrate the multifaceted aspects of nature's agency, showing us how that agency is revealed in numerous unexpected and negative as well as positive ways. From a certain perspective, however, his accounts, especially that of corn, offer a kind of zero-sum game in which these plants' successful, evolutionary drive to populate the world occurs at the expense of humans who spend increasing amounts of time, and create increasingly costly technologies, in order to support and promote the plant. What is missing in this account, of course, is that some humans shoulder far more of the costs of producing corn, while other humans reap more of the emergent benefits of these human-plant relationships.

Fuller's point about the lack of attention to power in Latour's work is well taken and much discussed elsewhere (e.g., Castree 2002; Haraway 1996; Ingram 2007b). His take on agency, however, leaves out the possibility that power and "mattering," which is what agency is all about, emerges through interaction. As I will describe, a main focus of much of the recent scientific research on microbes moves beyond visual representations of microbial bodies to investigations of communication and signals—ways that microbes interact with each other and with "us." A performative understanding of agency allows us to not only conceive of microbes as agents, but to take a relational view such that we can anticipate how, in part, it is through our own actions that certain microbes come to matter. This understanding of agency also allows us to assess the different technologies that mediate interactions between our bodies and microbial ones. The hope here is that once cognizant of the political implications of how we relate to microbes—who stands to gain and who to lose in different networks—we become more capable of evaluating those relationships and networks, and therefore more able to choose technologies that enable more equitable relations. I will now turn to some of the recent scientific research on microbes, before discussing some of the implications for policy related to managing microbes, and more specifically, for sustainable agriculture.

NEW RESEARCH ON MICROBES

Emerging technologies and methods of investigation into microbes and microbial behavior are challenging long-standing characterizations of the "silent majority" (Handelsman and Smalla 2003). Microbes are not only of many types but also participate in diverse communities and communicate with each other about how to interact with host macroorganisms. Microbial species do not always have set beneficial or negative characters, new research indicates, but change according to conditions and context. Even pathogenesis, for example, has been described as a "relationship" between host organisms and microbes, which cannot be grouped into strictly "virulent" and "nonvirulent" categories (Swerdlow and Johnson 2002). Microbes are social and even sentient subjects, citizens of complex little societies that communicate with each other (Bower 2004).

Microbes form biofilms, which have been described as "cities," heterogeneous assemblages of various bacterial species that create communities by shifting lifestyles from nomadic, unicellular individuals to sedentary, multicellular groups (figure 5.2). "We liken the multispecies bacterial biofilm to a city where bacteria settle selectively, limit settlements of new bacteria, store energy . . . and transfer genetic material horizontally all for the good of the many" (Watnick and Kolter 2000, 2678).

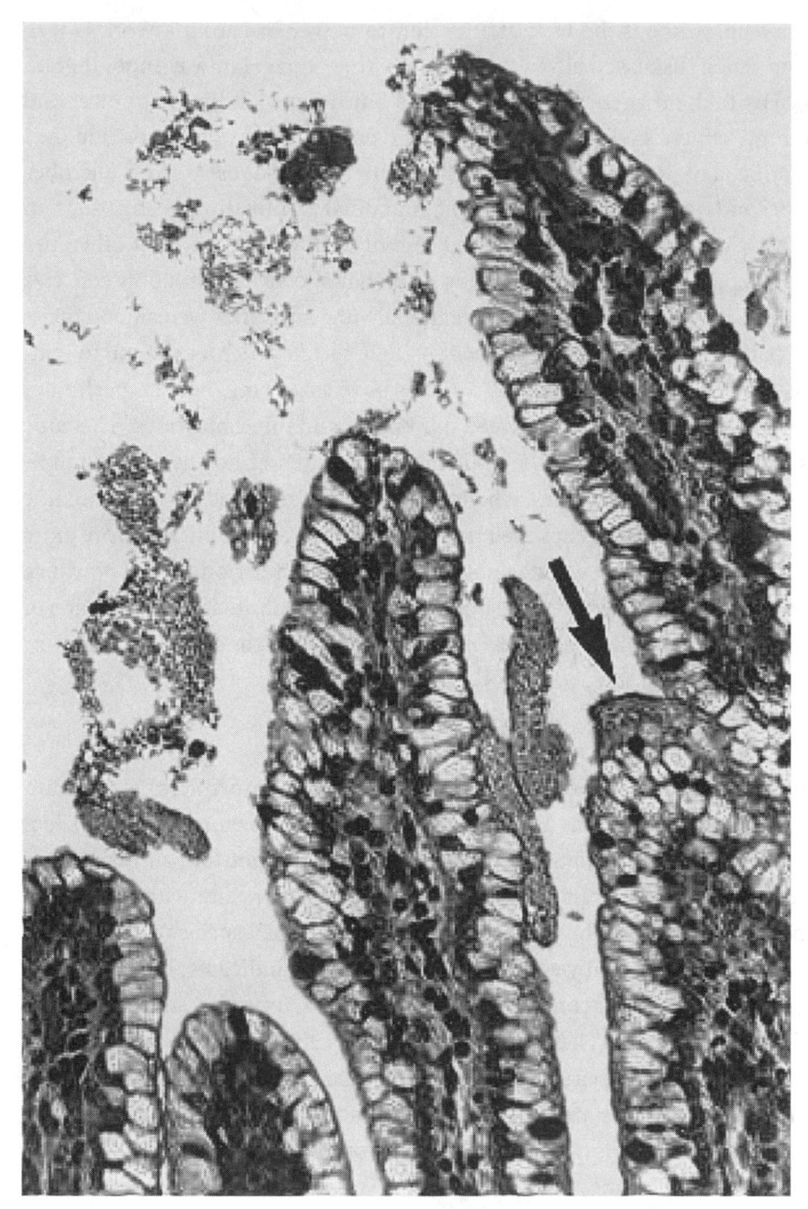

FIGURE 5.2 Biofilm (*arrow*), containing aggregating bacteria and mucus, in a piglet fed *E. coli* strain 042. This piglet did not contract diarrhea. Biofilms are sophisticated, diverse communities of microbes, and are implicated in activities as diverse as desert varnish (which thrives on rocks) to tooth decay and food poisoning. Reproduced by permission from Nataro and Kaper (1998).

Related investigations into microbial diversity has found that in contrast to the "everything is everywhere" maxim that has guided much thinking about microbiology, there are genetically distinctive microbial communities, resident only in particular places (Whitaker, Grogan, and Taylor 2003). The National Science Foundation's Microbial Observatories/Microbial Interactions and Processes supports research on microbial communities in wilderness areas as well as actively managed areas such as farms, with the long-term goal to develop a network of sites or "microbial observatories" in different habitats to study and understand microbial diversity over time and across environmental gradients.

While the idea of a microbe joining the endangered species list might seem far-fetched, medical research on microbe communities in the human gut has revealed that "indigenous" microbes, typically found living in the intestines of most people, are now disappearing due to the widespread use of antibiotics (Blaser 2005). Although the influence and role of these microbes has not been clearly determined, researchers believe they play a positive role in intestinal health. Other research suggests that each of us possesses unique microbial communities (Bik et al. 2006). These findings led one researcher to state, "It certainly means that human genomes alone are not enough to describe all the important biological variation found in humans" (Powledge 2006).[1]

CHALLENGING A "CONTAIN-AND-CONTROL" APPROACH TO HUMAN-MICROBIAL RELATIONS

Some of the scientists involved in this recent research are viewing this new information as an important call to change dominant thinking about microbes, and to appreciate the tremendous complexity and positive nature of human-microbial relationships. Ongoing problems with multiple drug-resistant bacteria, for example, has led some researchers to argue for a paradigm shift away from a war on microbes toward using improved understandings of our coevolution to "direct" microbial evolution away from virulence (IOM 2006; Levy 2002). In a 2004 proposal for a "symbiosis initiative" at the University of Wisconsin–Madison involving the funding for new microbial science faculty, two bacteriologists wrote that "within the last ten years scientists have come to recognize that microbes in the environment live as complex, social communities on all surfaces, living and nonliving. These communities . . . offer whole new realms of investigation and possible entries into microbial manipulation such as hormone signaling, nutrient exchange, and cellular differentiation." The scientists added that while microbes plague us, "conversely, and less well-recognized, microbes aid us by providing essential nutrients to plants and animals and blocking invasion by would-be pathogens. . . . Pathogenesis may be the less common relationship microbes develop with their hosts . . . rather than the dominant interaction."[2]

Problems in our food system reveal the urgency of the need to rethink our relationship with microbes. Large-scale food-poisoning incidents have resulted in high levels of concern among consumers and health agency officials. In 2006, for example, more than 200 people in twenty-six states became ill and three people died from eating bagged raw spinach contaminated by *Escherichia coli* O157:H7 bacteria. Over 183 people in twenty-one states reported salmonella poisoning from tomatoes, and 160 people in the Northeast became sick with *E. coli* poisoning from eating at a fast-food restaurant. Even without taking dramatic food-poisoning events like these into consideration, however, people frequently report poisoning from food. The Centers for Disease Control and Prevention estimates that 325,000 people are hospitalized and 5,000 people die every year because of something they ate (Mead 1999). Agricultural centralization magnifies the impact of any single contamination, and is a significant factor in the high-profile nature of many recent food scares. Eric Schlosser, who has detailed at some length the vulnerabilities of our industrial food-production system, reports that only thirteen slaughterhouses process the majority of beef consumed by 300 million Americans, and that a single Taco Bell distribution center, a possible source of *E. coli* contamination in 2006, served over 1,100 restaurants in the northeastern United States (Schlosser 2006).

In addition to the high-profile food scares, multiple-drug-resistant microbes (Brody 2006) are a growing problem, and current agricultural practice has been identified as a significant perpetrator. Research has shown that poultry farmers, laborers, and slaughterers develop antibiotic resistance merely as a result of working with poultry that are routinely fed antibiotics as a preventative (van den Bogaard et al. 2002). Other research has shown how antibiotic-fed beef was the source for "clusters" of drug-resistant urinary-tract infections in women (Ramchandani et al. 2005); how antibiotic drug resistance can be spread to workers through the dust created on confined animal-feeding operations (B. Harder 2005); and how manure from operations feeding antibiotics to animals (which contains about 90 percent of the antibiotics originally fed) can actually increase antibiotic resistance in microorganisms living in soil receiving the manure (Raloff 2005).

At the same time, however, other agricultural research has revealed how a more "ecological" concept of microbial life can be effective in fighting disease. For example, research has shown how finishing stockyard beef cows on grass (which they'd be eating naturally) helps to manage the threat of pathogenic bacteria in their manure (Diez-Gonzalez et al. 1998). Other research is investigating how spraying young chicks with a mix of microbes (which they'd get from their mother naturally) helps build their immunity and lowers death rates (Raloff 1998). In addition, numbers of studies have connected lower rates of asthma and allergies in people who are exposed to farms (Alm et al. 1999; Braun-Fahrlander et al. 1999; Kilpelainen et al. 2000; Svanes et al. 1999). "Instead of focusing research on killing microbes," authors of one recent article assert, "more and more scientists will search for ways to improve our relationship with them" (Swerdlow and Johnson 2002, 51).

THE END OF THE ANTIMICROBIAL AGE? AGRICULTURAL POLITICS OF MICROBIAL AGENCY

Might this new research herald the end of the antimicrobial age? When it comes to discussions about human-microbe relationships, it is useful to take the long view. As Latour described in his work on Pasteur (1988), the "war" on microbes has a long history. Ever since Pasteur and his colleagues connected the presence of specific microbes with incidence of disease, germ theory has supported the widespread use of antibiotics and a general view of microbes as risk-producing and potentially pathogenic. Contemporary societal commitments to the "antibacterial" and to containing and eradicating microbial bodies are deeply held. Emily Martin's work on metaphor (1987, 1994, 1998) has shown how our naming and narratives about bodies have great implications for how we approach aging, medicine, and disease. Microbes, of course, have been recognized as critical in making bread rise, beer ferment, and cheese blue. Understanding this role in culinary achievement has not, however, translated into a wider appreciation for microbial diversity and behavior and their critical role in our ongoing good health. In addition, antimicrobial policies are firmly embedded in our conventional food-production system, and the industrialized methods through which we produce meat and vegetables. The centralized nature of food distribution poses significant barriers to developing what we might call "promicrobial" practices.

Indiscriminant antibiotics are used liberally in agricultural production, for example, as a routine way to manage disease and promote growth in industrial production conditions. In contemporary livestock production, thousands of cattle or pigs, or 100,000 or more chickens, are produced in a single facility. These large confinement operations have been associated with increases in livestock disease and the emergence of new, frequently antibiotic-resistant diseases (Tilman et al. 2002). Antibiotic-resistant *Salmonella*, *Campylobacter*, and E. *coli* strains that are pathogenic to humans are increasingly common in poultry and beef produced in large-scale operations (Smith et al. 1999). According to the Union of Concerned Scientists, U.S. livestock producers annually administer almost 25 million pounds of antimicrobial drugs—the same types used in human medicine—for prophylactic use and growth promotion (Mellon, Benbrook, and Benbrook 2001). This estimate equals eight times the quantity of antimicrobials used in human medicine. Even at half of this amount, which is what has been reported by the Animal Health Institute and industry groups, routine antibiotic drug use in animal agriculture is remarkable. While the evidence against prophylactic antibiotic drug use in agriculture has been building for years, the livestock industry has dragged its feet on making changes, and U.S. policy makers have done little to require the industry to change (Denmark, a major exporter of pork, instituted a ban on antibiotic growth promoters in pigs and chickens in 1999).

Instead, food scares and fear of bacterial contamination of food has led to calls for increased microbial containment and control through scrutiny of food production, processing, and handling. Legislators have called for food safety regulations for vegetable production involving a number of "control points," including frequent testing of water and produce, the use of antimicrobial products, pasteurization, irradiation, and other techniques. The Food and Drug Administration and the U.S. Department of Agriculture both carry out food safety policy, the general direction of which has been to increase the requirements for inspection and surveillance. Fears of microbial contamination have been extended to places like fairs, farms hosting field days, and petting zoos (Steinmuller et al. 2006). The requirements of food safety regulations such as these can include everything from purchasing testing equipment, hiring the labor to perform the testing, and the mandatory use of antimicrobial products, to adding a phone line or improving a driveway to provide "proper" access for inspectors (Dreher 2003). Requirements like these favor larger, more centralized operations.

Impacts of food safety concerns include limits on practices associated with sustainable agriculture. Due to the ecological "realities," if you will, of the movements of manure, water, wind, soil, and plants, other animals besides cattle have been implicated as potential carriers of deadly *E. coli* bacteria, including wild pigs and raccoons as well as other farm animals. In Salinas Valley, California, vegetable processors have pressured growers to follow new "clean farming techniques," in the wake of the spinach scare, including clearing and mowing areas that might attract wild animals that could be carrying *E. coli* or *Salmonella*. As a result, farmers are removing grassy areas planted to reduce erosion, allow wildlife habitat, and filter chemical nutrients and pesticides that might otherwise move into the water system (Lochhead 2009). "When we plant hedgerows now, we have to use the bait stations or we lose our contracts," one farmer was quoted as saying. "Later you see birds of prey perched over the bait. They eat mice sluggish from the poison and get poisoned themselves. It kind of defeats the whole purpose of putting in the habitat" (G. Martin 2006).

Food safety concerns have also placed limits on composting by USDA certified organic farmers. The 1990 U.S. Organic Foods Production Act created a set of standards for organic production implemented in October 2002. Along with requirements for crop and livestock production, the National Organic Program (NOP) contains requirements for making organic compost. Historical and contemporary writing about organic agriculture reflects composting as an important piece of an organic farmer's toolkit, and a symbolic process in which farmers work with nature's "expert" microbes to recycle waste on-farm and to produce an enlivening, disease-combating and nourishing substance.

The regulatory language on composting in the organic federal law, however, was taken directly from practice standards developed for large municipal composting facilities by the USDA's Natural Resources Conservation Service, and is consistent with Environmental Protection Agency language for the production of sewage sludge. The regulation's requirements for turning and temperature tracking raised concerns among organic farmers because of the time, labor, and technologies involved and because of the narrow definition of the composting process. The NOP, despite repeated petitioning, has ignored several suggested practice standards that would allow for some additional flexibility in the implementation of the organic compost requirements while maintaining food safety standards. In public meetings about these issues, the NOP has made it clear that the risks of a food safety incident to a program relying so heavily on public concern over food safety are too great to allow any flexibility. According to several midwestern organic certifiers and an annual nationwide survey of organic farmers, the on-farm production and use of compost on certified organic farms has steadily declined since the implementation of the NOP (Ingram 2007a).[3]

Antimicrobial thinking can also be seen behind discussions over raw milk and raw-milk cheese (Enticott 2003; Fletcher 2005; Paxon 2008). Since 1924 the U.S. Pasteurized Milk Ordinance has banned the sale of raw milk in most states. Farmers have been able to get around the ordinance by selling their milk "for pets only" or through "cow share" programs, where consumers buy a part of the cow, but there is ongoing debate over whether even these activities should be allowed. As an Iowa farmer producing raw milk for customers once put it to me, "The state puts me in the same category as a meth lab."

Small-scale, limited-volume artisan cheese is one of the fastest-growing sectors of the dairy industry and offers economic viability to small farmers. This success has been limited by food safety regulations requiring the use of pasteurized milk in all cheeses that are not aged for sixty days at temperatures not less than

35° Fahrenheit. Thus, artisanal cheese producers in the United States are prohibited from using raw milk to make many types of traditional softer cheeses, for example, Camembert. Pasteurized milk alters (for the worse, according to experts) the taste and texture of these cheeses.[4] Much like the composting regulations designed for large municipal waste-handling facilities that restrict the practices of small organic farmers, these regulations were developed to provide safety in the context of large industrial cheese plants and place an unnecessary and restrictive burden on small-scale cheese producers (Knoll 2005). In addition, concerns have led to increasing restrictions on importing raw-milk cheeses from Europe (Fletcher 2005).

Given the current climate in U.S. food safety regulatory agencies, it is unlikely that the United States will soon allow artisan cheese makers to produce raw-milk cheeses. However, a step in that direction would be to require adequate funding for, and restructuring of, the food safety inspection service in a way that does not favor only the very large and industrial-type producers (Schlosser 2006). France offers a model of successful small and dispersed cheese producers using raw milk in cheese production in France, which relies, ironically enough, on a U.S.-developed system of hazard analysis and control points to regulate food safety in its cheese-production sector (Knoll 2005). The European approach to food safety involves a stringent set of sanitation procedures and adequate funding for inspectors to routinely examine small cheese factories and test for dangerous microbes. These things are not outside the realm of possibility in the United States—the rapid development of a nationwide network of organic certifiers and inspectors shows how quickly such programs can develop with adequate funding and economic incentive.

CONCLUSION: BREEDING FAMILIARITY, NOT CONTEMPT

The purpose here is not at all to minimize the potential threat of dangerous microbes. My goal is to encourage an appropriate political focus on the scales and the technologies that create contexts for negative human-microbial interactions, and to suggest that actively pursuing positive human-microbial interactions is a critical step for human health. It is not enough to say that some microbes are good guys. Even while recent medical and scientific scrutiny of microbial bodies recognizes more diversity, sophistication, and interaction than ever before, much of it occurs in the context of seeking ultimate control; the manipulation of one microbe through another, for example, "teaching microbes a new language," and "bioengineering" of "designer microbes" through genetic modification (Gerchman and Weiss 2004; Ferber 2004). Members of the pharmaceutical industry pay close attention to these discoveries, hoping to capitalize on the next breakthrough that will take advantage of microbial signaling technologies (Pollack 2001). Clearly, some people stand to gain a great deal from emerging microbial science and technology.

Furthermore, multiple species of bacteria once used in food production are disappearing. According to food writer Harold McGee (2004) over a dozen different microbial species once involved in making yogurt have now disappeared. Industrial yogurts, in contrast, are limited to only two or three bacteria. Traditional fermented milks such as kefir, which are no longer widely produced, contain multiple strains of bacteria, many of which have been identified as important for human health. If we do not understand the potential for positive human-microbial relationships and nurture the places in which these relationships can unfold, a status quo regulatory command-and-control approach may eradicate these opportunities for good. Beyond a sense of what doesn't work, we must articulate desired human-microbial relationships and practices, and eventually new laws.

Small, well-tended compost piles, artisan cheese-making sites, and even kitchen yogurt makers offer important examples of dispersed and diverse technologies of positive human-microbial relations. Because they are small-scale and with few barriers to entry, technologies like these are accessible to people. Even city dwellers can compost successfully, it turns out, cohabitating with numerous digesting microbes in a 300-square-foot apartment on New York's Fifth Avenue (Green 2007). In addition, these performances are manageable: unintended consequences, good or bad, have limited impact, unlike bacterial contaminations in a heavily centralized and consolidated production system.

Multiple-species settings and positive interactions like these offer an important correction to the flow of fear over contamination from invisible bodies. At root is recognition of our own hybridity, the impossibility of any fundamental separation. Equally important is recognizing the implications of the technologies that mediate our interactions with other species. A beautiful compost pile emerges from the working together of farmers, microbes, and other organisms. Delectable, ripe Reblochon is born out of the interactions of grass, cows, microbes, soil, farmer, and cheese maker. By focusing on how to nurture those performances, and by thinking about how to value them, we celebrate what we cocreate. Harold McGee (2004, 62) captures much of the emergence of these relationships where he writes eloquently about that eclectic and satisfying product of human-microbial performance, cheese: "So these are the ingredients that have generated the great diversity of our traditional cheese: hundreds of plants, from scrubland to alpine flowers; dozens of animal breeds that fed on those plants and transformed them into milk; protein-cutting enzymes from young animals and thistles; microbes recruited from meadow and cave, from the oceans, from the animals' insides and skins; and the careful observation, ingenuity, and good taste of generations of cheesemakers and cheese lovers. This remarkable heritage underlies even today's simplified industrial cheeses."

NOTES

1. The ubiquity of microbes across very different geographic contexts, the vast range of genetic diversity being discovered by new technologies, and the microbial trait of "horizontal" gene transfer whereby genes are transferred between organisms within a generation, are all challenging conventional definitions of what we mean by "species" and how they are created (Ogunseitan 2005).

2. http://wiscinfo.doit.wisc.edu/cluster/view_proposal_single_final.asp?id=147 (accessed February 2005).

3. The Organic Farming Research Foundation's farmer surveys published in 1999 and 2003 indicate that farmers regularly producing compost for their own fertility management has more than halved, from roughly 70 percent to 31 percent (http://www.ofrf.org/publications/survey/).

4. Even in France issues of food safety are threatening traditional Camembert-making techniques by outlawing the use of raw milk. Any change will not occur without a fight, however. As a mayor of a town in the Camembert region put it, "A camembert not made out of raw milk is like making love without sex" (Bitterman 2008).

CHRIS DUVALL

6 FERRICRETE, FORESTS, AND TEMPORAL SCALE IN THE PRODUCTION OF COLONIAL SCIENCE IN AFRICA

When observed over a human lifetime, the mantle of the Earth is as rigid as steel, but over thousands and millions of years it acts as a highly viscous fluid.
—*P. C. England, P. Molnar, and F. M. Richter,*
"Kelvin, Perry and the Age of the Earth"

In this chapter, I explore the role of apparently inert entities—things that appear unchanging from a human perspective—in the production of scientific knowledge. I focus specifically on ferricrete, or indurated, iron-rich soil that is functionally an inert rock to humans although it changes continuously, if minutely, through time. My concern with ferricrete arises from my interest in relationships between vegetation, soils, and land degradation, and how scientific knowledge of these relationships has been produced in African contexts. This chapter contributes to broader theoretical concerns by examining how perceptions of temporal scale affect how humans and nonhumans interact to produce scientific knowledge.

I must first consider how spatiotemporal scale shapes human perceptions of the biophysical environment. A core theme in political ecology has been the importance of scale in configuring human-environment relationships. Indeed, a principle in natural and physical sciences is that biophysical entities exist at specific spatial and temporal scales (Meyer et al. 1992; Wilbanks and Kates 1999). These scales range from minute to vast, from a human perspective: humans occupy intermediate scales whether considered as individuals, societies, cultures, or species (figure 6.1). Relative scale is crucial when assessing relationships between entities (Gibson, Ostrom, and Ahn 2000; Meentemeyer 1989). In particular, as the epigraph indicates, the human capacity to observe change in other entities decreases as the scale of these entities increases relative to humans.

Many natural entities change so slowly relative to human experience that they appear inert, as simply the biophysical context of human activities. However, inertness is a matter of perspective. If I am concerned only about events that happen over decades, many entities—like large, old trees—can appear inert; if I am concerned with change over centuries or millennia, I can understand that many apparently inert entities change over time. Individual perspectives on environ-

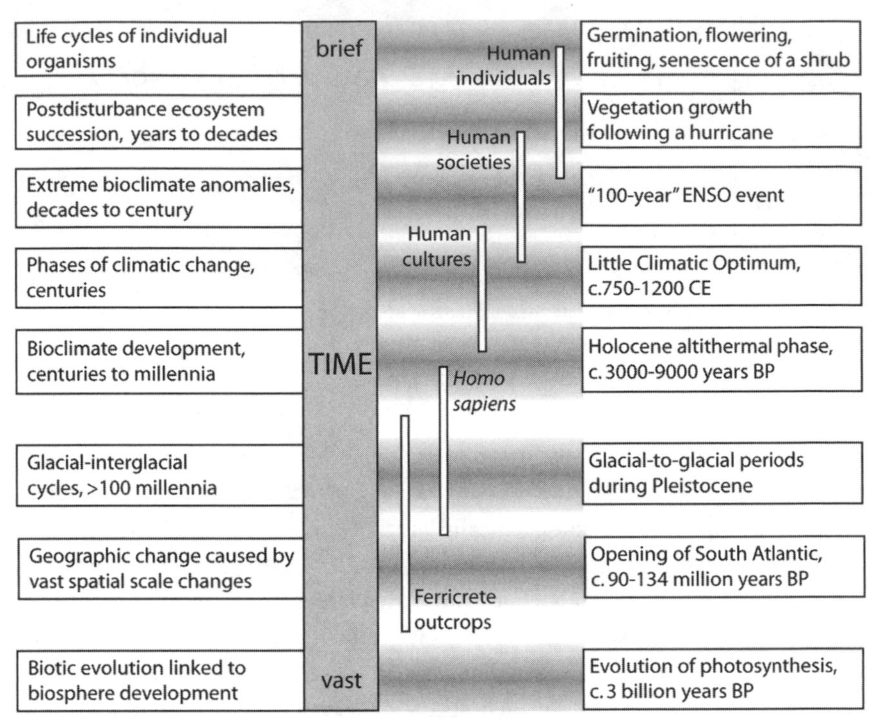

FIGURE 6.1 Conceptual representation of relative scales of human existence. Various natural processes are described at the left, and are arranged schematically along the time scale from brief to vast, relative to human existence. Examples of events characteristic of these processes are given at the right. All natural entities could be arrayed in the center portion of the figure alongside human individuals, societies, cultures, *Homo sapiens*, and ferricrete outcrops to represent relative temporal scale.

mental change reflect the epistemological traditions of specific knowledge cultures. Different scientific cultures have adopted different time frames of reference in studying particular objects, helping to create markedly different perspectives on singular entities, including ferricrete. Ferricrete occurs widely in the semiarid tropics, particularly western Africa, my focal region. Ferricrete horizons in soils form over, and persist through, tens to hundreds of thousands of years (Aleva 1994; Maignien 1966; Tardy and Roquin 1992). Yet ferricrete is also important in relatively short-term interactions. For instance, it constrains the growth and distribution of plants; where ferricrete lies near the ground surface, barren hardpans may exist (figure 6.2). Different scientific cultures have focused on limited portions of this range of timescales, reflecting discipline-specific interpretations of meaningful landscape change.

In this chapter, I show that epistemological differences between scientific

FIGURE 6.2 Ferricrete hardpan in southwestern Mali. The barrenness of ferricrete hardpans is accentuated—as in this photo—during dry seasons after annual grasses that grow on the ferricrete during rainy seasons have been burnt, and when juxtaposed with neighboring woodland vegetation.

cultures have led to multiple scientific knowledges of ferricrete in Africa. This is important because political-ecological analyses of scientific knowledge cultures in Africa have focused on how political-economic contexts influence science, and less on the methods and scope of inquiry scientists have used in producing knowledge. For instance, two groups of researchers have shown that ferricrete hardpans represent anthropogenic environmental degradation in science-policy discourse in Guinea (André, Pestaña, and Rossi 2003; Thomas, Van Dusen Lewis, and Dorsey 2003). These authors explicitly build upon political-ecological literature that challenges environmental degradation discourse in Africa (i.e., Fairhead and Leach 1996; Leach and Mearns 1996), because they are concerned that this discourse sustains unfair and ineffective resource management policies, and misrepresents "local realities" of resource use and landscape change (André, Pestaña, and Rossi 2003; Thomas, Van Dusen Lewis, and Dorsey 2003). Their analyses are accurate and contribute significantly to our understanding of how conflicting, state-versus-local, material interests affect environmental management in Guinea. Yet by overlooking epistemological differences between knowledge cultures, political-ecological analyses do little to clarify how the interweaving of scientific method, cultural change, *and* political-economic interests influence how knowledge is, or is not, produced in scientific institutions. In contrast, I de-emphasize, without ignoring, the political-economic contexts of ferricrete science in order to explore how different perspectives on relevant temporal scale can lead to different portrayals of human and especially nonhuman entities. In making my argument, I build upon theory in two specific research areas within political ecology and STS: colonial science, and nonhuman agency in knowledge production. I discuss these areas of research next.

NONHUMAN ROLES IN KNOWLEDGE PRODUCTION

Analyses that focus on the political-economic contexts in which scientists participate privilege human-to-human interactions and neglect how humans *and* nonhumans interact to produce scientific knowledge. Actor-network theory (ANT) provides an analytical approach that encompasses the range of interactions that produce knowledge. Actor networks exist as relations between entities, particularly between humans and the nonhuman objects they work with to create knowledge (Latour 2005). Scientists create scientific actor networks by assigning roles to other entities that circumscribe the relations these entities are supposed to have with other entities (Callon and Law 1989). If nonhuman entities do not conform to the roles assigned to them, they gain agency by altering relationships within the network (e.g., Whatmore and Thorne 2000).

Soil can have agency within scientific actor networks (Wynne 1992). The processes of soil formation cause soils to behave in particular ways, which may not conform to roles scientists assign them. Although ferricrete is technically a type of soil, it is as inert as rock. Apparently inert natural entities do not "behave" in the normal sense of the word, but only exhibit states of being. For instance, ferricrete is often barren, heats when exposed to the sun, holds itself together tightly when struck, and exists in certain locations but not others. The roles scientists have assigned ferricrete in actor networks have always accommodated at least some of its observable states of being. As scientific networks extend over time and space, scientists gain observational knowledge and can assign natural entities increasingly circumscribed roles in networks.

While observations are important in scientific knowledge production, scientific knowledge also arises from frameworks of symbolic meaning (Forsyth 2003; Proctor 1998). These two domains of scientific knowledge—observational/material and symbolic/metaphoric—are distinct yet complementary (Pratt 1995). Humans project symbolic meaning upon natural entities in order to materialize *ideas* about the biophysical environment. Symbolic projection is necessary to understand relationships between individual observations and fundamental ontological structures (Bhaskar 1979), but also problematic because it can serve to materialize politicized opinions about specific people and places (Forsyth 2003; Wolch and Emel 1998). Symbolic projection and material observation both contribute, though not necessarily equally, to the roles scientists assign entities in scientific networks. Thus, both symbolic and real behavior of nonhuman entities can convey agency to them because agency is developed only in relationship to human actors. Importantly, the symbolic meaning projected upon an entity can be in contradiction with its real behavior, although symbolic meaning is not necessarily revised based on material observations.

Scientific actor networks include nonhuman entities, but only human actors—scientists—propose and bring these networks into being. Nonhumans can shape science only if humans create roles for them that are both materially *and* symbolically meaningful. In scientific knowledge production, meaning is performative, arising through sustained interaction between scientists and natural entities, and thus potentially changing. This interaction is, from a human standpoint, observation. Observation is necessary for nonhuman agency to exist in scientific networks (Clark and Murdoch 1997; Whatmore and Thorne 2000; Wynne 1992), although the importance of observation has been obscured because scholars have focused on natural entities that exhibit clear activity because they exist at temporal scales similar to humans. For instance, Callon's scallops (1986; Callon and Law 1989) became agents because scientists observed them to be absent on the collection devices to which scientists proposed the mollusks would attach. It is easy to ignore, overlook, or generally underobserve nonhuman entities that do not actively and obviously interact with us; such underobservation can help dilute with symbolic projection the roles these apparently inert entities are assigned in scientific networks.

COLONIAL SCIENTIFIC NETWORKS

The way observation creates agency for nonhuman entities is crucial to understanding knowledge production in colonial scientific networks. Unlike the networks most STS scholars have studied—well-funded, technocentric networks that generate and rely on rich databases of observations—colonial scientific networks emerged (and persist) in times and places where scientific observations and funding were limited.

While some political ecologists have recognized that limited observational knowledge has helped create inaccurate views of African environments, scientific observation has been represented as merely an aspect of the "poor" science that has served to support the political-economic interests of states (e.g., Bassett and Crummey 2003; Fairhead and Leach 1996; Leach and Mearns 1996). From this perspective, research methodologies are a means to an end predetermined by the concerns of (colonial) states. True, many colonial-era scientists were formal functionaries of colonial institutions. A common example is André Aubréville, who, while serving in the forestry service of French West Africa, popularized the term *desertification* and argued that much of Africa had experienced anthropogenic deforestation (e.g., Aubréville 1949). By assigning other humans and nonhumans specific roles in scientific networks and conducting research in ways that preserved the proposed interactions, colonial scientists promoted state control of natural resources. Science and social order were co-produced (cf. Jasanoff 2004b) by colonial functionaries for the benefit of colonial states. Thus, although the colonial period ended decades ago in most of Africa, colonial scientific networks persist because they have remained an effective means of promoting state authority (e.g., Duvall 2003; Swift 1996).

This interpretation of colonial science is accurate but too simplistic. Knowledge production in colonial science resulted from the interplay of political-economic interests with discipline-specific epistemological constraints, within a broader cultural context in which "tropical" otherness was being constructed and refined. Research methods were not merely means to an end in colonial science, but represented increasingly ossified perspectives on meaningful spatiotemporal scale. Many colonial scientists were leading scholars within their disciplines, not just colonial functionaries. For instance, Aubréville was one of the foremost tropical botanists during the 1940s to 1970s, and many of his works remain important to botanists, biogeographers, and plant taxonomists in Africa, South America, Asia, and Oceania (Aymonim 1983; Leroy 1983). Different scientific disciplines have distinct ways of thinking about, and thus observing, vast-scale natural entities (Gibson, Ostrom, and Ahn 2000; Meentemeyer 1989); many colonial scientists helped construct the epistemological frameworks that continue to shape basic and applied research in several disciplines. The roles colonial scientists assigned nonhuman entities reflected discipline-specific perspectives on what constitutes meaningful environmental objects and processes (Turner 1993), as much as any desire to produce social order and state authority. Additionally, these framings were often intertwined with cultural constructions of "the tropics" that emphasized perceived differences with the world(s) of European and North American scientists (cf. Arnold 2006). Science and place were co-produced in ways that often helped stagnate perspectives on the rate and causes of "tropical" environmental change. And, of course, colonial political economies also shaped knowledge production, by privileging certain ways of knowing as well as certain conclusions about environmental change. The persistence of colonial scientific networks is an outcome of historic inertia in the epistemological and cultural bases of scientific institutions, as much as the continuity of state dominance.

The case of ferricrete in Africa particularly reveals linkages between epistemology, nonhuman agency in scientific actor networks, and colonial scientific knowledge production, because scientists have developed strong and divergent opinions about ferricrete's meaning(s) in different disciplines. In the remainder of this chapter, I provide background on ferricrete as a biophysical entity, then excavate how two differing perspectives on ferricrete have emerged in its scientific history in Africa.

FERRICRETE LANDSCAPES

In the semiarid tropics, distinctive landscapes exist where ferricrete occurs at the ground surface (figure 6.2). In discussing ferricrete, attention must be given to terminology, because it is one of many soil features associated with the term "laterite," whose use has been confused and confusing for nearly two hundred years (Maignien 1966; Tardy 1992). "Ferricrete" has had several synonyms, including "ironstone," the French *grés ferrugineux*, and some usages of "laterite." Most synonyms have clear meanings, though that of "laterite" is not; where sources use "laterite" but appear to mean "ferricrete," I quote the text and provide its apparent meaning in brackets. Multiple terms have also signified the process of ferricrete formation. In the Fulfulde language spoken in Guinea, *bowal* means "land cover associated with surficial ferricrete" (André, Pestaña, and Rossi 2003; Thomas, Van Dusen Lewis, and Dorsey 2003). From *bowal* is derived "bowalization," a hypothesized process of land degradation that has concerned some scientists in Africa since the early 1900s. Bowalization is an increased rate of soil formation following (anthropogenic) deforestation, leading to the creation or lateral expansion of ferricrete horizons (Aubréville 1947). This hypothesized process reflects a particular understanding of the soil-forming process often called laterization, the enrichment of soil in certain clays and metal oxides (hereafter sesquioxides). However, "laterization" and similar terms, such as "lateritization" and "latosolization," have sometimes served as synonyms for "bowalization." It is important to distinguish these concepts because "bowalization" implies human agency, whereas "laterization"—in its current, technical sense—does not. Where sources use "laterization" or similar terms, I quote the text and provide its apparent meaning in brackets.[1]

Ferricrete forms if sesquioxide-enriched soil horizons desiccate (Tardy and Roquin 1992). Sesquioxide accumulation is primarily a function of landscape position: low-lying areas receive material transported by water from adjacent uplands (Maignien 1966; Ollier 1991). Erosion-resistant ferricrete hardpans form over tens to hundreds of millennia (Goudie 1973; Trendall 1962), and are exposed if overlying soil erodes away (Ollier 1991). Vegetation on surficial ferricrete may consist only of sparse annual grasses and stunted shrubs growing in cracks (Goudie 1973). Ferricrete retains almost no moisture; vegetation on hardpans desiccates quickly and burns early in dry seasons. Bare ferricrete rapidly absorbs and reradiates solar heat, so that during dry seasons hardpans are extremely hot and barren (figure 6.2).

The broadly tropical distribution and marked barrenness of ferricrete outcrops have been salient to many scientists. Ferricrete is unusual because it is hard like rock but is distributed more like a soil, being associated with certain climates and topographies rather than geologies. Its durability has led people to use it as a building material throughout recorded history, and colonial and postcolonial states have used ferricrete widely in road construction.

SOIL, PEOPLE, FORESTS, AND FERRICRETE IN THE TROPICAL WORLD

Scientists have studied ferricrete for nearly two centuries. During the nineteenth century, scientists observed fairly obvious features, such as how ferricrete responds when hammered, and its different textures and shapes. I will concentrate on the twentieth century, when scientists studied the physical appearance of ferricrete outcrops, changes in the shape of ferricrete masses over time, and the distribution of ferricrete, within soil profiles, across landscapes and regions, and globally.

Ferricrete's global distribution has contributed to a subtext in ferricrete science that portrays it as characteristic of the "tropical" world. Early on, ferricrete served to represent the unhealthiness that is part of the social construction of "the tropics" (cf. Arnold 2006). For instance, in 1903, the British geologist Thomas Holland, working in India, hypothesized that an unidentified organism is responsible for chemically altering bedrock to form "laterite [various substances including ferricrete]." "If this fancy turns out to be well founded we must add laterisation [sesquioxide accumulation] to the long list of tropical disease[s], against which even the very rocks are not safe" (Holland 1903, 63). Though evocative for researchers working in Africa (e.g., Chautard 1905, 137), representation of ferricrete as an environmental illness did not explicitly emerge in the African context until the 1940s, when it was described as "this leprosy of soils and vegetation" (Aubréville 1947, 340–41). In this latter case, calling upon ferricrete's barrenness served to emphasize the otherness of tropical (African) places and people relative to the (European) scientists who, as part of the colonial project, sought to impose European resource management practices in Africa. Long cast as part of the "tropical" world, ferricrete's scientific history reflects many of the epistemological changes that have shaped scientific knowledge of Africa's natural resources.

In the early twentieth century, scientists assigned ferricrete roles in two growing bodies of knowledge, which had distinctly different time frames of reference. First, within the field of African plant ecology, scientists assigned ferricrete a role based mainly on observation of the barrenness of ferricrete outcrops. The time frame of reference in this field was decades to centuries, because participating scientists were concerned mainly with the ecological interactions that affected the distribution of particular plant species. Botanists and foresters found that "trees are dwarfed, gnarled, and widely scattered" on "stiff laterite formations [ferricrete hardpans]" (Thompson 1911, 136–37), and that the grass-dominated vegetation of hardpans burns nearly every dry season (Chevalier 1909; Garrett 1892; Pobéguin 1906). This is the core of the first role scientists assigned to ferricrete: it forces vegetation to adopt certain characteristics by impeding root growth and resisting water absorption. Other scientists working in the field of African geomorphology assigned it a role based mainly on the observation that ferricrete occupies predictable positions in landscapes—the tops of plateaus. These scientists were concerned with the long-term evolution of landscapes, and thus adopted a referential time frame of dozens to hundreds of millennia. Geologists, physical geographers, and soil scientists found that ferricrete outcrops do not change appreciably over decades by either growing or disintegrating (Falconer 1911; Lacroix 1913; Shantz and Marbut 1923). This is the core of the second role assigned to ferricrete: it creates specific landforms by resisting erosion. The cores of these two roles have persisted, but less central features have emerged and declined for various reasons—the complex set of reasons that have contributed to the emergence and persistence of colonial scientific networks.

FERRICRETE IN PLANT ECOLOGY

Political-economic interests have shaped the role proposed for ferricrete in the broad field of African plant ecology through much of the twentieth century. In the early 1900s, many botanists and foresters were concerned that African farmers and pastoralists had degraded the continent's vegetation and soils, and that the degradation was continuing at an alarming rate because of inappropriate, indigenous resource management practices (e.g., Bassett and Crummey 2003; Fairhead and Leach 1996). These ideas dominated natural resource policies in colonial-era Africa, because of the close and often symbiotic relationship that existed between scientists and policy makers—the production and application of knowledge is difficult to separate.

Yet this political-economic relationship was but one interconnection in a scientific network that also included African farmers and pastoralists, and various biophysical entities. The quality of these other interconnections had less to do with political-economic factors, and more to do with the cultural and epistemological perspectives embedded within the relevant scientific institutions. Scientists proposed roles for each entity in this network, and revised these roles if observations showed that an actor did not perform its role. This network has centered on processes of environmental degradation occurring over decades to centuries, and particularly the deforestation most of Africa was supposed to have experienced very recently—whether the date of observation was 1908, 1949, or 1992 (Fairhead and Leach 1998). As shorthand, I will call this network "anthropogenic deforestation." Significantly, within the anthropogenic deforestation network African farmers and pastoralists were assigned a static role that was primarily symbolic. These people, like "the tropics," were represented as being without history, timelessly managing resources in anachronistic ways that European observers considered unsustainable in contemporary social and demographic settings (Benjaminsen and Berge 2004). For European observers, this timelessness was materially evident in the "primitiveness" of African resource practices (Aubréville 1949; Chevalier 1906; Schonck de Goldfiem 1936). By simply farming or keeping cattle in ways foreign to European observers, Africans passively conformed to the role imposed upon them in the anthropogenic deforestation network. Simply by being different from their observers, African people confirmed pre-established notions European scientists brought about both the "tropics" and environmental degradation in Africa.

Since African farmers and pastoralists were assigned a static role within the anthropogenic deforestation network, the network's interactions could be sustained only if the natural resources that were supposedly degraded and degrading exhibited change. Environmental objects were seen as evidence of recent environmental change, so that the formative processes for these features were necessarily conceived as sufficiently rapid for observation, even if such observations had not yet been recorded (Duvall 2003; Fairhead and Leach 1996). The sparseness of observations strengthened the symbolic meanings attributed to many environmental objects, eventually despite contradictory knowledge of material reality. In the early 1900s, botanists came to believe that areas with surficial ferricrete had been recently deforested, so that surficial ferricrete represented the supposedly destructive character of African land-management practices (Chevalier 1909; Pobéguin 1906; Schonck de Goldfiem 1936). For instance, Chevalier (1928, 85) argued that African use of fire in agriculture destroys forest vegetation and "brings about greater and greater soil degradation. It becomes hard at the surface . . . its *latérisation* [transformation to ferricrete] is accelerated." This representation was based on observations that ferricrete outcrops are generally barren and burn early in dry seasons, but the representation of ferricrete as damaged soil also materialized the idea that European resource management techniques must be imposed to halt the environmental degradation supposedly caused by indigenous land management (Hubert 1920; Renner 1926; Stebbing 1935). Importantly, scientists in this network had no explanation grounded in material knowledge of soil formation of how humans could create ferricrete through deforestation—the existence of a link was explicitly the "impression" (Aubréville 1947, 355) they had in viewing sparsely vegetated hardpans. In its role as "damaged soil," ferricrete symbolized associations scientists sought to impose between other actors—particularly farmers, wildfires, vegetation, and soils—to help sustain the science-policy network of anthropogenic deforestation.

FERRICRETE IN GEOMORPHOLOGY

Political-economic concerns and cultural othering have not strongly shaped ferricrete's role in the field of geomorphology, which has instead reflected changing ideas about rates of tropical soil formation. Geologists, soil scientists, and physical geographers assigned ferricrete a role in a second, distinct scientific network, centered on landscape evolution. Although the ferricrete geomorphology network coalesced only in the 1950s, several aspects of ferricrete's role in this network were proposed earlier in the century. In the 1930s, ferricrete symbolized ancient climate change for scientists in this network, materializing belief that Africa's climate has changed between humid and arid in geological time. Such representation of ferricrete arose from various observations: ferricrete outcrops showed no growth during dozens of years of measurement; sesquioxide-enriched soils mainly occur today in humid tropical areas; sesquioxide-enriched soils may harden if air-dried (J. Campbell 1917; de Chételat 1938; Falconer 1911; Lacroix 1913; Shantz and Marbut 1923); and vegetated sand dunes occur in semiarid areas, while many dry lakes occur in the Sahara (Chudeau 1921; Falconer 1911; Gautier 1928; Lahache 1907; Mission Tilho 1910–11). The timescale of reference for these scientists was tens to hundreds of millennia, far greater periods of time than considered by their colleagues in botany, forestry, and cognate disciplines. As a result, humans did not have an expressed role in this network (Lahache 1907, 150).

During the 1930s, the decades-to-centuries timescale became increasingly interesting to soil scientists, physical geographers, and geologists working in Africa, who observed the American dust bowl from afar (Swift 1996). Many feared that human activities could amplify natural climate variation. Awareness of climatic change during the Quaternary and Holocene was increasing across tropical

Africa following research at Lake Chad and dry lakes in the Sahara, and in the Great Lakes region (Gautier 1932; Scaëtta 1937; Tilho 1928). In this context, a handful of scientists began to question the rate of tropical soil formation, asking whether this may be much faster than previously accepted due to seasonally intense precipitation and the high available solar energy (Aufrère 1932; Scaëtta 1938, 1941). No new observations of soil or ferricrete formation underpinned this thinking, which was instead an inchoate (and ultimately aborted) change in the epistemology of ferricrete geomorphology. Knowledge of tropical soil formation (e.g., Hardy 1935; Shantz and Marbut 1923) also did not contradict the new *idea* that soils might evolve rapidly. Furthermore, the Italian agronomist Hélios Scaëtta (1938, 1224; 1941) recognized that this idea meant that drastic changes in vegetation characteristics—including human "destruction of the vegetation"—could replicate the effects of climate change by increasing rainfall intensity and exposure to solar energy at the ground surface, and thus potentially lead to rapid ferricrete formation. Scaëtta simply recognized that humans—whether African or European—could be significant if soil formation occurs rapidly, and did not express judgment about indigenous agriculture.

THE BRIEF MERGING OF SCIENTIFIC NETWORKS

By suggesting a new way of thinking about tropical soil formation that associated soil, ferricrete, forests, and people over a decades-to-centuries timescale, Scaëtta allowed scientists in the anthropogenic deforestation network to assign a more precise role to ferricrete, and helped briefly bring together this network and the emerging ferricrete geomorphology network. In 1947 the French botanist André Aubréville coined the term "bowalization" and, citing only Scaëtta, explained his belief that surficial ferricrete was forming at an increasing pace because Africans had deforested and "abusively cultivated" so many areas (1947, 353). Elsewhere, Aubréville (1949, 332) argued that bowalization was a primary cause of desertification throughout Africa. Botanists, foresters, and some soil scientists immediately accepted Aubréville's view of deforestation and ferricrete formation. Many attendees of the First Conference on African Soils (held in Goma, Belgian Congo, in 1948) cited his 1947 paper. However, none presented observations of rapid ferricrete formation following forest clearing, but just confirmed their beliefs that this was true. The (fearful) idea of rapid soil change was clearly widespread, and Aubréville's presentation of ferricrete hardpans as evidence for such change materialized the idea for these authors. Most citations of Aubréville (1947) were linked to value judgments about African agriculture, and supported the conference's primary theme: the imperative of replacing these "primitive practices" with "modern techniques" (Jungers 1949, 28). By identifying a link between supposedly degraded landforms, processes of degradation, and African agriculture, the conferees provided the colonial administrations that they represented a clear political imperative to advance their views and methods of natural resource management.

Soon after the 1948 conference, however, these two networks diverged. Through the 1930s and 1940s, many soil scientists still believed that soil development requires much time even if intense tropical weathering might marginally speed the process (Hardy 1935). Most believed that the formation of "laterite [various substances including ferricrete]" required "long quiet periods in earth history" (E. Harder 1952, 35). Ferricrete remained symbolic of ancient climate change and active in long-term landscape evolution. Yet the ferricrete geomorphology network coalesced in the 1950s only after detailed, new observations that ferricrete's distribution contradicted its proposed role as a product of human deforestation. Ferricrete exists in predictable forms and locations in soil profiles, landscapes, and regions, and it is not associated in space or time with deforestation (e.g., Alexander and Cady 1962; de Swardt 1964; Maignien 1958; Trendall 1962). Their observations allowed these scientists to understand that ferricrete changes significantly only over a millenarian timescale.

The evolving land surface on which the ferricrete geomorphology network centered existed meaningfully only over periods of tens to hundreds of millennia. Thus, the only other entities ferricrete interacts with in this network are also apparently inert objects like landforms, soil horizons, and planation surfaces. Humans are not explicitly assigned a role in this network, because it exists at timescales far beyond human experience. At the Second Conference on African Soils (held in 1954 in Léopoldville, Belgian Congo), the Belgian soil scientist Jules d'Hoore directly rejected the idea that humans are significant in ferricrete formation (1954b), concluding elsewhere that "the nefarious influence of man, and especially of primitive man . . . seems . . . to have been somewhat exaggerated" (1954a, 90). The networks of anthropogenic deforestation and ferricrete geomorphology ceased to overlap: Aubréville's 1947 paper was not cited in any papers presented at the 1954 conference, nor subsequently by scientists in ferricrete geomorphology.

In contrast, some botanists, foresters, and others have maintained the role Aubréville assigned ferricrete, because this continues to materialize belief that much of Africa has recently experienced major environmental degradation. As the formal colonial period closed, ferricrete symbolized the supposed destructiveness of African agriculture and the need to rationalize resource use through the application of exogenous knowledge (e.g., Jaeger 1956; Pitot 1953; Sillans 1958). More recent sources that cite bowalization as an active form of land degradation also link this process to perceived shortcomings of African agriculture, and argue for improved conservation through increased state control of resources (Kimball 2005; Ministère de l'Environnement et de l'Eau 2000; Schmitz, Fall, and Rouchiche 1996; Timofeyev, Dembele, and Danioko 1988). The political-economic interests of states continue to encourage representation of ferricrete as degraded soil (André, Pestaña, and Rossi 2003; Thomas, Van Dusen Lewis, and Dorsey 2003).

As important, however, has been epistemological inertia in how associations between people, plants, soil, and ferricrete are conceptualized within the anthropogenic deforestation network. Recent authors who have confirmed the role assigned to ferricrete in this network rely only on Aubréville (1947), whether directly or indirectly through citation chains. As a result, scientists in this network have maintained the decades-to-centuries frame of reference for understanding soil formation, and continue to see ferricrete as an entity produced in the recent (but unspecified) past. Ferricrete continues to have a primarily symbolic role in the anthropogenic deforestation network, representing recent, unobserved environmental change. This metaphoric meaning is supported circumstantially by observation of the barrenness of ferricrete outcrops, which seems to belie the potential for vegetation development in the semiarid tropics. The epistemological framework underpinning the anthropogenic deforestation network has inhibited detailed observations of ferricrete landscapes (cf. Turner 1993). Belief in recent, rapid environmental change has prevented scientists in this network from assigning ferricrete a role that reflects the vast temporal scale of its development.

CONCLUSION

Our interactions with and understanding of apparently inert natural entities feel very different from those we have with entities that exist at our same spatiotemporal scale. The challenge for scholars is to integrate the entire range of natural entities—from obviously active to apparently inert—in understanding how nonhumans may contribute to knowledge production. The case of ferricrete shows that it is human observation that enables nonhumans to have agency in actor networks. Advances in observational knowledge represent moments of association between scientists and the entities they observe, in which the activity or the state of being of an observed entity does not conform to a role previously imposed upon it. Thus, d'Hoore's observations (1954a, 1954b) that ferricrete does not necessarily exist in deforested places advanced knowledge because it was an instance in which ferricrete's distribution contradicted the role scientists had assigned it in the anthropogenic deforestation network.

Why did d'Hoore's observations not affect associations in the anthropogenic deforestation network, but instead helped coalesce the ferricrete geomorphology network? The anthropogenic deforestation network has persisted in part because it continues to support state-dominated natural resource management policies. Yet this network has persisted not only for political-economic reasons. The epistemologies of component disciplines have often allowed scientists to underobserve the apparently inert entities enrolled in the network. The decades-to-centuries time frame proposed for many forms of environmental degradation in Africa has meant that many entities—from ferricrete hardpans to forest patches—cannot actively interact with humans, because these entities exist at far greater timescales. Apparent inactivity allows scientists to project symbolic meaning upon natural entities with a low likelihood that these entities will contradict the roles assigned them. Thus, Latour's view (2005, 185) that "scale is the actor's own achievement" seems correct. The received wisdom of colonial scientific networks has persisted in part because their epistemological frameworks have inhibited knowledge production, helping to perpetuate particular ideas about African environments.

NOTE

1. A technical disclaimer: this chapter is concerned only with autochthonous ferricrete formation from illuvium, and no other processes, including erosion that exhumes buried ferricrete and ferricrete formation from sesquioxide-rich colluvium.

Part 2

PAUL NADASDY

APPLICATION OF ENVIRONMENTAL KNOWLEDGE

THE POLITICS OF CONSTRUCTING SOCIETY/NATURE

All of the case studies in part 1 begin their explorations of environmental politics by focusing on the activities of scientific experts and their efforts to produce scientific knowledge about the environment. Each, however, demonstrates the inadequacy of focusing on scientific knowledge producers *in isolation*. The authors of part 1 were all compelled to cxamine, in varying degrees, how the knowledge produced by scientific experts is taken up and applied in situations apparently far removed from its production. This was not simply out of curiosity to see what eventually becomes of scientific knowledge once it is released into the world. Rather, all these authors show that scientific knowledge producers—as well as their knowledge-producing activities—are all intimately caught up in the politics surrounding the circulation and application of the knowledge they produce. Indeed, they demonstrate clearly that the production of environmental knowledge by scientific experts is merely one aspect of a much larger process. To examine the production of environmental knowledge without at the same time taking into account its application and circulation is to generate an impoverished view of knowledge production itself.

The chapters in this second part of the book approach the study of environmental knowledge and practice from a slightly different angle. All the authors of part 2 take as their starting point an examination of the politics at a particular site (or sites) where environmental managers are attempting to apply environmental knowledge. Just as the authors in part 1 realized they could not examine scientific knowledge producers in isolation, however, all the authors in this part quickly expand their analyses beyond a narrow focus on the politics of knowledge application in a particular place. They demonstrate clearly that one cannot hope to understand the politics of environmental knowledge application without taking into account how that knowledge is produced and the circumstances of its circulation. Indeed, all the chapters in this part help to disabuse us of the oft-held notion (in resource management, if not critical academic, circles) that knowledge somehow exists in a pure state, fully formed, merely awaiting its application in particular places. The application of environmental knowledge, as a social—and deeply political—activity, simply cannot be distinguished from the social activity that constitutes its production and circulation. Nevertheless, a focus on the politics surrounding the application (or attempted application) of environmental knowledge at particular sites can provide an important window into the broader sociopolitical processes in which such efforts are embedded.

This part's focus on the politics surrounding specific attempts by scientific managers to apply environmental knowledge in particular situations is an especially powerful way of drawing attention to the existence of multiple, sometimes incommensurable, ways of knowing the environment. In so doing, the chapters all highlight problems associated with the notion of *expertise*. In my analysis of wildlife management in Canada's Yukon Territory, for example, I describe how wildlife managers' knowledge of animals and their ideas about how to manage them, structured as they are by metaphors of agricultural production, differ profoundly from those of aboriginal hunters. Mara Goldman, in her analysis of efforts to introduce wildlife corridors in East Africa, shows that a similar dynamic exists between Maasai pastoralists and conservation biologists, each of whom have very different understandings of wildlife and their relation to humans and livestock. In a similar manner, Karl Zimmerer shows how scientific models of water resources used to manage water in the Cochabamba region of Bolivia differed substantially from local farmers' understandings of water use in the same region. Because the environmental sciences are for the most part applied sciences, the knowledge they produce is necessarily knowledge *for* a particular use. This renders the distinction between the production and application of knowledge particularly problematic and highlights the political nature of knowledge production. As Peter Vandergeest and Nancy Peluso put it in their analysis of scientific forestry in postcolonial Southeast Asia, "forestry is concerned not just with scientific knowledge about these naturalized objects [forests], but also with *transforming* . . . ecologies through forest management . . . to achieve social, economic, or political ends." Thus, the knowledge-production process—especially (or at least most obviously) in the applied ecological sciences—is necessarily shaped by those social, economic, or political agendas; and ideas about proper resource use (or, indeed, about what constitutes a resource in the first place) are themselves the products of particular sociocultural histories. Zimmerer shows especially clearly how intended use structured the collection of information for scientific water-management models. It should not be surprising, then, if aboriginal hunters in the Yukon, Maasai pastoralists in Kenya, Runa farmers in Cochabamba, and Orang Asli in newly designated Malaysian forest reserves produce very different kinds of knowledge about the nature of wildlife, water, and forests than do scientific resource managers employed by the state, who have very different ideas about how these resources should be used. Thus, each of the chapters in this part exposes quite clearly the flaw in quantitative understandings of expertise (i.e., that some people are experts because they *know more* than others) and makes the case, instead, for an approach to expertise that can accommodate multiple ways of knowing. Each chapter attends to the production of multiple forms of knowledge, the different weights assigned to each by various participants in the management processes under examination, and the political consequences of these epistemological encounters.

The point, however, is not simply that there exist multiple ways of knowing a single given environment. "The environment" itself cannot be known apart from our engagement with it. "Nature," in this sense, is always socially constructed. Such constructions are always only partial and never politically neutral: one person's soil erosion is another person's soil fertility; one person's home is, for another, a collection of commodities. Much work within political ecology has striven to uncover the multiple perceptions at play in defining nature and the various politics involved in creating, managing, and policing it. While less overtly focused on the politics of nature, work within STS has drawn attention to the nuanced and intricate ways in which nature and society are connected and interact, and how that leads to very specific (and political) outcomes. STS has also highlighted the various ways categories, names, and systems are created, maintained, and manipulated to produce "nature" and "society" as separate realms, and to "manage" nature effectively. The chapters in this part all draw from the insights of both political ecology and STS to explore the co-production of nature and society. The authors are particularly interested in how the application of scientific knowledge leads to the construction of certain kinds of natures (at the expense of others).

Goldman shows how wildlife corridors, theoretical constructs emerging from conservation biology, come to appear to their proponents as natural entities, components of ecosystems in urgent need of protection; and she notes that the naturalization of the corridor concept serves to limit debate and foreclose consideration of other possible conservation strategies. In their chapter, Vandergeest and Peluso show that in an important sense forests in Southeast Asia (as opposed to "jungles") did not precede scientific forestry, but were constituted by it. Indeed, they argue that the production of forests and forestry have been part and parcel of postcolonial counterinsurgency and nation-building projects throughout the region. Zimmerer, too, describes the spatial naturalization of particular watershed models, which ignored social complexity and heterogeneous land use to "create" areas of water abundance and scarcity. These new "natures" were then used to justify water projects, sometimes on a massive scale. For my part, I show how the agricultural metaphor creates a world in which wildlife is croplike, thus naturalizing assumptions about human ownership and control over animals and authorizing particular management interventions while precluding others.

In all four cases, the application of scientific knowledge entails the production and imposition of one nature/society and the erasure of others. Thus, environmental management must be viewed as an inherently political process. Contributors to this part are all interested in the power and politics of such processes. How are "natural" categories/zones created and what does it mean for the natural and political landscape in question? How are such categories contested, within science, with local people, in the lab and on the ground? These questions are all deeply political and engage simultaneously with a politics of knowledge (Whose knowledge counts when, where, and why?); a politics of nature (Which natures are to be protected, by whom, for whom, and in which ways?); and a politics of economy (Is nature a resource or commodity, in abundance or scarce, and for whom?).

In each of the case studies in this part, a critical mechanism for the imposition of one particular society/nature at the expense of others is the state's assertion of and control over property rights. Goldman shows us that the establishment of wildlife corridors, however "natural" they may appear to their proponents, is wholly dependent upon the Kenyan state's appropriation of Maasai land—a fact not at all lost on the Maasai who vigorously opposed their creation on precisely these grounds. I show that wildlife managers' conception of animals and how to manage them are intimately bound up with the state's assertion of property rights over wildlife. Similarly, Vandergeest and Peluso show that the assertion of state jurisdiction over forests (often accompanied by the forcible relocation of populations within them) was a prerequisite for the creation of state forests and the practice of scientific forestry in Southeast Asia. State jurisdiction over water is essential for the scientific management of that resource as well, as Zimmerer shows us, though sometimes there are limits to what the state can do. This became abundantly clear when the Bolivian state attempted to privatize water, setting off the Cochabamba Water War.

The application of scientific knowledge necessarily also implies its circulation. Although scientific knowledge, methods, and standardized models for management do circulate globally, it is difficult to generalize about how those knowledge artifacts will be received, understood, and acted upon in particular places. By analyzing specific efforts to apply scientific knowledge in particular situations, contributors to this part provide an important perspective on how those knowledge artifacts are packaged and transported. They all show that a simple diffusion model of science is inadequate for understanding how scientific artifacts travel. In every case of attempted knowledge application, different people attach different meanings and understandings to the knowledge, methods, and models to be applied. The authors in this part all use analytical tools they draw from STS (e.g., boundary objects, standardized packages, metaphorical overlap) to make sense of the multiple meanings that get attached to seemingly agreed-upon but in fact essentially contested scientific terms and models, the illusion of agreement that can result from the use of these seemingly standardized terms and models, and the ways in which people negotiate among and struggle over these various meanings. Vandergeest and Peluso show that although the legal and institutional framework of scientific forest management was transferred to Southeast Asia from abroad (mostly during the colonial period), these were quickly transformed, given new meanings, and put to new uses in pursuit of regional and national agendas, agendas that transformed forests and the people living in them on a massive scale. Zimmerer describes the application of standard water-management models and how these were received and reacted to by local people, which in turn led to a revision of the management models and sometimes open rebellion. Goldman and I show that the terms "wildlife corridor" and "wildlife management," respectively, can and do mean different things to different people. Use of such essentially contested terms can facilitate communication and serve to connect scientists, local people, activists, state and donor agencies, and policy makers, but it can also lead to misunderstandings and catalyze resistance to scientific management, as occurred in both the Kenyan and Yukon cases. Taken together, the cases in this part show that we must not view efforts to scientifically manage particular environments as a passive process of importing and applying knowledge artifacts that were produced elsewhere. Rather, scientific management always entails a negotiation and refashioning of meanings, a struggle not only over control of resources, but also over the very nature of those resources. In other words, it is always an effort construct a particular form of society/nature.

PAUL NADASDY

7 "WE DON'T *HARVEST* ANIMALS; WE *KILL* THEM"

AGRICULTURAL METAPHORS AND THE POLITICS OF WILDLIFE MANAGEMENT IN THE YUKON

"We don't *harvest* animals. When a bear gets one of us it doesn't *harvest* us. It kills us. And we kill them too. We don't *harvest* animals; we *kill* them." With these words, Mary Jane Johnson, a citizen of Kluane First Nation (KFN),[1] urged participants at a 1996 salmon-management meeting to stop using the term *harvest* when referring to human hunting and fishing. Like other subarctic Athapaskans, the Kluane people of Canada's southwest Yukon are hunters, not farmers. Mary Jane objected to the term *harvest* because it implies ownership and control; people harvest crops that they themselves plant, so they *expect* to harvest their whole crop every year (e.g., you reap what you sow). She felt this to be a very dangerous mindset when it comes to wildlife and urged everyone to use words like *hunt* and *kill*, rather than *harvest*. For the rest of the day, everyone attempted to follow Mary Jane's advice, but this proved no easy task. Throughout the day, everyone—First Nation people included—kept on catching and correcting themselves. Nor did Mary Jane's intervention have a lasting effect; at subsequent meetings everyone lapsed back into old habits.

Avoiding agricultural terms in wildlife management meetings is so difficult because such terms are ubiquitous in the field of wildlife management. Not only does the verb *harvest* regularly stand in for less obviously metaphorical terms such as *shoot* or *kill*, but biologists also refer to the overall number of animals killed by hunters each year as the *annual harvest* (e.g., the "annual salmon harvest"). In addition, a host of other agricultural terms, such as *crop*, *cull*, *husbandry*, *seed*, *brood stock*, *yield*, *fallow*, and *transplant*, are all commonplace and essential in the field. This suggests that Euro-American wildlife managers view what they do as somehow analogous to the production of crops and domesticated animals. While most Euro-Americans are hardly even conscious of the metaphorical nature of such language, it can be jarring to Yukon First Nation people who have never been farmers. Indeed, as we have seen, they are sometimes quite explicit in their rejection of the agricultural metaphors of wildlife management.[2] Instead, as I will show, they tend to subscribe to a different conception of wildlife management altogether. Given the centrality of agricultural metaphors in the discourse and practice of scientific wildlife management, it is worth attending to the effects the metaphor has had upon the discipline and to the political consequences of its application in a cross-cultural context like the Yukon.

In contrast to early studies that viewed metaphor as little more than rhetorical flourish, more recent scholars have argued that metaphors actually play a critical generative and structuring role in the production of all human thought and practice (Black 1962; Lakoff and Johnson 1980; Ortony 1979). Building on these insights, others have explored how metaphors structure scientific knowledge and practice in particular (Hesse 1980; E. Martin 1987, 1991; Merchant 1980; Todes 1989). Few scholars, however, have analyzed the dynamics of cross-cultural interactions in which various participants subscribe to different, perhaps even incompatible, metaphorical systems. Anthropologist Colin Scott (1996) paved the way for such an analysis by pointing out that the root metaphors structuring human-animal relations among Cree hunters in northern Quebec are radically different from those underlying scientific understandings of wildlife. I build on Scott's insights by examining two root metaphors that structure wildlife management in the Yukon: WILDLIFE MANAGEMENT IS AGRICULTURE, subscribed to by biologists, and WILDLIFE MANAGEMENT IS THE MAINTENANCE OF SOCIAL RELATIONS, which structures many First Nation people's concept of wildlife management. Although these two root metaphors are incompatible with one another in many ways, we shall see that there is, in fact, enough overlap between them that wildlife biologists and First Nation people can talk to one another and sometimes even agree on particular management strategies. This can create the illusion that the two parties share a mutual understanding of the situation—and even of wildlife management more generally—thus obscuring the fact that their perspectives are in fact profoundly different, structured as they are by different root metaphors.

Lakoff and Johnson (1980, 97) describe metaphorical overlap of this sort as resulting from "complex coherences between metaphors," and they note that "a metaphor works when it satisfies a purpose, namely, understanding an aspect of the concept. When two metaphors successfully satisfy two purposes, then overlaps in the purposes will correspond to overlaps in the metaphors. Such overlaps, we claim, can be characterized in terms of shared metaphorical entailments and the cross-metaphorical correspondences established by them."

Key to their argument is the observation that all metaphorical understandings are always partial (10). Just as a metaphor highlights (and perhaps even creates) certain aspects of a concept (e.g., the agriculture-like aspects of wildlife management), it simultaneously obscures or effaces others. Thus, it makes little sense to ask whether a metaphor is "true," or even which of two partially overlapping metaphors is "more true." Instead, each metaphor, providing it "works" in Lakoff and Johnson's sense, merely creates and/or highlights different aspects of the concept and has its own set of metaphorical entailments.[3] For the purposes of this chapter, the important work is to understand how each root metaphor structures thought and practice by creating or highlighting certain aspects of wildlife management and obscuring others. In a cross-cultural context, where there is no agreement about the root metaphors themselves, the question of power becomes especially critical. We need to ask which metaphors are accepted (at least implicitly) and acted upon, in what circumstances, and by whom. And, in such situations, what is the role of partial metaphorical overlap of the sort described above? Does it foster cooperation between parties *despite* their different understandings of and approaches to wildlife management? Or does it serve to mask unequal power relations between them?

With these questions in mind, I begin by examining the agricultural metaphor and its role in structuring the knowledge and practice of scientific wildlife management. I then turn to the social-relations metaphor that underlies First Nation notions of wildlife management; I examine its entailments and show that they partially overlap those of the agricultural metaphor. Finally, I examine the political consequences of all this by analyzing a particular case of wildlife management in the Yukon.

METAPHOR ONE: WILDLIFE MANAGEMENT IS AGRICULTURE

THE AGRICULTURAL METAPHOR AND THE ORIGINS OF WILDLIFE MANAGEMENT

WILDLIFE MANAGEMENT IS AGRICULTURE is a clear example of what Schön (1979) refers to as a "generative metaphor" in the field of wildlife management. It played a crucial role in the development of scientific wildlife management and has continued to structure knowledge and practice in the field ever since. One useful way of getting at the metaphor's generative role in the field is through a close reading of wildlife management textbooks. Thomas Kuhn (1970, 136–38) argued that scientific textbooks are products of the prevailing scientific paradigm—so much so that they need to be rewritten with every paradigm shift. As such, textbooks enmesh students in the prevailing paradigm, inculcating in them a particular way of seeing the world. Emily Martin (1987, 1991) has demonstrated the value of analyzing medical textbooks as a means for getting at the metaphorical underpinnings of biomedical approaches to human reproduction. In a similar way, I trace the use of the agricultural metaphor in wildlife management textbooks to show how it has structured the field. In the process, I pay particular attention to one set of the metaphor's entailments: namely, those objected to by

Mary Jane Johnson in this chapter's opening vignette: human ownership and control over land and animals.[4]

In 1933 Aldo Leopold, widely regarded as the father of scientific wildlife management in North America, wrote *Game Management*, the first textbook in the newly emerging field. In it he defined wildlife management in explicitly agricultural terms: "Game management is the art of making land produce sustained annual crops of wild game for recreational use. Its nature is best understood by comparing it with the other land-cropping arts. . . . Like the other agricultural arts, game management produces a crop by controlling the environmental factors which hold down the natural increase, or productivity, of the seed stock" (Leopold 1933, 3). It is clear that Leopold did not view the relationship between agriculture and wildlife management as metaphorical at all. For him, wildlife management was simply one of many different forms of agriculture, and the common thread linking all the various "land-cropping arts" is the exercise of human control. Indeed, for Leopold, control was the essence of wildlife management: "any practice may be considered as entitled to be called game management if it controls one or more factors with a view to maintaining or enhancing the yield" (4), and he structured *Game Management* around what he saw as the emerging discipline's primary objective: to control the factors that affect the production of wildlife.[5]

Leopold noted that "in game, as in forestry and agriculture, there is no sharp line between the practice which merely exploits a natural supply, and the practice which harvests a crop produced by management" (1933, 4). He then went on to describe a spectrum of activities characterized by differing amounts of human control over animals: from no control at all (exploitation of a natural supply), through game management, to game farming, and then to the raising of livestock, which, he noted, are "incapable of survival in the wild state, much less of perpetuating themselves as wild populations" (4). He argued that by the beginning of the twentieth century, however, there no longer existed any "natural supplies" of wild animals on the continent—that, in effect, the low-control end of the exploitation spectrum by then existed only in theory. As he put it: "Every head of wild life still alive in this country is already artificialized, in that its existence is conditioned by economic forces" (21). In other words, wildlife populations had by then become every bit as incapable of perpetuating themselves as livestock populations. As a result, he concluded that "hunting is the harvesting of a man-made crop, which would soon cease to exist if somebody somewhere had not, intentionally or unintentionally, come to nature's aid in its production" (210). Like it or not, then, humans were *already* "in control" of wildlife populations; and as a result, he argued, we have an ethical responsibility to exercise that control so as to ensure the continued existence of these already artificialized populations (19). For Leopold, this meant applying the scientific methods that had by that time already been adopted in other forms of agriculture.[6]

WILDLIFE MANAGEMENT, AGRICULTURE, AND PROPERTY

Underlying the notion of control that is so critical to Leopold's agriculturally inspired vision of wildlife management is the concept of ownership. Given the importance of property rights in the practice of other forms of agriculture, this should hardly come as a surprise. Leopold himself noted quite explicitly that the kind of control required for wildlife management "can only be accomplished by the landowner" (21), and his account of the evolution of wildlife management (5–18, 408–10) deals in large part with the evolution of property rights, in both animals themselves as well as in land; and, indeed, the different sorts of management interventions he described depend variously upon ownership rights in animals and land. As Leopold (409) was aware, there is a long tradition in Euro-American property law, which holds that no one owns wild animals until they are killed or captured (e.g., Locke 1947). As some scholars have noted (Asch 1989; McCandless 1985), however, the common-law principle that wildlife are unowned has in effect meant that they are owned by the state, and this is born out—explicitly or implicitly—in the language of North American wildlife statutes and aboriginal land claims agreements. Indeed, state ownership of wildlife was essential to Leopold's vision of management because it justifies the state's restriction of hunting through the imposition of hunting seasons, bag limits, and the like (Leopold 1933, 409).

As Leopold noted, however, the state's ownership of wildlife—and hence its ability to exercise the control necessary to manage wildlife populations—is not absolute. The existence of private property limits the state's ability to manage wildlife, particularly when it comes to implementing strategies that protect or improve wildlife habitat on private lands (409). Indeed, Leopold adopts the Lockean position that it is only landowners (whether individuals or the state) who would ever be willing to invest in such "improvements" to the land. Because private landowners are in a unique position to "produce" wildlife crops cheaply, one of the primary goals of state-employed wildlife managers should be convince them to do so (398).

Thus, the agricultural metaphor has two entailments that are critical for the purposes of this chapter. First, humans can and should control animal populations and the factors that affect them so as to maximize the wildlife crop for human benefit. Second, humans (collectively or individually) own both the wildlife crop and the land on which it is grown. This ownership not only justifies human control over and manipulation of wildlife; it also provides the necessary incentive for humans to "improve" animal habitat and so increase production.

PARADIGM SHIFT: TRANSCENDING THE AGRICULTURAL METAPHOR?

We have now seen that the agricultural metaphor played a critical role in shaping Leopold's conception of wildlife management. It is reasonable to ask, however, whether the metaphor continues to shape the discipline eighty years later. And in fact, in their 2003 textbook, *Wildlife Ecology and Management*, Eric Bolen and William Robinson acknowledge the importance of the agricultural metaphor to the history of their discipline but suggest that it has now been transcended: "Since its inception, wildlife management developed in the context of an 'agricultural paradigm' that stressed the production and harvest of a few species. . . . However, a shift to an 'ecological paradigm' marks the continuing maturation of wildlife management. . . . With this shift comes the broader recognition and appreciation for the conservation of all species and the functions and services of healthy ecosystems" (188).

This leads them to propose a definition of wildlife management that differs significantly from Leopold's: "wildlife management is the application of ecological knowledge to populations of vertebrate animals and their plant and animal associates in a manner that strikes a balance between the needs of those populations and the needs of people" (2), and they link this ecologically based concept of wildlife management to the rise of *adaptive management*, an approach that treats the making of management policy as part of the scientific enterprise, in which "policy options are tested repeatedly, with each option thereafter given more or less weight based on comparisons of predicted and actual results" (187).

There have certainly been important changes in wildlife management since Leopold wrote his seminal text. It is by no means clear, however, that the paradigm shift described by Bolen and Robinson really frees wildlife management from the agricultural metaphor that spawned it. Although these authors do make considerably less use of agricultural language than did Leopold, some agricultural terms, such as *harvest* and *yield*, have become so deeply entrenched in the field that they, too, use them quite unselfconsciously. For instance, it is in a section entitled "Harvest and Hunting" that they introduce the basic principles of adaptive management, supposedly one of the centerpieces of the new paradigm. Their continued focus on managing *harvest* levels (adaptively or otherwise) suggests that the new paradigm has not fully transcended the discipline's generative metaphor.

Indeed, although other recent wildlife management textbooks also invoke the notion of a "paradigm shift" in the field, there is considerable variation in how they characterize this shift. In his 2002 textbook, *Introduction to Wildlife Management*, for example, Paul Krausman describes the shift from the "traditional" paradigm of managing for multiple use and sustained yield of resources, in which "habitat protection for a diversity of wildlife is often viewed as a constraint to realizing management goals," to the new paradigm of *ecosystem management*, "which is concerned with preserving the complex interactions that drive ecological processes and individual species" (354–55). Significantly, he never identifies the agricultural metaphor as a problematic part of the old paradigm. Indeed, his entire text—itself presumably a product of the new paradigm—is redolent with agricultural language. What is more, Krausman explicitly adopts Leopold's definition of wildlife management, cites *Game Management* extensively, and even adopts the general structure of Leopold's original text.

Despite the degree to which contemporary wildlife management texts vary in their use of agricultural terms, all of them continue to take for granted the entailments of the agricultural metaphor: ownership and control. Whether the focus is on managing for a few game species or overall ecosystem health, wildlife managers seek to control human activities, predator and competitor species, and habitats. Indeed, many of the approaches espoused by contemporary textbooks are quite consistent with—if not foreshadowed by—Leopold himself. For instance, although *Game Management* predated the elaboration of the ecosystem concept, Leopold was well aware of the complex interrelationships among species, and he explicitly argued that failure to take such complexity into account in designing management strategies could lead to disaster (Leopold 1949, 129–33). Nor would Leopold likely have been very surprised by the "invention" of adaptive management, since he himself advocated an experimental approach to policy making.[7] Even if we allow that the ecological insights informing contemporary wildlife management bring something new to the field, this does not necessarily translate into radically new practices, since wildlife managers continue to be constrained by the political economy of resource extraction, which is what gave rise to the need for scientific wildlife management in the first place (Nadasdy 2007a).

METAPHOR TWO: WILDLIFE MANAGEMENT IS THE MAINTENANCE OF SOCIAL RELATIONS

I turn now to Yukon First Nation people's understanding of wildlife management. As we saw, First Nation people are sometimes explicit in their rejection of the agricultural metaphor. Instead, they subscribe to a very different set of beliefs about the nature of human-animal relations and so have very different ideas about what it might mean to "manage" wildlife.

The key to understanding this alternate vision of wildlife management is First Nation hunting, which is—and has long been—the fundamental organizing principle governing social relations not only among Yukon First Nation people, but also between them and animals. Like other northern hunting peoples, many Yukon First Nation people conceive of hunting as a reciprocal social relationship between humans and animals. In this view, fish and animals are, to use Irving Hallowell's phrase (1960), "other-than-human persons" who give themselves to hunters in exchange for the hunters' performance of certain ritual practices. These practices vary across the North—as well as by animal—but they commonly include the observance of food taboos, ritual feasts, and prescribed methods for disposing of animal remains, as well as injunctions against overhunting and talking badly about, or playing with, animals. Hunters who fail to live up to the obligations that they incur through hunting risk the possibility of retribution from the animals, who may withhold themselves in the future or visit sickness or even death upon the hunter and his or her family.[8]

It should hardly be surprising, then, that northern First Nation people generally do not view wildlife management as a process whereby humans tend to an annual "crop" of wild animals. Rather, they see it as being about the management of social relations—among humans as well as between humans and animals. I suggest, then, that we might view First Nation people's concept of wildlife management as structured by the root metaphor WILDLIFE MANAGEMENT IS THE MAINTENANCE OF SOCIAL RELATIONS.[9]

Yukon First Nation people do see themselves as intimately involved in the ongoing production of animal populations. By observing hunting rituals and maintaining proper social relations, First Nation hunters play a critical role in the continuous renewal of animal populations (e.g., proper disposal of animal remains is critical if the animals are to be reborn). Although First Nation hunters may resort to trickery and even coercion in their conduct of social relations with animals (Nadasdy 2007b), they do not generally subscribe to the view that they *control* animals, who may abandon their human partners at any time if they do not live up to the social obligations incurred through hunting. Indeed, many Yukon First Nation people find the assumption of control inherent in the agricultural metaphor absurd, perhaps even offensive to the animals. At wildlife management meetings one KFN hunter regularly objected to use of the term *wildlife management* itself. Humans cannot "manage" wildlife populations, he said. Animals manage themselves; they make their own decisions about when to reproduce and where to go, decisions that are quite independent of any human desires. Wildlife management, he said, is not about managing animals; it is about managing people (see also Natcher, Davis, and Hickey 2005).

The agricultural metaphor's entailment of ownership is equally incompatible with Yukon First Nation people's ideas about appropriate human-animal relations. A case in point is the concept of aboriginal hunting rights, viewed in Canadian law as a form of usufructuary property right possessed by people of aboriginal descent. However useful this notion has been to Yukon First Nation people in defending their interests against the Canadian government, it is nevertheless largely incompatible with the social-relations metaphor. Recall that animals give themselves to hunters only if they prove themselves worthy; if hunters don't live up to their responsibilities, then the animals will cease giving themselves. In such a world, one does not possess a "right" to kill animals simply because one was born of First Nation parents. In fact, the expectation accompanying someone's aboriginal right to hunt is contrary to the proper attitude of respect and humility that a hunter must have if he or she is to be successful in the hunt (Nadasdy 2003, 242–47).

Much like the field of wildlife management, Yukon First Nation people's beliefs and practices regarding animals have changed significantly over the past eighty years, largely in response to contact with Euro-Americans. Some of the old practices regarding disposal of animal remains, ritual feasting, and food taboos have fallen into disuse.[10] These changes, however, have not altered the social-relations metaphor. Many Kluane people continue to conceive of animals as intelligent, social, and spiritually powerful other-than-human persons with whom they are engaged in ongoing social relations; and they view these relations as vital to their own survival. For them, wildlife management is about the careful maintenance of these relations, not the production of a crop they own and seek to control.

Having laid out these two quite different concepts of wildlife management, structured as they are by different root metaphors, it is important to note that there is some common ground between them. It makes sense from both perspectives, for example, to talk about managing for some optimum-size animal population, though the explanations for *why* animal populations change and the kinds of "management" interventions that are effective and permissible differ markedly from one perspective to the other. Like scientific wildlife managers, Yukon First Nation people subscribe to the notion that human hunting affects the size of animal populations. Thus, on the surface it seems that their beliefs and practices are at least partially in line with those of wildlife managers, but this is something of an illusion. Though Kluane people and biologists agree that overhunting is "bad," they differ fundamentally in their understandings of *why* it is bad. As far as at least some Kluane people are concerned, overhunting and meat wastage affect the animals not so much because they reduce the number of animals in the total population, as wildlife biologists would have it, but because they offend the animals, making it less likely that they will give themselves to hunters in the future. Indeed, many Yukon First Nation people view the prohibition against overhunting as simply one facet of their complex relationship with animals, a relationship that also entails many other responsibilities not so easily classified by biologists as "wildlife management," including prohibitions against talking badly about animals, "playing" with them, laughing at them, and so on (Nadasdy 2003, 83–94; 2005).

The illusion of agreement created by the partial overlap of metaphorical systems can lead to misunderstandings and political difficulties, as we shall see in the case of the Aishihik wolf kill, to which I now turn.

PARTIAL METAPHORICAL OVERLAP AND THE CASE OF THE AISHIHIK WOLF KILL

In 1993 the Yukon Territorial Government began a five-year moose and caribou recovery program in the Kluane and Aishihik areas. One of the main components of the program involved reducing wolf populations, principally by shooting them from helicopters. In the press and everyday conversation in the territory, this program came to be called the Aishihik "wolf kill"—much to the consternation of wildlife biologists, who preferred the program's official moniker, Aishihik-Kluane Caribou Recovery Program, and who were quick to point out that the program involved more than just the killing of wolves. Although wolf populations in the Yukon are not endangered, the species' status as a charismatic large predator combined with its near-extirpation from much of the rest of the continent—largely as a result of just this kind of predator control—virtually assures that any such program will be extremely controversial. This one was no exception; it drew protesters from as far away as Germany.

Despite all the controversy, however, territorial wildlife managers and local First Nation governments—all of which came out in support of the program—appeared to be in complete agreement with one another and to share an understanding of the program. As I will show, however, this was an illusion, a product of partial metaphorical overlap of the sort discussed above. This illusion was threatened a few years into the program when wildlife biologists introduced a new management technique: sterilization. In this section, I first discuss scientific wildlife managers' and then First Nation people's perspectives on the Aishihik wolf kill, with attention to the different understandings that underlay their respective positions in support of the program. I then examine the wolf-sterilization initiative and the disagreements it sparked between biologists and First Nations.

Predator control has a long history in Euro-American society (long predating the rise of scientific wildlife management). Leopold himself devoted a chapter of his seminal textbook to predator control, identifying it as one of the standard techniques in the wildlife manager's toolbox. As a predator-control program, then, the Aishihik wolf kill was simply the latest application of a time-honored wildlife management technique. Because of its controversial nature, however, government biologists were under intense pressure to design an ecologically sound program that would not threaten Yukon wolf populations in the long term.[11] Taking an adaptive approach, they framed the wolf kill as part of a broader "large mammal recovery" experiment designed to take into account not only valued ungulates, but also the health of the wider ecosystem (Hayes 1992). Not surprisingly, the agricultural metaphor shaped the way they conceived of the wolf kill. Participants at an October 1992 meeting to design the program,[12] for example, regularly used the term *harvest* to refer to human hunting, and they referred to the wolf kill as a *cull* (YFWB 1992). The experimental-design document that resulted from the meeting also uses *harvest* (though it generally refers to wolf population *reduction*, rather than a *cull*). That biologists designing the program also subscribed to the underlying assumptions of ownership and control entailed by the metaphor will become apparent shortly.

Although the wolf kill was designed and carried out almost exclusively by territorial biologists, the political climate in the territory at the time was such that the territorial government could not have proceeded without the support of local First Nation governments. As noted above, these did support the program, but it is also worth noting that individual First Nation citizens were far from uniform in their support (Nadasdy 2005; YFWB 1992, 8). Although they held a wide range of views on the program, most had quite ambivalent feelings about it. On one hand, many agreed that because of a decline in trapping in recent decades, the wolf population had grown too large and was having a devastating impact on the ungulate populations upon which First Nations depend. On the other hand, many were troubled by the methods the government used and would have preferred bounties or trapping subsidies to a capital-intensive helicopter hunt (Buckley 1993a; *Yukon News* 1993).

First Nation support for the wolf kill puzzled—and angered—some of its non–First Nation critics, who had difficulty squaring First Nation people's assertions about the sacredness of wolves with their support for the wolf kill (Nadasdy 2005). Assertions about the "sacredness" of wolves are largely based on First Nation people's view of wolves as other-than-human persons who possess particularly potent spiritual power (McClellan 1975, 135–39).[13] Yet Yukon First Nation people have *always* killed wolves, and their "spiritual relationship" with these animals—as with all animals—is *predicated on* their need to kill them (Nadasdy 2003, chap. 2; 2007b).[14] Indeed, the view that a belief in the sacredness of wolves is incompatible with the practice of killing them is the result of an inappropriate projection of Christian conceptions of "the sacred" onto First Nation practice. As one First Nation supporter of the wolf kill stressed to me, wolves *are* sacred, but when a native person shoots a wolf, "it's not the same as shooting St. Peter." It is not a sin to shoot a wolf. On the contrary, killing animals has *always* been a critical aspect of the long-term social relationship between human- and animal-people in the Yukon. As long as First Nation hunters behave properly toward wolves and their remains, killing them is a perfectly respectful—indeed, sacred—act.

Thus, the social-relations metaphor is not necessarily incompatible with predator-control programs such as the Aishihik wolf kill, and territorial wildlife managers and First Nations did indeed appear to be in substantial agreement about the program—at least for the first few years. Both groups wanted to reduce the number of wolves in the region to limit their impact on ungulate populations. The alliance between them, however, was a fragile one, because it was based on the partial overlap of different metaphorical systems. This apparent agreement broke down in the third year of the program, when wildlife biologists began sterilizing wolves in the study area.[15]

PERSPECTIVES ON THE STERILIZATION OF AISHIHIK WOLVES

Yukon biologists hoped that the sterilization of wolves, because it is nonlethal, would make the program more palatable to the general public. They also had reason to believe that sterilization would improve the program's long-term effectiveness and affordability (Ralston 1996b; Spence 1998, 4–5). Shooting wolves by helicopter is a costly endeavor that produces only short-term results.[16] Every year new wolves moved into the Aishihik region to fill the vacuum left by the previous year's wolf kill. So to reduce the wolf population to the target level each year, wildlife managers had to kill off these incomers as well as deal with any natural increase in the population. Sterilization offered a potential solution to this problem. The plan was to sterilize the dominant breeding pair of each wolf pack in the area and then kill off the rest of the pack. The remaining wolf pairs would kill significantly fewer ungulates; and, because wolves are territorial, the hope was that they would defend their territories against incursion by outside packs.[17] If this worked as hoped, biologists would have to kill fewer total wolves over the course of the program.[18]

First Nation governments in the region decided not to oppose the sterilization program, because they did not want to jeopardize the wolf kill as a whole. Nearly every First Nation person with whom I spoke, however, expressed to me his or her profound discomfort with it.[19] This discomfort was fully in keeping with their concept of wildlife management. In contrast to the killing of wolves, there is no way to reconcile forced sterilization with the ongoing social relationship that First Nation people seek to maintain with wolves as other-than-human persons. There is simply no precedent for such behavior; indeed, it explicitly violates many of the beliefs about the respectful treatment of animals discussed above. As a result, many viewed the program as dangerous because it exposed the entire community to the possibility of retribution for failing to live up to their social obligations to wolf-people.

By contrast, sterilization is predicated on notions of ownership and control that have their roots in an agricultural worldview. Indeed, sterilization is a standard practice in the management of livestock populations. That Burwash residents explicitly viewed sterilization as an agricultural practice was evident in one man's objection to wolf sterilization at a meeting in Burwash Landing on April 18, 1996. He interrupted a formal update on the sterilization project by asking the biologist who was making the presentation if she knew what happens to a horse after it is sterilized. When she replied negatively, he said simply, "It's good for nothing" (AKCRSC 1996a). Horses are the only livestock common to the region; in fact, many First Nation people have become skilled with them as a result of their experience working as horse wranglers and hunting guides for outfitters, and they're quite used to the idea of gelding horses. So it is not surprising that they viewed wolf sterilization through the lens of their experience with horses. But horses, relatively recent imports closely associated with the arrival of Euro-Americans, are anomalous (as are dogs, for different reasons); Yukon First Nation people relate to them differently than they do to most other animals. In this case, what is appropriate for horses is not appropriate for wolves. As a result, biologists' sterilization of wolves caused even First Nation people who had been in support of the wolf kill to begin distancing themselves from the program, which they now saw as a threat to the web of social relations upon which they depend.

CONCLUSION

In this chapter, I have shown that notions of wildlife management in the Yukon are structured by two different root metaphors. These metaphors grow out of very different views of the world, and they lead to very different conceptions of wildlife management, its goals, and appropriate methods. We also saw, however, that there is considerable overlap between these metaphors, such that First Nation people and wildlife biologists sometimes appear to share common interests in and understandings of particular wildlife management situations. Partial metaphorical overlap of this sort can facilitate cooperation between First Nation people and wildlife biologists. This is what we saw in the first three years of the Aishihik wolf kill, when First Nation people and wildlife biologists, despite their different conceptions of wildlife management, seemed to agree on what needed to be done to restore ungulate populations. Cooperation of this sort, however, is extremely fragile, based as it is on the accidental overlap of different metaphorical systems. The illusion of agreement engendered by such overlap is liable to disintegrate as soon as members of either party can no longer make sense of one another's' actions through their own metaphorical lens—as occurred when biologists began sterilizing Aishihik wolves.

The real danger, however, is not that the illusion of common interests and understandings might disintegrate, since the resulting recognition of radical difference might actually lead participants to try to understand and appreciate one another's different positions. Rather, the danger is that, despite legitimate disagreement among the parties resulting from their different worldviews, the illusion of common understanding will *not* disintegrate. If the parties do not appreciate the extent to which their understandings actually differ from one another—*despite* the apparent convergence of interests due to metaphorical overlap—then they are likely to also misunderstand any disagreements that subsequently arise between them as resulting from ignorance or bad faith rather than from legitimate cultural differences. To those who take for granted the agricultural metaphor, for example, First Nation objections to sterilization may appear to be nonsensical, the result of childlike anthropomorphizing. As a result, they become easy to dismiss. By masking cultural difference in this way, the illusion of agreement engendered by metaphorical overlap also serves to obscure the politics underlying environmental knowledge production and application in management settings.

Wildlife biologists and First Nation people are equally likely to misunderstand and judge one another in this way, but because wildlife biologists are more closely linked to state management institutions, it is their understandings of wildlife management that tend to prevail in the making of management decisions.[20] It is thus the entailments of the agricultural metaphor, rather than those of the social-relations metaphor, that form the basis for most wildlife management decisions. Indeed, what may on the surface appear to be a case of wildlife biologists cooperating and compromising with First Nation people (e.g., by taking into account First Nation people's beliefs and interests when it comes to killing wolves, but not to sterilizing them) is, in fact, no compromise at all. Every aspect of the wolf control was consistent with wildlife managers' agricultural view of management, whereas First Nation perspectives were respected and acted upon only insofar as they happened to be compatible with the agricultural metaphor. This suggests that attention to metaphorical overlap of the sort discussed here may provide a critical perspective on the politics of knowledge production and application in what may appear—on the surface—to be successful efforts to cooperatively manage environmental resources.

NOTES

To begin with, I would like to thank my co-editors, Matt Turner and Mara Goldman, and all the contributors to this volume for helpful comments, criticism, and feedback that enabled me to improve this chapter considerably. I also received helpful comments from Ken George, Kurt Jordan, and Marina Welker. As always, I acknowledge the help and friendship of the people of Burwash Landing, Yukon, without whom this chapter could never have been written. Thanks, too, to the members of the Aishihik-Kluane Caribou Recovery Steering Committee for allowing me to attend their meetings. Finally, the research on the Aishihik wolf kill was carried out with the support of the National Science Foundation (grant 9614319) and the Wenner-Gren Foundation for Anthropological Research.

1. *First Nation* is the accepted term in Canada for referring to aboriginal people and their governments.

2. In fact, it is the implicit comparison with large-scale *capitalist* agriculture, with its focus on maximizing production and attendant assumptions about ownership and control, to which they object. Small-scale noncapitalist forms of agriculture may be more compatible with First Nation ideas about human-animal-land relations (see Feit 2001).

3. Max Black's interactive model of metaphor (1962) would suggest that use of the agricultural metaphor not only shapes its users' understandings of *wildlife management*, but also alters the meaning of *agriculture* itself. It is beyond the scope of this chapter to explore the full range of semiotic interactions created by the metaphors I examine, but they would be worth exploring further.

4. Elsewhere (Nadasdy 2008) I have explored a different set of the agricultural metaphor's entailments.

5. In part 1 of the book, "Management Theory," Leopold (1933) set out to describe the properties of wildlife populations and identify the factors that influence their size. In part 2, "Management Technique," he focused on the specific techniques wildlife managers can use to control those factors and so effect change.

6. It was to scientific forestry in particular that Leopold turned for inspiration (Leopold 1933, 21; Kosek 2006, 91–92).

7. "The detail of any policy is an evanescent thing, quickly outdated by events, but the experimental approach to policy questions is a permanent thing, adaptable to new conditions as they arise. . . . There are conflicting theories on how to bring the land, the means of payment, and the love of sport into productive relationship with each other. No one can confidently predict which theory is 'best.' The way to resolve differences is to bring all theories susceptible of local trial to the test of actual experience" (Leopold 1933, 411).

8. For the classic description of hunting as a reciprocal relationship between human- and animal-people, see Hallowell (1960). For more on the specifics of the Yukon case, see Nadasdy (2003, 2007b).

9. Elsewhere (Nadasdy 2007b) I have argued that we must not dismiss First Nation people's beliefs about human-animal relations as "just metaphor." In suggesting that the social-relations metaphor structures First Nation people's understandings of wildlife management, then, I am not denying the "truth" of these understandings—any more than I intend to deny the "truth" of the beliefs about animals that underlie the agricultural metaphor. As noted above, it makes little sense to ask whether or not a metaphor is "true." Rather, the agricultural and social-relations metaphors both clearly "work," in Lakoff and Johnson's sense (i.e., they each create or highlight different aspects of the concept *wildlife management*), and each has its own attendant set of metaphorical entailments.

10. For a detailed discussion of such changes and their significance, see Nadasdy (2003, 88–94).

11. To this end, the Yukon government formed a citizen's action committee to consider the issue. The result of the committee's deliberations was the Yukon Wolf Conservation and Management Plan, which developed a set of criteria governing the implementation of any wolf-control program. The plan was widely endorsed by those on all sides of the wolf-kill debate, but there was significant disagreement over whether the wolf kill was consistent with the plan. Critics, including some nongovernment biologists, objected to it on the grounds that it was *not* consistent with the plan (Buckley 1993b, 1996; Jickling 1994; Theberge 1993; YCS 1994). Notably, First Nations were not represented on the committee (Buckley 1993b).

12. The meeting was attended almost entirely by biologists, and there was no KFN representative in attendance.

13. Along these lines, wolves appear in Yukon First Nation songs and stories, often providing assistance to human hunters (see, e.g., Smith 1982). In addition, wolves have totemic significance throughout the region.

14. Because First Nation people see wolves as competition for moose and caribou, they have historically killed wolves not only for their pelts but also in a conscious effort to keep wolf populations down (McClellan 1975, 135–37).

15. One wolf pair was actually sterilized in 1994, but sterilization was not introduced as a general management strategy until 1996.

16. By the end of 1996, the program had already cost nearly $2 million, with one more season to go (Ralston 1996a).

17. According to the master's thesis of the biologist who designed the study (Hayes 1996), however, a pair of wolves can kill as many moose as a pack of six or seven.

18. In keeping with the biologists' adaptive management approach, the sterilization program, too, was designed as a scientific experiment (Spence 1998).

19. KFN decided not to oppose sterilization at a meeting in Burwash Landing on February 2, 1996. At a meeting of the Aishihik-Kluane Caribou Recovery Steering Group two days later, wildlife biologists acknowledged that although First Nation people "were not wholeheartedly behind sterilization," they had "decided not to interfere with it" (AKCRSC 1996b,

3). It is worth noting that until this time KFN had been largely uninvolved in the wolf kill. At the February 2 meeting, however, many Kluane people complained about not having been involved in the process. Thereafter, KFN began sending representatives to AKCRSG meetings; and, at the First Nation's insistence, the steering group began holding alternate meetings in Burwash Landing.

20. As Lakoff and Johnson (1980, 157) put it, "People in power get to impose their metaphors."

8

PETER VANDERGEEST AND NANCY LEE PELUSO

POLITICAL VIOLENCE AND SCIENTIFIC FORESTRY

EMERGENCIES, INSURGENCIES, AND COUNTERINSURGENCIES IN SOUTHEAST ASIA

How do political violence and war contribute to the co-production of scientific forestry and forests as *stabilizing* components of national states? In this chapter, we argue that the major episodes of what government officials called jungle-based political violence in all Southeast Asian countries in the decades after World War II contributed to the co-production of postcolonial states, scientific forestry, and forests as natural objects or "artifacts" (Braun 2002, 3) of scientific forestry. We focus on Indonesia, Malaysia, and Thailand, as they are the nation-states we have studied in our larger project on the making of professional forestry in Southeast Asia (Peluso and Vandergeest 2001; Vandergeest and Peluso 2006a, 2006b). However, we believe that our key findings have relevance to other sites where jungle-based political violence has led to increased state or state-authorized transnational control over forests for production and conservation in the name of national development.

Our work on the making of professional forestry in Southeast Asia has been located primarily in political ecology, but it intersects with science studies through our consideration of forestry as an applied science. Elsewhere we have argued for qualifying what postcolonial "technoscience" scholars have labeled the "diffusionist hypothesis" (W. Anderson 2002, 648) that describes science as spreading outward from center to periphery. In the case of scientific forestry, this approach is still common in accounts that explain forestry in Asia largely in terms of its spread from Germany and France during the colonial period, and from international development agencies during the postcolonial period (for a contrary view, see Vandergeest and Peluso 2006a, 2000b). We have instead emphasized how scientific forestry practices were also produced, circulated, and transformed through interaction with the politics, economies, and ecologies of specific Asian sites. This account focuses on two aspects of scientific forestry: first and most important, the legal demarcation of what we call political forests, which provides the legal basis for forestry departments' claim to exclusive control of forests necessary for the practice of scientific forestry. Secondarily, we have taken up in less detail the silvicultural models through which forest departments attempted to reshape forest ecologies to achieve certain ends, including but not limited to timber production. We do not directly take up logging and other extractive enterprises in this chapter. During the period under consideration, logging companies were usually required to adhere to practices prescribed by state forestry agencies in selecting trees to cut, replant, or otherwise manage, although in practice there were many ways around these rules.

In the language of science studies, these legal and management practices might be called "boundary objects" (Forsyth 2003, 141; Goldman, chapter 10 in this volume), as they span, move through, and link different networks, including those concerned with advancing knowledge of forest ecologies and those implicated in the actions of state agencies such as the civil administration, police, and military. State forestry departments may similarly be understood as "boundary organizations" (Forsyth 2003, 145–47), in that they link the worlds of applied forestry science and these other state agencies. The notion of a boundary organization thus draws attention to how postcolonial scientific forestry was implicated in the production of both national states and political forests.

A political ecology perspective that accounts for the constitutive role of political violence in the making of postcolonial states can also contribute to science studies' interest in the co-production of scientific knowledge and social order (Jasanoff 2004b; Forsyth 2003, 104). Science studies provides analytical tools that show how natural or boundary objects are produced, stabilized, moved around, and transformed through scientific and government networks. Political ecology can in turn show how political violence enables the materialization, establishment, and maintenance of boundary objects and institutions. Even the political ecology literature, however, has only minimally explored the ways that land-management institutions—including conservation organizations—are caught up in the kinds of violence that threaten state stability, such as civil wars and insurgencies. We suggest that insurgencies challenging the very basis and ideology of central state power and property rights are a unique kind of political violence, one that has historically made scientific forestry easier to justify and maintain.

Our earlier work has shown how "the forest" as an object of knowledge and practice did not precede scientific forestry, but was constituted through the practice and recognition of this discipline (cf. Mitchell 2002, chap. 2). Neither forests nor nation-states were preconstituted templates for the application of state forestry, nor were they teleologically determined endpoints. Rather, they were open-ended and incomplete outcomes of often-violent projects to intensify and stabilize state rule. What we call "colonial forest practices" produced territorial jurisdictions as objects understood to comprise certain species and configurations of vegetation (Peluso and Vandergeest 2001). Forestry as a science involves "field-based" knowledge about the ecologies of these extensive terrestrial spaces, usually dominated by woody species and particular animals, as well as laboratory-based knowledge (e.g., about the composition and characteristics of various timbers, nontimber products, soils, and insects). As an *applied* science, however, forestry is not just concerned with scientific knowledge about these naturalized objects, but is also about remaking these forests now understood as artifacts (Braun 2002, 3; Latour 2005), through forest management practices (silviculture, harvesting, and other on-site management techniques) to achieve social, economic, or political ends; for example, enhancing timber or biomass production, regulating watersheds, and producing pulp, resins, or latex.

To achieve these ecological transformations, foresters must be able to predict, control, eliminate, or stimulate certain ecological processes depending on their management objectives. This can only be achieved if foresters have exclusive management rights over forest territories. In other words, the making of forests as objects of scientific knowledge has always been a political act of allocating jurisdiction to agents charged with forest management, usually state forestry departments. Further, territories are more than spatial arrangements. When a state intentionally creates a territory, it has to be understood as constituting a claim. A state agency's "jurisdiction" is a way of laying out a territory for realizing property rights. Thus we call the territories allocated to forestry agencies "political forests" (Peluso and Vandergeest 2001).

As many political ecology writers have shown, the making of political forests often involves violent displacement. We go further than most political ecologists, however, in our claim that we need to understand the specific ways that state-making violence shapes scientific forestry under different conditions. During the post–World War II insurgencies in Southeast Asia, the "jungles" of Indonesia, Malaysia, and Thailand were transformed ideologically, materially, and institutionally to "forests" (see, e.g., Slater 1995; Peluso 2003; Sioh 2004). This process involved the violent undoing and reconstituting of the spatialized society-nature boundaries inherent in the term "jungle," where relatively unassimilated people lived in undisciplined spaces of what was understood as wild nature. Turning jungles into forests further distinguished *forests* from *agriculture*, both spatially and in terms of agency jurisdiction (see also Dove 1992; Sivaramakrishnan 1997; Sivaramakrishnan and Agrawal 2000).

A dimension of state formation that we have found to be particularly important to the making of political forests is the racialization of forest residents. State responses to insurgency helped produce racialized categories of minorities and majorities in Southeast Asia, categories now taken for granted as constitutive of national populations. We have described elsewhere how the colonial model of making political forests often involved the settlement of local claims to forests and their component resources through the creation of "Customary Rights," and how these settlements were often racialized through colonial classifications (Peluso and Vandergeest 2001). Here we show how the racialization of counterinsurgency shaped the making of forests through programs for moving people either to or away from areas where insurgents were active. In general, racial and ethnic groups that were believed by state authorities to be more loyal than others to new nation-states were more likely to be moved to jungles, and those that were considered likely to support insurgents were moved away from jungles. These population movements organized where and how forests and agricultural zones were created out of jungles.

In the remainder of this chapter we illustrate in more detail some of the specific and diverse ways that a specific form of political violence—counterinsurgency— contributed to the co-production of national states and scientific forestry through political forests. We provide background on jungle-based insurgencies and counterinsurgencies. We then focus on how counterinsurgency helped create political forests through the removal of agriculture from jungles. Finally, we give some brief examples of how the intensification of military access and surveillance capacity sometimes strengthened, and sometimes weakened, the political control of state forestry departments over vast territories.

JUNGLE-BASED EMERGENCIES AND INSURGENCIES IN SOUTHEAST ASIA

Although we do not focus on the Second World War in this chapter, the Japanese occupation (1942–45) was a key moment in the emergence of jungle-based political violence in Southeast Asia, as people from occupied countries took to the jungles and mountains to launch resistance to Japanese wartime occupation throughout Southeast Asia. Both subsequent anticolonial struggles in some parts of the region (1945–63) and many of the insurgencies that we discuss here involved the same peoples and were partially based on their experiences of wartime resistance to the Japanese (e.g., see Phatharathananunth 2006).

We have drawn our account from the following cases of jungle-based insurgencies and emergencies: the 1948–57 Malayan Emergency in peninsular Malaya; the Communist Party of Thailand's insurgencies from the mid-1960s to the early 1980s; Communist and Islamist insurgencies in Indonesia from the late 1950s through the early 1970s; and the "Emergency" years of counterinsurgency in Sarawak (approximately 1963–70).[1] Most of these insurgencies were inspired by Maoist revolutionary ideas about peasant-led revolutions, such as the call to "let the countryside surround the cities" (Marks 1996). But not all jungle-based insurgencies were Maoist, nor were they all called "jungle" insurgencies. The regional Islamic rebellions in Indonesia (1950–57), particularly in western Java, parts of

Sumatra, and Sulawesi, launched by militants and disaffected soldiers wanting an Islamic state in Indonesia and supported internationally by the United States, Britain, and other capitalist powers, took place in montane forests (*gunung*). These jungle-based rebellions included the *Darul Islam* movement and its military arm, *Tentara Islam Indonesia* (the Islamic Army of Indonesia), referred to hereafter as DI/TII.

While insurgent groups enlisted, attracted, or forced local people living in these forests to engage in violence or the physical support of their jungle-based fighters, organized political violence did not generally emerge primarily out of the forest villages. Rather, violence was often initiated by Communist Party leaders, students, and other people of urban origins, who "went down" to the countryside, into the mountains, or into the jungles, intending to stage resistance.[2] Once there, they often connected with or took over leadership of existing agrarian movements or tried to attract local people to their cause. The initial association of communism with urban labor and intellectuals, and the adoption of Maoist revolutionary strategies, all contributed to the idea in Malaya and Thailand, and later in Indonesia's Kalimantan, that communism was not primarily a "Thai" or "Malay" or Dayak ideology, but more often a political ideology embraced and propagated by Chinese or other alien groups. This idea, which local people knew was a myth at best and at worst a lie, also served the purposes of counterinsurgency operations.

Among communist parties, the Chinese Communist Party's success had provided a model that was emulated in Vietnam, Malaya, and Thailand. Southeast Asian forests and mountains were ideal bases for the Maoist revolutionary strategy, given the recent histories of resistance to the Japanese and the colonizing forces returning to these areas, the weak reach of nascent nation-states in forests and mountains where insurgencies were being fought, plus the presence of what Kunstadter and his colleagues called "farmers in the forests" (Kunstadter, Chapman, and Sabhasri 1978). *Political* forests had yet to be created in most of these difficult-to-access areas.[3] The territorial and political threats posed by insurgents to decolonizing and postcolonial states motivated state authorities to devote huge resources to strengthening their control of these areas, both by "winning hearts and minds" (creating national subjects) and by turning "wild jungles" into zoned and controlled state territories, with newly created formal property rights for forest departments, other state agencies, or private property holders.

Central to counterinsurgency strategies were the civilizational discourses used by central states to discredit the insurgents. Jungle-based insurgencies understood themselves as alternative civilizing projects, but state powers depicted them as uncivilized and opposed to the modernizing goals of the urban-based states, which included the demarcation and scientific management of forests for capitalist development. Characterizations of forests as "jungle" and insurgents as wild and recalcitrant subjects lent legitimacy to state projects to control these regions and their people through the elimination of areas they called jungles. It is crucial to recognize that these alternative civilizational projects were not only imaginable but eminently possible during this period, as demonstrated in the outcomes of insurgencies or revolutions in China, Vietnam, and Laos. In "jungles," therefore, counterinsurgency strategies included massive projects to redistribute people, and to remake settled areas and "nature" by changing forest and agricultural ecologies. The next section considers how population resettlement contributed to the making of forests by taking agriculture out of the jungle.

MOVING PEOPLE

During the insurgencies, millions of people were forced or encouraged to resettle because their spatial locations were crucial to counterinsurgency operations. This movement also helped create political forests, where livelihood practices not explicitly permitted by forestry departments were criminalized. Resettlement also contributed to the remaking of the silvicultural practices by which foresters reshaped forest ecologies—by giving foresters unpeopled spaces in which to model, experiment, and manage the tropical vegetation.

Insurgents sought to take advantage of rural people's presence either by enrolling them (recruitment) in their alternative state visions or by using them to provide food, supplies, and shelter. State authorities countered by moving people in order to separate insurgents from the rural people who might support them. In general, suspect subjects were moved away from conflict zones or consolidated into closely controlled settlements, while loyal subjects were encouraged or helped to move into conflict zones. More then one strategy was often pursued in any given site, and in some places, such as West Kalimantan and Malaya, all three were deployed. The intention was always to divide forests and agriculture into separate territorial domains, isolating insurgents from the cover (physical and social) provided by jungle.

The postcolonial state's assessments of loyalties thus played a major role in determining which specific strategy was adopted, helping to produce the racialized landscapes of contemporary Southeast Asia (Vandergeest 2003; Sioh 2004).[4] Racialized identities were refashioned out of colonial-era ideas about ethno-racial characteristics, people's presumed territorial histories and claims, and their presumed associations with insurgent and counterinsurgent forces. The new national context made possible the idea that some populations were national minorities, especially in relation to other citizens whose racial or ethnic identities were in the majority, or most closely associated with the national identities (e.g., Javanese in Indonesia, Malays in Malaysia, and Thais in Thailand). Minorities could be either tribal (e.g., hill tribes, Orang Asli, or Dayak) or alien (e.g., Chinese in Malaya, Sarawak, and Indonesia and Vietnamese in Thailand). They were acted upon as if specific ethnic and racial identities carried political-economic characteristics.

In many areas of the "jungle," forest farming "tribals" such as Dayaks in Kalimantan and Sarawak, Orang Asli in Malaysia, and Hmong and other "hill tribes" in Thailand were consolidated into more easily controlled spatial settlements. "Tribal" peoples in general were represented as backward, because they lived far from the nation-states' centers and practiced swidden agriculture. Some tribal peoples were considered innocent and in need of development aid and, at the same time, primordially fierce and violent with the potential of using their tribal warfare skills to support the "wrong side" in political violence. Their alleged backwardness in fact fed this representation; unruly "savage headhunters" and "wild settlers" had to be tamed as part of the broader project of producing scientifically managed forests. Complicating these representations were tribal peoples' associations with and knowledge of the difficult jungle terrain, which often made them important participants (or pawns) in counterinsurgencies as well as insurgencies (e.g., Leary 1995; Endicott 1997; Harper, 1997; Peluso and Harwell 2001). During the insurgencies these consolidated villages were often abandoned as nonviable and counterproductive after a few years (e.g., A. Walker 1983, 468; Tapp 1989), but all states in the region have since continued their efforts to limit access of tribal peoples to jungle/forests, especially for agriculture.

Not only ethnic minorities were resettled. Where foresters were relatively strong in their ability to control territory and influence state policy, and where violence threatened the integrity of international borders (Peluso and Vandergeest 2001), even majority ethnic groups were moved out of contested zones. For example, in the Malaysian state of Kedah, forestry maps confirm oral-historical accounts of upper-watershed hamlets[5] being forced to move to lowland sites during the early 1950s. These hamlets included people previously classified as Malay, Siamese, and "Sam-Sam" (Siamese-speaking but Islamic). Where the emptied upland areas were not already gazetted as forest reserves, this resettlement was accompanied by forest reservation, consolidating the territorial control of the forest department.

A more widely known example was the resettlement of half a million Chinese forest "squatters" into camps called "New Villages" during the emergencies in Malaya and Sarawak (Stubbs 1989, 262). Similarly, in Kalimantan, even long-resident rural Chinese were forced either to move to resettlement areas or to become refugees in urban areas (Coppel 1983; Somers-Heidhues 2003; Peluso 2003;

Davidson 2008). In Kalimantan, the Indonesian state forcefully engaged Dayaks against the alleged Chinese communists (Peluso and Harwell 2001; Davidson and Kammen 2002). The same year that 50 to 100 thousand rural "Chinese" were evicted from their homes, 1967, was the year the first *national* Indonesian Forest Law provided for a *national* forest estate. Not long after that, lines on maps were drawn to allocate forest concessions in West Kalimantan and elsewhere. The military forces received logging concessions along contested borders—a twenty-kilometer-wide swath of territory along Kalimantan's thousand-kilometer border with Sarawak. No recognition was made on these maps of the Dayak (or Malay, or mixed) villages enclosed within these forests—nor of their pre-existing land and resource claims. With the construction of these new national forests the national government excluded loyal locals, again remaking the jungles, their people, and the state.

While suspect populations were moved out of conflict zones, trusted majority populations such as ethnic Thais and Javanese were encouraged or sponsored to move into these zones to clear forest for permanent agriculture, because it was assumed they were less likely to support insurgents.[6] The resulting agricultural expansion was represented as an alternative landscape solution to jungle-based insurgency. A well-known case is the rapid expansion of cash cropping in Thailand as millions of migrant farmers were encouraged to occupy reserve forests as a counterinsurgency strategy (Vandergeest and Peluso 1995; Uhlig 1984), particularly in the northeast, where Isan people (Phatharathananunth 2006) were understood to be potentially loyal subjects who could be incorporated through development projects, roads, and the promotion of cash crops. Although these occupations violated reserve forest laws, they were condoned by civil authorities who saw them as a way of decreasing jungle cover for insurgents, and as a means of retaining the loyalties of land-poor farmers vulnerable to insurgent propaganda.

The redistribution of people was also tied up with the remaking of forest management or silvicultural practices. One set of examples concerns how programs providing temporary cultivation rights to poor farmers in exchange for establishing tree plantations were refashioned during the insurgencies in Indonesia and Thailand. These programs—called *taungya* or *tumpang sari*—emerged first in Burma and Indonesia during the nineteenth century as a compromise between the German forestry model that assumed a forest without people, and the actual presence of forest-based peoples in the teak forest areas of Java and lower Burma (Peluso 1992; Bryant 1997). In the latter countries taungya was formulated by foresters as a way of forcing resident people to plant valuable teak forests in exchange for limited access (two to three years) to land for farming in the short time span before the seedlings' canopies shaded the ground.

Ironically, perhaps, in West Java, where Islamic rebels hoped to establish an alternative Indonesian state, taungya or tumpang sari had its own local form, called *talun*. This rotational agriculture system for farming forests, used extensively throughout West Java's hilly terrain, was a field-to-forest cultivation system that was the envy of many foresters, and entirely controlled by West Java's farmers. During the DI/TII insurgency in the late 1950s and early 1960s, foresters destroyed much more forest cover than insurgents did, ordered by the military to burn down the forests and reveal the mountain jungle hiding places of insurgent cells. In the aftermath of this period, forest areas were governed by the West Java Forest Service (Dinas Kehutanan), which had a very small budget and almost no real presence in the region. However, in 1978, forest-dwelling villagers felt the heat turn up when the administration of these forests was turned over to Perum Perhutani, the same State Forestry Corporation (SFC) known for its repressive and violent management tactics in Central and East Java. Just as the SFC used labels implying antistate subversive affiliations—"communists" or Communist Party Members (PKI) to threaten allegedly illegal loggers and forest farmers in Central and East Java—they called people "ex-DI" or "DI" in West Java, turning their claims in the forest into crimes against the state. Through labeling, counterinsurgency and forest management were inextricably linked together.

In postwar Thailand, the taungya model was adapted by the forestry department for the forest village program initiated in the northern mountains in 1967. This program consolidated farmers who were previously dispersed in conflict zones into controlled forest villages, and sought to rehabilitate forests believed to have been degraded by swidden agriculture. Farmers were given temporary cultivation rights contingent on participation in reforestation activities, including the clearing of existing "degraded" forests and the planting of a specific area and number of trees, usually teak, on a presumed forty-year rotation following specific instructions as to spacing and so on (Boonkird, Fernandes, and Nair 1984). When the program was expanded to other regions of Thailand, other species, mainly fast-growing trees, were also added (Watanabe, Sahunalu, and Khemnark 1988). In 1982, as the insurgency came to an end, the forest village program became the basis of the National Forestland Allotment Project, which attempted to prevent future settlement in reserve forests by providing temporary five-year land usufruct certificates to more than 700,000 farmers in reserve forests as a kind of amnesty (Vandergeest and Peluso 1995), while in theory keeping any new settlement illegal.

In Malaya the "Malayan Uniform System" was developed after the war as a technique for regenerating forests cut by the Japanese during the occupation. It was a reformulation of a prewar Malayan silviculture model (the regeneration improvement fellings, RIF), which was in turn based on Clementsian successional theories of ecology. The underlying notion that tropical forest management should be about managing complexity was already a departure from the German approach to forest management based in radical simplification and predictability (cf. J. Scott 1998). In the context of the Emergency, the forest department revised the RIF into a model for regenerating forests that had been cleared for cultivation by ethnic Chinese who had occupied forest areas during the Emergency and who were suspected of supporting Malayan Communist Party insurgents (Wyatt-Smith 1947, 1949; Ali 1966). This silviculture model can be described as a boundary object[7] that spanned forestry and counterinsurgency networks; the system was simultaneously a form of scientific knowledge about recently depopulated forests, an applied management technique, and an argument for removing squatters so that forests could be regenerated using ecological and forestry science.

SURVEILLANCE AND ACCESS

Our final argument is that the vast resources devoted to increasing state access to formerly inaccessible areas, and to counterinsurgency surveillance technologies, often were transferred across institutional boundaries into state forestry departments. Efforts to stabilize capitalist postcolonial states during the Cold War drew huge international and national resources for enhancing surveillance and access.

Counterinsurgency activities called for intensified monitoring and mapping of forest areas, particularly around international borders. In Thailand, the Royal Survey Department became in many ways an arm of the U.S. military, which used aerial photos to produce the well-known series of 1:50,000 topographic maps of all forested areas starting in the 1950s, periodically updated based on new aerial photographs. Although public access to these maps was restricted into the 1980s, they were shared with other government departments, including the Forestry Department, where they became the base maps for forestry work. In particular, they were the basis for the demarcation of reserve forests, the boundaries of which were drawn along the contour lines and vegetation zones shown in these military maps (Vandergeest 2003). By the early 1970s over 40 percent of the terrestrial area of Thailand was demarcated as reserve forest, largely through drawing lines on these maps, with minimal ground-truthing of local forest uses or claims.

In Malaya, aerial photographs of the entire peninsula were taken during the Emergency for military purposes. Forestry officials and anthropologists subsequently used them to map information on Orang Asli and other forest residents (Harper 1997, 21). We have mentioned above how in Kedah upland areas were mapped and demarcated as reserve forests during the Emergency while residents were forced to move to the lowlands.

In West Kalimantan, the insurgent areas in the 1960s were primarily (almost exclusively) the same as those called the "Chinese Districts" under colonialism. These areas had the most technologically sophisticated and detailed maps of anywhere in Borneo, maps that dated back to Dutch colonial times and were much more comprehensive than those for other regions that the U.S. Army had made in the early 1950s.[8] The reasons for the existence of these maps paralleled those we discuss here, although they were of an earlier century and an earlier kind of political violence. These areas had been contested by Java-based Netherlands East Indies colonial authorities who resented the political autonomy and territorial power of Chinese *kongsis* mining and farming in the western part of today's West Kalimantan. They were glossed as insurgents and eventually forced to recognize Dutch colonial authority in the mid-nineteenth century, after over a century of unsuccessful attempts.

Another significant "technology of jungle access" for both counterinsurgency and forestry purposes was roads. From a security perspective, roads had two purposes: to provide easier access for military operations, and to draw existing populations closer into the sphere of central state rule, away from insurgents' alternative civilizing projects. Roads also facilitated access for logging, both legal and illegal. For example, in Thailand, the Accelerated Rural Development (ARD) scheme, supported by USAID (Muscat 1990, 164–66), was primarily a program to build roads. This scheme was active in provinces designated as most in need of development, which meant those where the insurgency was most active.

Even after insurgents no longer posed a serious challenge to national states, security concerns continued to shape the practice of professional forestry. The fear of further insurgencies helped motivate the reshaping of legal access to forest lands and products, the allocation of concessions, and the use of retired or current military personnel as security guards for forest enterprises. These continued to put foresters, militaries, and big extractive businesses into close connection. The new physical and legal changes significantly transformed local people's modes of access to and control over forest products and territories.

CONCLUSION

This chapter shows how the political violence of insurgency and counterinsurgency articulated with the deeply territorialized form of state forestry that characterized forest-agrarian politics during the long historical moment that was the Cold War. Jungle-based insurgency challenged the legitimacy of state power at its very core and its most metaphorical and physical peripheries. Our study compares three Southeast Asian countries that were not the region's "First Front,"[9] thus illustrating some of the ways forest territories and forest-based populations were imbricated by the tactical actions taken by national militaries, civil administrations, and forest departments in counterinsurgency operations.

The jungle-based challenges to state power that we present here resonate with similar conflicts in other South and Southeast Asian countries with significant areas of forest. Even today, the discourses of jungles and the practices of counterinsurgency remain alive in the Philippines, Nepal, and India—as well as in other parts of the world such as Colombia, Peru, and Bolivia—even as the tactics and concepts of desert wars and international terror reshape Western imaginaries of spatialized threats.

Like actors caught up in the antipolitics machinery of development (Ferguson 1990), state foresters represented themselves as professional practitioners whose work was predicated on territorial control of natural objects or artifacts called forests, so that they could manage them according to scientific methods. As many critical (political ecology) accounts of colonial and postcolonial forestry have shown, the sanitized and rational accounts of forestry's contributions to national development through provision of forest products and forest revenues ignores the ongoing violence of state enclosures and territorializations that are preconditions for the practice of scientific forestry. This chapter goes further, however, in that we take up not just the violence implicit in the expansion of scientific forestry, but also the relationship between broader political violence in the making of postcolonial states, scientific forestry, and forests. This broader approach helps us to augment narratives organized around the advance of forestry against the resistance of local people, with one that gives more room for seeing how scientific forestry was simultaneously strengthened and limited through its participation in projects to stabilize state rule in the immediate postcolonial context.

Our research has been framed largely through a political ecology framework, but participation in this volume has highlighted how it also resonates with current themes in science studies. Most political ecology scholars are aware of how science studies has already provided additional tools to help explain the politics of knowledge inherent in constructions of nature or socionature(s). The work of Bruno Latour (1993) on nature, he and his associates on actor-network theory (Latour 2005; Law and Hassard 1999), and Donna Haraway (1988) on the gendering and racialization of science have all been influential in political ecology (see, e.g., Forsyth 2003). Our chapter further highlights how the co-production approach in science studies is engaging questions similar to those we take up here, specifically, the co-production of states, political power, and scientific knowledge (see, e.g., chapters in Jasanoff 2004b). In our specific focus on forests and forestry, we have found the idiom of co-production a useful way of ensuring that we attend to the question of how scientific forestry also helps produces states, not just how states produce scientific forestry.

This chapter and our work more broadly is an extension of these research themes into terrains not often covered in science studies. First, like much work in political ecology, our point of departure is the global South. Science studies arguably remains largely oriented to the global North, with science in the global South often framed as derivative of or related to global North frameworks (W. Anderson 2002; Lowe 2006). We are not suggesting that there is some kind of fundamental epistemological difference between research in the South and North, but this chapter, and political ecology more broadly, points to how a global South orientation raises new and different questions about understanding the co-production, movement, and hybridizations of applied science in the context of empire. In this chapter, we show how in the immediate postcolonial context in the global South, the production of forestry science was tied up with international political projects to stabilize pro-Western postcolonial states in the face of challenges from alternative state-building projects launched from jungles, mountains, and forests. Second, science studies has generally not ventured into the empirical terrain we traverse in this chapter, namely, the ways that war, insurgency, and other forms of political violence co-produce political order and scientific practice. Even within political ecology, there has been relatively minimal exploration of how specific types of war and political violence have shaped scientific forestry.

We finish with a brief comment on the implications for scholarship on contemporary debates in scientific forestry or the science of conservation. The period of political violence that we outline here set the stage for the post–Cold War's "global conservation era." Our work unpacks some of the common assumptions that frame contemporary debates, and suggests new research questions. For example, we suggest that one outcome of Cold War political violence was a dramatic expansion of forests produced as "natural objects" administered by professional foresters and conservationists today. The significance of these ideological and material shifts is indicated by the way that "jungle" discourses of insurgency have now largely disappeared in contemporary Southeast Asia, resurrected mostly when a militant or oppositional group is perceived as taking refuge in or launching attacks from forests.

In addition, we have briefly indicated how the contemporary racialization of these landscapes was produced *in part* through counterinsurgency. Political violence induced, forced, or enabled massive movements of people across these landscapes, transforming territories, property rights, and claims along racialized lines. This violent transition has ultimately enabled the production of political forests in today's Southeast Asia. The creation of political forests was in turn a precondition to conservation discourses that imagine "ideal forests" (both pristine and imaginary!) as those subject to minimal human disturbance, and are predicated on the assumption that conservation should be primarily about reducing human impact.

Nation-state formation and conservation in political forests continue to be open-ended and incomplete projects. Many Southeast Asian forests are still used by local farmers, usually without legal sanction, although new surveillance technologies have made it increasingly difficult for farmers to hide farming practices rendered illegal under postcolonial regimes. Under these conditions, new challenges to forest-farm boundaries have emerged, framed in the more scientific and disciplining language of agroforestry and community forestry. These are often presented as acceptable alternatives to visions of forests without people, acceptable because these political-forests-as-boundary objects bring local forest management practices into the realm of state law and national development, rather than invoking the law of the jungle—perhaps a necessary adjustment given the ongoing intensification of forest surveillance.

The implications are broader than the impacts on conservation practice. The idiom of co-production directs our attention to consider how forest conservation practices remake political order. Most significant may be the way that influential nonstate organizations (e.g., WWF, IUCN, Forest Stewardship Council) and movements (e.g., indigenous and forest farmer organizations) have become increasingly important in making forest management practices visible, in setting normative standards for managing ecologies, and in participating directly in administering territories. The effect has been to open up state sovereignty in new ways, as many of these organizations make claims that challenge the exclusivity of state jurisdiction, and institute new forms of extraterritoriality.

NOTES

1. The latter occurred when the second president of Indonesia criminalized communism and was hunting down both West Kalimantan and Sarawak members of the guerrilla forces trained by Sukarno and his army inside Indonesia (and by the Indonesian army). For a detailed account of these low-impact wars, see, e.g., Mackie (1970); Coppel (1983); Dennis and Grey (1996); Poulgrain (1998); Davidson and Kammen (2002); and Davidson (2008).

2. Not all these insurgencies took place in forests, but the lowland agricultural conflicts are beyond the scope of this chapter.

3. It needs to be noted that the major political forests created during the colonial period, in what became the nation-states of Thailand, Malaysia, and Indonesia, were in Java and on the Malay Peninsula (Peluso and Vandergeest 2001). Only there were colonial forest departments able to gain formal legal control of large territories. In the other colonial areas we examined,

such as Siam, Sarawak, and Dutch Borneo, most forested or swiddened areas were not yet legally allocated to the jurisdiction of state or professional forestry agencies.

4. We do not assume that the various "ethnic groups" we discuss pre-exist the counterinsurgency, although this was a contention or a conception utilized by military, intelligence, and other government strategists.

5. Discussed in Cheah Boon Kheng's account (1988) of social banditry on the border during 1900–1929.

6. Ethnic majority populations were also recruited into armed village self-defense units in all three countries, another indicator of their presumed loyalty.

7. See Goldman, chapter 10 in this volume, for a more thorough discussion of how a conservation model can be boundary object.

8. The army essentially copied the Dutch colonial maps of the Chinese districts and redated them.

9. This term was reserved for Vietnam, Laos, and Cambodia for much of that time.

9

KARL S. ZIMMERER

SPATIAL-GEOGRAPHIC MODELS OF WATER SCARCITY AND SUPPLY IN IRRIGATION ENGINEERING AND MANAGEMENT

BOLIVIA, 1952–2009

PREVIEW: EXPERTISE AND EXPERIENCE IN IRRIGATION DEVELOPMENT

July 18, 1991, "Campamento Proyecto Laka Laka," near the town center of Tarata, Bolivia: The Canadian director and his Bolivian NGO counterpart reviewed the sheets of hydrological calculations. Both irrigation engineers looked approvingly at the estimates. The new Multiple-Use Laka Laka Project (Proyecto Múltiple de Laka Laka), financed through the Canadian government and an international NGO channeling lunch-money donations of Montreal schoolchildren, could count on sufficient water inputs and delivery. The project's aim of impounding upland runoff at the Laka Laka dam was designed to bring modern improvements to the "water-scarce" valley agricultural area and adjoining provincial capital. Technical expertise was vital. Deciphering rainfall, runoff, topography, and other catchment conditions, the engineers estimated water flows to the reservoir that would form behind the hundred-meter-high dam under construction. If soil erosion caused problems—a scenario given little attention at the time (described below)—presumably they could be addressed through reforestation and physical structures.

February 14, 2009, Guillermo's peach field, Arbieto (2.5 kilometers downstream of Tarata): The president of the Association of Irrigators (Asociación de Regantes) looked over his three fields of peach trees laden with fruit. Revenues appeared promising. Still a cloud of uncertainty hung over the irrigation landscape. The president himself was having a new well drilled to augment water supply. The *regantes*, now in full control of the Laka Laka project, had spent nearly their entire treasury on legal disputes with the Tarata townspeople. Many sides complicated this dispute, which had included hearings in a national court in La Paz, violent confrontations, and vandalized waterworks. One problem is sediment buildup. It has forced redesign of the original project several times at costs of several million dollars and is severely reducing the reservoir's storage capacity. Technical expertise had been marshaled to tackle this problem beginning shortly after the Canadian ambassador inaugurated the project in 1993. A British

Columbia–based firm, the Bolivian government, and various NGO institutions and individuals guided numerous and ongoing mitigation measures. Yet not one of these institutions or individuals doubted per se that the upland was integral to the watershed. Flows encompassed the water and sediment, to be sure, and the people and products of one dozen large peasant communities passed to and through the town of Tarata and the peach fields below. Integrated watershed management received widespread and recurrent publicity. Indeed this concept was not foreign to the Canadian irrigation engineer and his Bolivian counterpart nearly twenty years earlier, while it did not prevent the unfolding saga of social and environmental problems associated with the Laka Laka project.

INTRODUCTION: WATER RESOURCE MANAGEMENT AND SCIENTIFIC MODELS

Water resource management—which ranges across scientific, engineering, and environmental projects to access and use rights of groups and individuals—is a paramount social-environmental challenge. Importance of water resources has escalated as a result of intensifying human impacts that include agricultural growth, changing management that includes privatization and community projects, global climate change, and expanding connections to resource sectors and environmental concerns such as energy and biodiversity conservation (Lovejoy and Hannah 2005). Scientific and engineering knowledge is increasingly used to guide policies on water resources, both in global frameworks and in national, regional, and local applications (Bakker 2000; R. Phadke 2002; Budds 2009). It is often aimed at conditions that are central to the livelihoods and resource use of large numbers of people. As a result, water resource management, along with its connections to other resource issues, is well-suited to the combined perspective of science and technology studies and political ecology.

Bolivia and the other Andean countries are critical to global water resources. The tropical and subtropical landscapes of the Andes Mountains, along with adjoining foothills and lowlands, are the headwaters of most of South America. Andean landscapes of Bolivia are inhabited by an estimated five million people, a majority of whom belong to Quechua, Aymara, and other indigenous groups. Livelihoods in these critical Andean watersheds involve resource use among populations ranging from the primarily indigenous areas of rural uplands to mixed social groups of urban areas and valley lands of intensive agriculture. Access to water, along with other determinants of well-being, is the result of resource governance (regulation of environmental activities) and scientific and technological models applied to the biogeophysical and social conditions of Bolivia's diverse Andean landscapes.

During the past sixty years in Bolivia, water resource policies and models, especially irrigation, have been created and implemented amid widely varied national governments and political, economic, and policy orientations. In the 1950s and early 1960s Leftist populist regimes, which nationalized Bolivia's major industries and enacted far-reaching agrarian reform, also shaped and supported major twentieth-century irrigation projects. Beginning in 1985 neoliberal governments in Bolivia designed and implemented (under guidance of the World Bank and International Monetary Fund) a series of drastic economic policies in the form of massive cutbacks, privatization, and globalization. Policies of Bolivia's neoliberal governments led to the well-known Cochabamba Water War in 2000—discussed below—while they also led, incongruously at first glance, to a host of well-intentioned community-based water-management projects. Indeed the neoliberal shocks had triggered the wave of earnest technical experts and welcomed expertise described in my opening vignette. This ongoing legacy has not been simply "good" or "bad" for resource environments and livelihoods in Bolivia. Most recently water development is being reconfigured, albeit complexly, amid the transitions to the government of Evo Morales. Characterized as Left neopopulist and revolutionary nationalist, the Morales government is still unfolding its environmental resource policies, including those aimed at water (see also Zimmerer n.d.).

Evolving expert knowledge about irrigation and water management in Bolivia is set in the context of development and agricultural projects undertaken since the mid-twentieth century. Initially this knowledge rested primarily on scientific and engineering models derived from irrigation technology, hydrology, climatology, and agronomy. Examples abound of irrigation knowledge and expertise in Bolivia. By 2005 the country was covered by an estimated 1,000 irrigation projects covering an area greater than four thousand square kilometers and affecting directly the lives of several hundred thousand people (see table 9.1 for estimated irrigation in the early 1990s; Montes de Oca 1989, 1992). This chapter is concerned with the broadly defined spatial and geographic dimensions of scientific and engineering models in Bolivian irrigation. It addresses explicit space-related parameters (such as distance and area) as well as geographic considerations in the interconnected environmental, livelihood, social, and political characteristics of irrigation areas. The study focuses on the vital role of knowledge and power in the combined spatial configuration in irrigation models of places of targeted water delivery (characterized typically as water scarce) and those demarcated for supply (surplus water).

This study's findings demonstrate the increased significance of spatial-geographic frameworks in the scientific models of water resources in Bolivia.

TABLE 9.1. *Irrigation projects in Bolivia in the early 1990s*

	Functioning Projects			Implemented Projects			Planned Projects			Total	
Department	Projects	Households	Area (hectares)	Projects	Households	Area (hectares)	Projects	Households	Area (hectares)	Households	Area (hectares)
La Paz	9	10,866	14,977	7	1,030	1,936	11	12,890	44,372	24,786	61,285
Oruro	134	5,205	9,210	26	1,239	693	14	1,166	11,828	7,610	21,731
Potosí	203	7,087	2,738	8	190	43	2	2,200	2,800	9,477	5,581
Cochabamba	24	15,211	20,845	7	1,579	1,730	26	38,140	60,627	54,930	83,202
Chuquisaca	10	481	860	3	1,945	3,800	3	1,863	6,000	4,289	10,660
Tarija	19	1,352	2,524	4	802	3,530	27	5,607	38,252	7,761	44,306
Santa Cruz	2	320	2,200				4	1,206	5,150	1,526	7,350
Total	401	40,522	53,354	55	6,785	11,732	87	63,072	169,029	110,379	234,115

Source: Data from Montes de Oca (1992).

Initially implicit, these frameworks have been given an important role in water resource sciences and engineering. During the mid-twentieth century Bolivian governments created spatially explicit nation-scale assessments and resource inventories, based mainly on biogeophysical information, to guide the selection, design, and siting of irrigation projects. Subsequently the models evolved to encompass both social and environmental factors. Methods and criteria of economics, in particular, have furnished the broadly social information for the spatial parameters of Bolivian irrigation planning in the 1990s and early 2000s. For example, the pricing of irrigation water has been used to indicate areas of relative availability. Similarly, the location-specific estimates of economic costs and returns are used to justify the selection and design of irrigation projects. Economic arguments applied to community-based irrigation (typically small and medium-scale projects, often in poor rural areas) frequently encompass income improvement and poverty alleviation, which were principal rationales in the Laka Laka project mentioned in the opening vignette.

A geographic shift has occurred along with this evolution to combined economic-environmental models of irrigation. The "standardized package" of irrigation models and planning (explained below) has been spatially reconfigured. The geographic areas of uplands and valley slopes, previously excluded, have increasingly become imagined and integrated as spaces of irrigation sourcing and water surplus. Spatial configuration of combined areas or zones of *both* water surplus (uplands, slopes) *and* water scarcity (valley bottoms) has not gone unnoticed. Marginalized peasant and indigenous people in the Bolivian Andes, including both *cholo* (and *chola*) peasants and those persons self-identified as "Indians" (*indios*), have challenged the geographic assumptions of the standardized package of water development, both in the earlier version (pre-1985) and in the post-1985 neoliberal period. Geographic issues in these disputes highlight the importance of multiscale understandings of water resource science and engineering needed to supplement the single fixed scale of the project area or watershed. This multiscale approach explicitly recognizes interconnected geographic areas of water flows and uses. This alternative approach highlights a need for conjoined analysis of Bolivian upland-slope space together with the valley bottomlands (as well as combined interconnected rural *and* urban water areas).

SCARCITY IN RESOURCE SCIENCE MODELS AND ENGINEERING

Environmental scarcity is a multifaceted and powerful concept. Defined as relative sparseness or inaccessibility, scarcity serves as *both* an important theme of ecological science critical to influential approaches in environmental management *and* one important root of the modern discipline of economics and, more generally, the modern "will to improve" that motivates development attempts of many stripes. (On the "will to improve" and rooting in "identifying problems"—albeit not the role of resource scarcity per se—see Li 2007.) Scarcity is the glue that bonds pricing in environment-related markets—including for example the expanding exchange mechanisms of ecosystem goods and services—with conservation goals. Environmental management and environmentalism in general have used the concept of scarcity as a foundation of claims for the current global water crisis. This study's perspective does not detract from the reality of lessening supplies of vital resources such as water to certain people, in particular those disempowered in their capacity to acquire goods and services. Rather, it stresses the need for careful social-environmental study in order to examine and engage the resource models that evoke, incorporate, and create the notion as well as the nature of environmental scarcity. (On nature as outcome of technological projects, see Mitchell 2002.)

Power and persuasion of the idea of environmental scarcity help drive prevailing albeit flawed accounts of environmental politics and resource conflicts. Increased scarcity of natural factors (water, pasture, agricultural land), often taken together with presumed rapid population growth (the so-called Malthusian specter), is seen as fueling the surge of so-called resource wars (Homer-Dixon 1993). The eruption of the Cochabamba Water War in 2000 has been interpreted in this view as an example of political conflict resulting from environmental scarcity. Yet political ecology offers trenchant critique and alternative framing of the social-environmental interactions, rather than singular environmental causation or determination, of the role of scarcity in these types of resource conflicts and issues (Bakker 2000; Birkenholtz 2009; Peluso and Watts 2001; Turner 2004). Such alternative framing has generated convincing counter-explanations to the overly simplistic claim of environmental scarcity per se as the cause of the Cochabamba Water War and other resource conflicts in Bolivia (Assies 2003; Hindery 2004; Perreault 2006).

Scarcity-based models have become more common and influential in conjunction with worsening environmental conditions and mounting challenges. "Stabilization" is frequently cast as the solution to environmental scarcity. Yet this now predominant logic was not prevalent even a few decades ago. In Bolivia, my analysis shows that this scarcity-based perspective in water resource science and engineering displaced a former based on rational use. The case studies demonstrate that scarcity-centered interpretations were then further expanded in the wake of the Cochabamba Water War. Ironically, the widespread reaction reversing neoliberal governments in Bolivia tended initially to reinforce, rather than reduce, the scarcity-centered perspective on water resources, at least for several years. Present-day government emphasis on water scarcity suggests a potentially new sort of environmental sensitivity within the administration of Evo Morales. Inclined to reject the neoliberal approach to market-based environmental solutions, the Morales government appears poised to shift national policy from a neoliberal economic to ecological understanding of "scarcity."

The role of scarcity in water resource issues often intersects with the issue of naturalistic versus social factors in scientific and engineering models. Naturalistic explanations of water scarcity, focused solely on physical-environmental deficiencies (e.g., climate), remain globally powerful, for example, in applications to water and irrigation management, such as the Sardar Sarovar Narmada dam complex in India (Mehta 2007). In Bolivia, by contrast, this unidimensional view, focused purely on physical parameters, was previously more common than at present. During the past fifteen to twenty years, it is increasingly subsumed in multifactor models combining environmental and economic elements of water scarcity. A standardized package—referring to repeated scientific procedures that make "doable" the undertaking of projects and research in a variety of settings (Fujimura 1987)—has evolved in the combined economic-environmental modeling, planning, and management of water resource development in Bolivia, especially irrigation for agricultural use. While this spatial-geographic framework was implicit in the early version of the standardized package of irrigation development (pre-1985), it has subsequently become more explicit, partly through the influence of enhanced technical renderings including, but not limited to, geographic information systems (GIS; on "rendering technical" problem solving in development, see Li 2007).

SPATIAL-GEOGRAPHIC FORMULATIONS IN IRRIGATION MODELS AND PLANNING

The case studies following this section argue that the standardized package of irrigation engineering and application has been grounded in a recurring spatial-geographic framework. Spatial-geographic framing of irrigation is especially vivid in Bolivia, given recent water conflicts, management crises, and related political strife. At the same time, it represents a largely overlooked core component of the standardized package of irrigation widely applied in other similar regions of the Andean countries and elsewhere in Latin America, Asia, and Africa. This study argues that spatial-geographic framing of water scarcity has gained salience in new scientific and engineering models of resource management in general. Integrated water management, for example, relies on basinwide models of areas of high-value use and hence scarcity (e.g., industrial water use) that are interconnected spatially with areas of low-value use and hence surplus (e.g., agricultural water use; see, e.g., Ward, Booker, and Michelsen 2006). Similarly, various models of climate change impacts on water availability are designed to identify spatial areas of scarcity and surplus within watersheds (e.g., Matondo, Peter, and Msibi 2005). Geographic technologies, especially GIS and remote imaging, have vastly improved capacities for collecting and compiling information. As seen in the following case studies, social-environmental data has become available at unprecedented levels of detail and spatial density, even in the relatively information-poor landscapes of Bolivia.

Scale illustrates the increasing role of the spatial-geographic dimension of scarcity in water resource management. The predominant scale has shifted from the national to the global in most water models and management. Earlier models for water management were conceptualized and implemented mostly at the national or regional scale in the mid-twentieth century. Prominent examples range from the Tennessee Valley Authority (TVA) in the south-central United States to similar projects in Latin America and the Andean countries, perhaps most notably the Cauca Valley Corporation of Colombia. The role of scarcity in this management fit neatly into a much desired though mostly imaginary calculus of rational planning and administration at the national level. Water scarcity in this national perspective required the efficiency-maximizing control of the central government over water resources (Alatout 2006; J. Scott 1998). By contrast, global and multi-country scales now predominate in water resource planning and management. Recent modeling of scenarios of water scarcity, for example, is often based on groups of countries or regions (see Mehta 2007). Economic markets now provide the principal unifying mechanism (along with hydrology and climate) at the cross-country scale.

My study argues that spatial naturalization is increasingly at work in contemporary models and management of water resources, especially irrigation. Spatial naturalization is defined here as the routinized identification and prefigured representation of paired areas of water scarcity and surplus. In practice spatial naturalization relies on (1) the compilation of spatially explicit information on environmental and economic factors involving water, enabling both direct calculation and inference; (2) visual and cartographic representations, which include plotting onto mathematically ordered surfaces; (3) scientific reasoning that can be defined as both deductive (derived from general principles) as well as inductive (based on scientific information); and (4) justification based on working with the best existing sources and information. The importance of spatial naturalization in water and resource management arguably applies also to relatively low-data landscapes, such as Bolivia. (In such cases, as discussed below, spatial naturalization in resource modeling is associated with the narrative of geographically differentiated scientific and social progress.)

IRRIGATION IN COCHABAMBA, BOLIVIA

NATIONAL PERIOD, 1952–85

Spurred by worker and peasant organizations, the 1952 Bolivian Revolution initiated the modern period of political rule and resource management with the rise to national power of the Movimiento Nacionalista Revolucionario (MNR; National Revolutionary Movement). The MNR leader Victor Paz Estenssoro pursued a distinct populist and reformist political agenda—rather than revolutionary per se—during his three successive terms as president (1952–64). Paz Estenssoro initiated extensive plans for Bolivia's agricultural and resource modernization, including the 1953 agrarian reform that was among the most ambitious in Latin America. The reform's slogan, "The Land Belongs to the One Who Works It," belied the widespread reliance of Bolivian agriculture on irrigation. Development of irrigation became a major goal for modernizing agriculture and the Bolivian nation under the Paz Estenssoro–led MNR governments as well as the military dictatorships and civilian governments that followed.

The National System for Irrigation Number One (Sistema Nacional de Riego Número Uno), which delivered water to Cochabamba's Central Valley (Valle Central; see figure 9.1), was the showcase of Bolivia's irrigation projects in the mid-twentieth century. The Bolivian government used the name México for the dam, which was designed by Mexican irrigation engineers and technicians. Paz Estenssoro and fellow MNR leaders, old-line left-leaning populists, looked to these countries for both ideological inspiration and technical and financial assistance. Locally, Cochabamba residents referred to this irrigation project as Angostura (narrows), named for the gorge where the Río Sulty was dammed and thus formed an expansive shallow reservoir known as Laguna Angostura (figure 9.1). (On the role of dams—and the Aswan dam in particular—as demonstrating the strength of the modern state as a technological and economic power, see Mitchell 2002; see also Swyngedouw 1999 on modern Spain.) As a local place name, Angostura referred also to the nearby hacienda estate, dating to the colonial period, that had been expropriated in the 1953 agrarian reform. Selection of the Angostura site thus redoubled the symbolic significance of the National System for Irrigation Number One project as the future of Bolivia's modern water management.

The Angostura project provided irrigation for approximately ten thousand hectares of agricultural lands belonging to small- and medium-size mestizo farmers in the central valley that immediately surrounds the city of Cochabamba

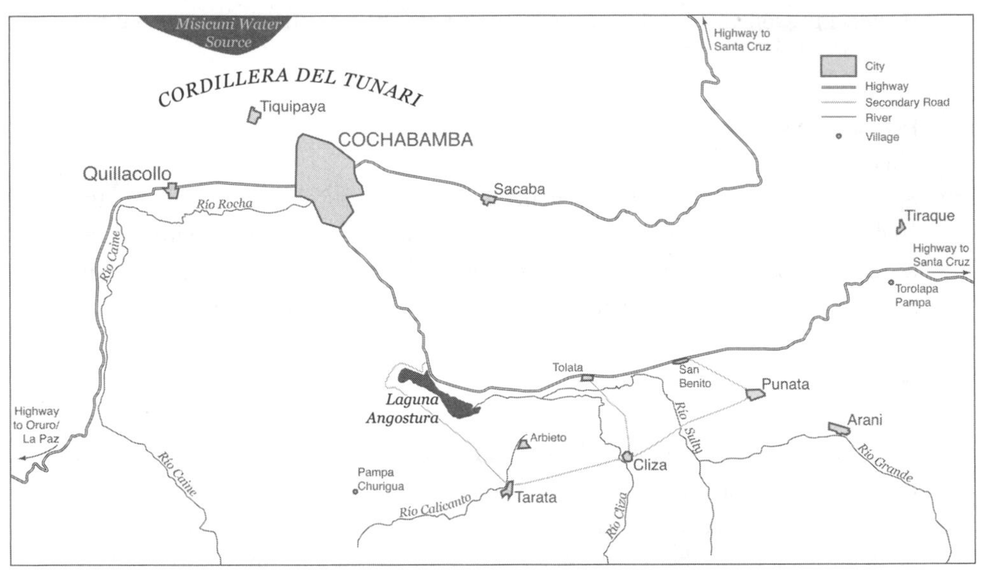

FIGURE 9.1 Cochabamba, Bolivia, with the Central Valley (Valle Central), High Valley (Valle Alto), and surrounding uplands.

(figure 9.1), which occupies a slightly lower elevation than the High Valley (Valle Alto). Documents of the Angostura project described these valley farmlands as "naturally" deficient in water and vulnerable to recurrent drought (Montes de Oca 1989, 1992). Technicians, agronomists, and hydraulic engineers relied on the geographic space of the valley bottomlands as demarcating conditions of water deficit. Representation of topographically defined chronic water shortages and recurring drought served the purpose of the Angostura project, although this depiction offered an inaccurate description of the existing agricultural landscape. Agriculture in the Central Valley was already subject to extensive use and development of irrigation systems that had enabled these farmlands to function as important breadbaskets of the Inca and Spanish colonial rulers—until the early twentieth century when national railroad connections began funneling the importation of inexpensive foodstuffs (Zimmerer 2000b). Second, agriculture and the use of irrigation extended well above the valley floor, yet only the floor became the spatial emphasis of modern irrigation projects beginning with Angostura.

Spatial naturalization of irrigation pioneered in the modern period has become integral to the standardized package of water-development models and management in Bolivia. Spatial naturalization here refers to this recurring spatial-geographic circumscription as characteristic and integral to depiction of development "problems" urging technical solutions. My argument is that this spatial-geographic reasoning is central to modern irrigation, yet it is often implicit and underrecognized relative to technical scientific and engineering approaches. In the case of the Angostura project, for example, data and models from climatology, hydrology, and irrigation engineering converged on the topographic unit of the valley bottomlands of the Central Valley, notwithstanding the counterevidence mentioned above. Data availability reinforced the role of geographic circumscription. The models, planning, and management of Angostura irrigation relied on the relatively extensive climate and soils data that were available only for the core of the Central Valley modeled as a zone of water scarcity on the basis of low annual precipitation (between 450 and 550 millimeters, or 18–22 inches), high interannual variability, and evapotranspiration rates. By contrast, the similar environment of the High Valley was characterized through volumetric estimates of flow and hence water supply. Cochabamba-based development organizations active in rational resource management encapsulated the above geographic description of water scarcity and supply areas in a series of regional studies (e.g., CIDRE 1984, 1985).

Similar spatial naturalization of the standardized package also distinguished plans for water resource development at the so-called Misicuni project starting shortly after completion of the Angostura dam and irrigation works (Oporto Castro 1999; Vera Varela 1995). Never launched and yet extensively studied and debated, the Misicuni project was first proposed in 1957 as a multipurpose project for generation of electricity as well as for supplies of drinking and irrigation water to the Central Valley. In planning the Misicuni project, the national and regional irrigation and development authorities, as well as regional Cochabamba counterparts, drew on existing ideas, models, and management schemes. Plans for Misicuni were anchored in scientific and engineering models of geographically circumscribed water deficits in the Central Valley—thus closely similar to the standardized package of the Angostura project. The standardized package planned at Misicuni relied on spatial naturalization, including cartographic depictions (figure 9.2) illustrating the interconnection of the water-deficit area (i.e., the Cochabamba valley) to the water-surplus area (see also figure 9.1). These twin areas of "sink" and "source" have served as geographic foundations of Misicuni modeling and planning. Spatial naturalization of the water-surplus area, while less publicized, was integral to concepts and calculations in this water resource modeling and proposed management. The Misicuni project map (figure 9.2) imputed water surplus in designating the mountain slopes as empty spaces except for the depiction of floods (*torentas*) while the valley base is displayed as entirely irrigated.

Scientific and engineering contrasts were characteristic of the spatial asym-

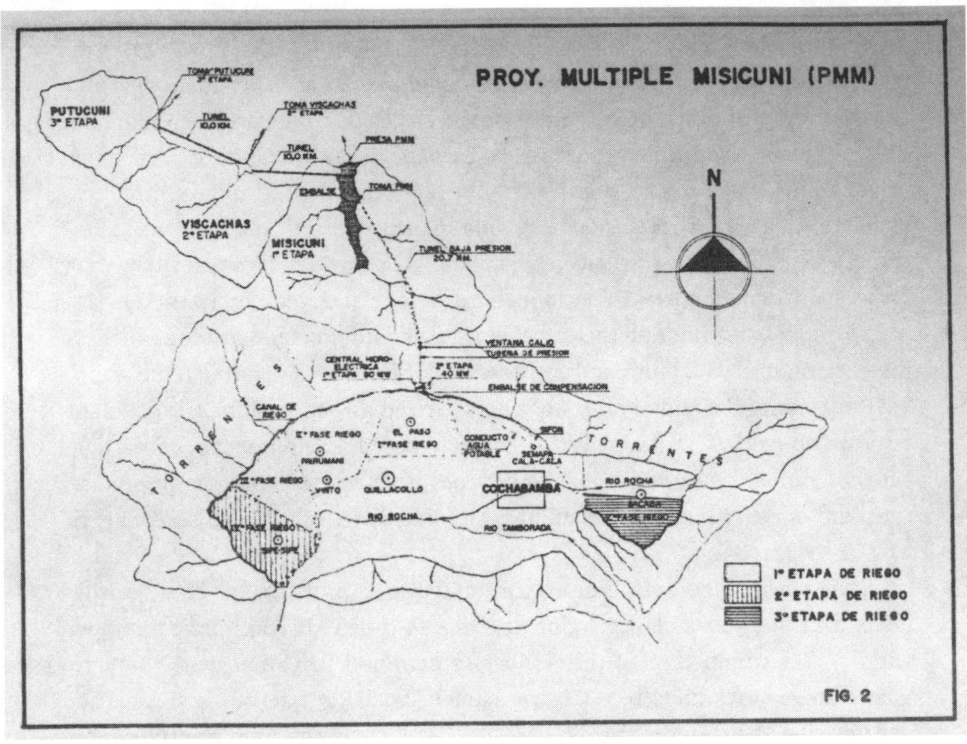

FIGURE 9.2 Map of the proposed Misicuni project (Proyecto Múltiple Misicuni), with the Central Valley (Cochabamba). From Vera Varela (1995).

metry of water-receiving places and water-supply spaces. Supply areas tended to be viewed through the estimates and modeling of the *inputs* to the place or places of scarcity (e.g., the river systems in the uplands in the Angostura project, or the "zone of floods" in the Misicuni proposals). Irrigation management and other consumption uses in areas of water supply were inferred to be inconsequential. These largely empty spaces of modeling and mapping did not, however, correspond to actually existing landscapes. Indeed the places and people of these areas of water sourcing become centers for the social and political contestation of resource development resulting from the standard modeling. Small-scale peasant farmers surrounding the Angostura project epitomized the vigorous albeit partial challenging of the standardized model of water resource development in Bolivia, because they wished their lands, otherwise labeled as a source area, to become a center of irrigation development. These High Valley agriculturalists numbered in the thousands and were among the most politically organized in the country. Recognizing that runoff from their fields and pastures filled the Angostura reservoir,

High Valley agriculturalists embarked on prolonged campaigns for the modernization of their moderately extensive yet rustic irrigation works utilizing flows of Río Cliza, Río Pucaramayu (near Punata), and Río Calicanto (near Tarata). The potential threat that High Valley farmers would stem flows to the Angostura project helped contribute in subsequent decades to political realization of irrigation projects benefiting their own farmlands.

NEOLIBERAL PERIOD, 1985–2005

Water resource management in Bolivia and in the Cochabamba region in particular hit the world-press headlines with eruption of the Water War in April 2000 (Assies 2003; Perreault 2006). The Bechtel Corporation, new owner of the privatized regional water utility (Aguas Tunari), had suddenly imposed a severalfold increase on the price of drinking water in Cochabamba city. Bechtel, which was expanding rapidly into the global water business, attempted to justify the price increase as a result of growing water scarcity in Cochabamba and what it would take to overcome this problem. Bechtel also cited the region's urgent need for extensive water infrastructure, including the possibility of reviving the Misicuni project. Massive well-organized protests against privatization took place in Cochabamba city. Violent oppression by government forces led to several deaths and scores of injuries. By November the coalition of protesting groups, uniting urban and rural social movements, succeeded in overturning the government's decision to privatize Cochabamba's water delivery.

Two decades of neoliberal rule in Bolivia transformed water resource development and irrigation in particular. Small- and medium-scale projects, many organized into participatory water users' groups and community-based resource management (CBRM), became the predominant form of water management beginning in the mid- and late 1980s in Cochabamba and elsewhere in rural areas (Perreault 2005, 2008; Zimmerer 2009). CBRM initiatives accounted for nearly all the growing number of projects following establishment of the country's first neoliberal government in 1985 (table 9.1). Many CBRM water projects were incorporated into a reaction ("soft neoliberalism") against the harsh conditions following the deep neoliberal cutbacks of services and support across multiple social and economic sectors that included agriculture and productive activities in rural areas (e.g., livestock raising, small-scale industry and manufacturing). While urban water privatization under neoliberal policies has been viewed as deregulation and reregulation (Bakker 2003; Laurie and Marvin 1999), a similar sort of transformation, largely unstudied from this perspective, occurred through CBRM irrigation projects in many rural areas (see below for details).

Rapid expansion for small- and medium-scale irrigation in Bolivia was supported through political decentralization initiatives, especially the 1994 Law of Popular Participation (Ley de Participación Popular, LPP) that served as a centerpiece of the neoliberal government's overarching Plan for Everyone (Plan de Todos). Decentralization under neoliberal policies provided an apt institutional setting for the staging and proliferation of participatory small- and medium-scale irrigation, including numerous projects framed as CBRM, as well as more general modes of social regulation (on the relation of irrigation development to the latter, see Birkenholtz 2008).[1] The standardized package of irrigation development entered a new phase during the neoliberal period in Bolivia. On one hand, water resource science and engineering occupied a somewhat less visible niche in the CBRM projects, which were typically smaller in size and scope (areal as well as financial) than the water projects of the preceding period. On the other hand, the general authority of the water sciences and engineering, along with the importance of market economics, expanded in conjunction with the decline of the national government as a legitimizing power for projects and policies (see also M. Moore 1989). Consultancies multiplied manyfold in irrigation modeling and management projects and programs in the Cochabamba region. Experts, both national and international, guided community irrigation projects as this form of water development became characteristic of the new CBRM initiatives.

The Multiple-Use Laka Laka Project (described in the opening vignette) became one of the most well known CBRM-type irrigation projects in Bolivia. It demonstrated the continuity of certain key concepts in irrigation planning and development after 1985. Environmental scarcity and spatial naturalization continued to figure prominently in the main models. Located in the western corner of the High Valley, near the town of Tarata, the irrigated area of the Laka Laka project was depicted in project documents as a problematic farm landscape prone to water shortages and recurrent drought (CIDRE 1985). This depiction was at odds with the extensive irrigation works already existing at the site of the Laka Laka project (Zimmerer 1995, 2000a, 2010). Spatial naturalization of Laka Laka also reflected a new phase in the context of Cochabamba water projects. Whereas the High Valley had served as the geographic source area for the earlier Angostura project, these farmlands were now depicted as the center of water scarcity. Canadian sponsors of the Laka Laka project, including the Montreal schoolchildren whose lunch-money donations provided chief financing through the Canadian International Development Agency, were presented the opportunity to solve, at least partly, the water deficit of the High Valley.

A similar geographic shift in the contours of spatial naturalization occurred in the Totora-Punata project that was completed in 1991. It covered a sizable irrigated area of several thousand hectares located mostly in the farmlands surrounding Punata at the eastern end of the High Valley (figure 9.1). The Totora-Punata project evidenced the influence of High Valley political organizations of mostly peasant farmers—massed into a regional peasant league (a *central*) with subgroups (*sub-centrales*)—whose agrarian agendas increasingly focused on irrigation and challenged the geographic template of the earlier standardized package of Cochabamba irrigation. One element of their success was shifting the geographic delimitation of the areas of water deficit to include the High Valley. While supported through new irrigation models and planning, this shift stemmed also from effective political pressure. High Valley political leaders and peasant organizations were able to gain necessary support of the Bolivian government. At the same time, and vital to financial backing, these High Valley residents successfully gained the support of regional, national, and international aid agencies (including Canada, Germany, and the Netherlands) and myriad NGOs.

Community-based water management at the Laka Laka project drew on a spatial design that consisted of numerous irrigating communities (joined into the special-purpose water-user groups, the Asociación de Regantes) plus the town residents of Tarata who were slated to receive potable water (Hoogendam and Vargas 1999a, 1999b). Water resource modeling and management focused mostly on the area of these communities of relatively more well-to-do irrigators—the area of water scarcity—where copious socioeconomic and environmental data were collected and compiled for irrigation modeling. By contrast, only somewhat detailed environmental information was collected and compiled in the Río Calicanto watershed, as illustrated by a map of geomorphic surfaces, distinguished by slope and substrate properties, in the sixty-four-square-kilometer Calicanto watershed (figure 9.3). The pair of consultants directing the Laka Laka project—described above—designed this map in order to calculate surface water runoff and hence water supply in the catchment area.

The Laka Laka project's scientific information about the uplands consisted mostly of physical environmental factors that influenced the supply of water in the Río Calicanto watershed (e.g., mapping of the geomorphic surfaces, as well as climate data collected in a portable weather station). As a result, the geographic space of the source area—the watershed and adjoining areas of the Río Calicanto—became populated, in effect, with physical scientific information. The source area was not subject, however, to information gathering or awareness about the sizable number of places and the people who have resided in the upper Calicanto landscapes. In fact, the Laka Laka project mostly overlooked the livelihoods and land use of an estimated two thousand people inhabiting and using the areas of water supply in the adjoining upland hills and mountain slopes.

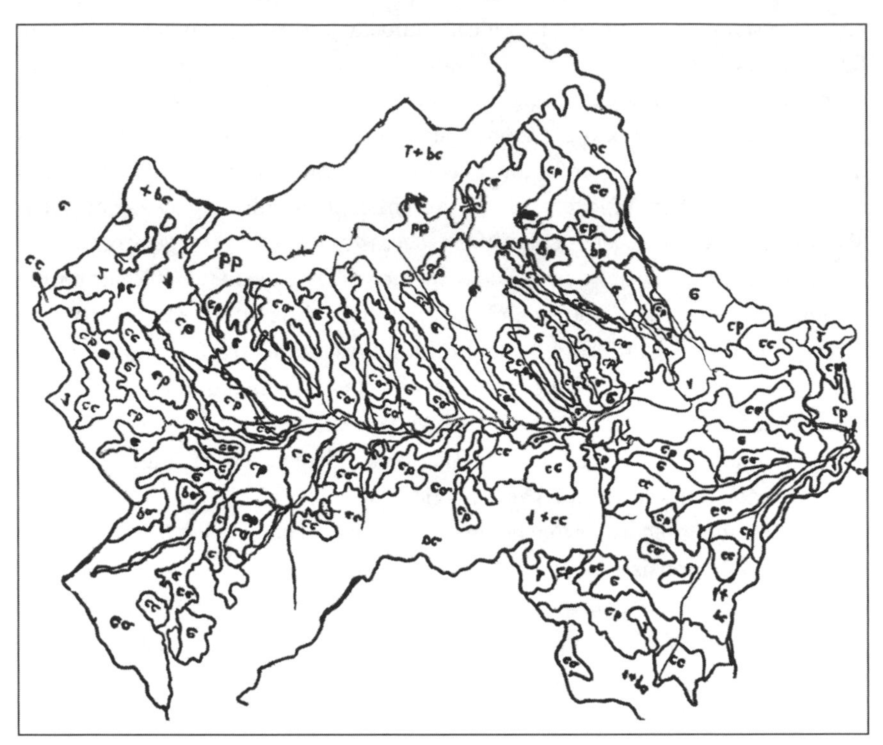

FIGURE 9.3 Geomorphic surfaces of the Río Calicanto watershed (drawn for runoff calculations by one of the head engineers of the Laka Laka project, June 1991).

This unaware overlooking of the upland inhabitants and their landscapes and livelihoods reflected divisions within Cochabamba and Bolivian society. Most upland dwellers are separated ethnically and racially as well as physically in the Bolivian landscape. People with homes in the uplands are significantly poorer and racialized as "Indians" (*indios*). Their recognition as "Indians," derived from their earlier identity as "hacienda Quechua" (*estancia runa*), stands in contrast to the identities of High Valley dwellers. The latter group comprises people recognized mostly as peasants (*campesinos*), agriculturalists (*agricultores*), and lower-class merchant/workers (*cholos*). Their mixed-blood or mestizo identity is predominant in the High Valley and is upheld as a mainstay of modern Bolivia. Yet while the upland "Indians" and their communities have been nearly completely cut off from irrigation projects, their personal movements and water flows make a lie of this separation. Asked to draw a community map, the members of Pampa Churigua chose to emphasize extralocal connections of the uplands (figure 9.4; see also figure 9.1). Their sketch map includes roadways and the "River to Laka Laka"

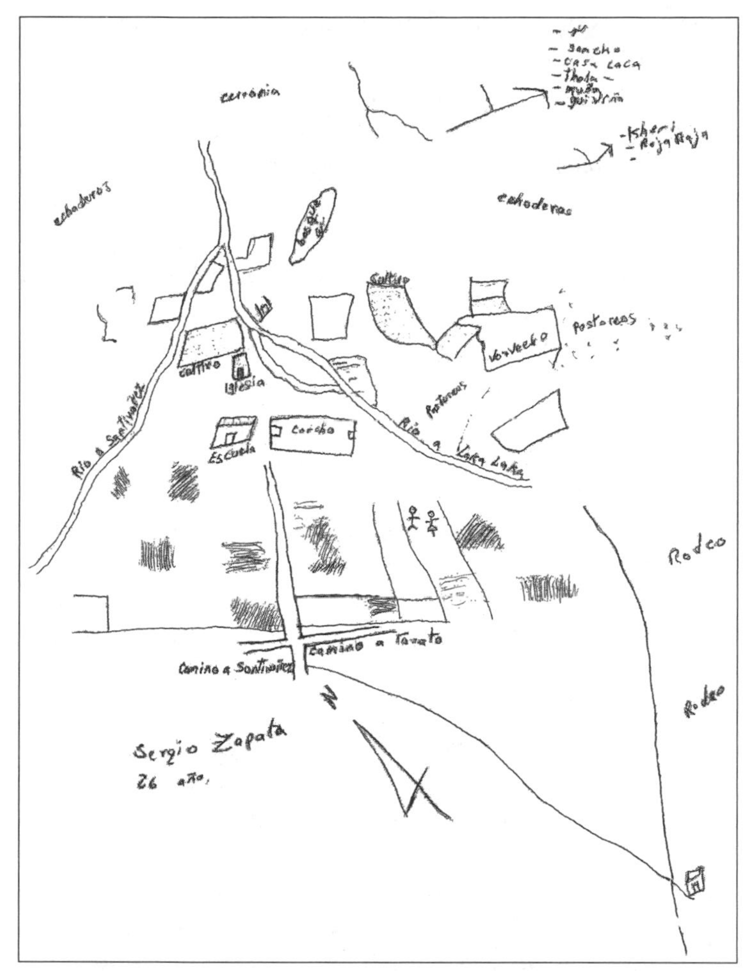

FIGURE 9.4 A sketch map of Pampa Churigua community (drawn by a group of several community members, including Sergio Zapata, May 1998, located approximately twenty kilometers upstream of the Laka Laka project).

(Río a Laka Laka), whose designation, reflecting the new irrigation project, was an alteration of local river geography (typically named for the town that the river flowed toward, or, alternatively, for the source area).

Notwithstanding these connections, it is the opposite phenomenon of fragmentation that has plagued the Laka Laka project. As described in the vignette, sedimentation in the water supply quickly became and has remained a huge problem, while it was overlooked in project design as a result of emphasis in the models and data collection focused primarily on the irrigated area. Serious sedimentation problems, which now threaten to reduce irrigation next year (2010–2011) by more than two-thirds, have contributed to occasional violent clashes. Opposing factions have caused major damage to project infrastructure, since the aqueduct and the primary feeder canal were severely damaged and have functioned below capacity. Spatial-geographic fragmentation, and the potential for water-related conflict, is not restricted to the Laka Laka project. While newer models of water resource development have embraced community-level planning and management, modeling such factors alone may fail to enable the important connections, linkages, and flows between source areas and delivery areas, such as occurred between the Río Calicanto watershed and the main valley site of the Laka Laka project.

CONCLUSIONS: CRITICAL UNDERSTANDING AND ENGAGEMENT WITH IRRIGATION DEVELOPMENT

This study demonstrates how scientific and engineering models have shaped geographically dichotomous configurations of social-environmental landscapes in water resource management. This demonstration is rooted in understanding how the concept of scarcity became widely adopted and standardized in the models and especially the management of irrigation resources in Bolivia (with emphasis on the Cochabamba region). Reasons for widespread adoption included the scientific credibility of the concept and models of water scarcity. Social and political persuasiveness also promoted the use of this concept in the context of stabilization approaches to irrigation and other forms of environmental management under prolonged neoliberal policies and governments in Bolivia (1985–2005). Spatial-geographic frameworks were integral to increased use of scarcity-based standardized packages of water resource management, particularly in irrigation engineering, planning, and management. Effectiveness of the scarcity concept within the standardized package relied increasingly on spatial analysis (biogeophysical and, recently, economic and demographic). Viewed historically, the standardized irrigation package involving spatial naturalization and the scarcity concept can be seen as having evolved during the span of the entire study period (1952–2009). At the same time, these spatial models did not arise de novo in Bolivia, either under the rule of the modern technological rationale-use state or under the economic-environmental neoliberal state (on colonial legacies in modern spatial-geographic models of irrigation development and technopolitics, see Birkenholtz 2008; Zimmerer 2000b). The evolving foundations of irrigation in Bolivia have reflected the spatial-geographic framework of scientific and engineering models of irrigation in many "developing" countries worldwide. Enhanced technical and technological capacities have furthered the spatial-geographic dimension of the scarcity concept in irrigation modeling and management, both in Bolivia and elsewhere. Yet expanding use of geographically referenced information and analysis (such as GIS and remote sensing) may contribute to countercurrents to the predominant two-part spatial-geographic model of water scarcity and surplus.

One recommendation of this study stems from overwhelming evidence of the spatializing assumption and effects of scarcity-based management on irrigation policy and planning. These dynamics tend to revolve around the physical-environmental (and, increasingly, spatialized economical rationales) contrast between geographic areas of scarcity and those of abundance or surplus. To be sure, such differences correspond to the important variation of environmental factors. Yet the models and management approaches often reinforce, and thus in a sense "construct," the sharpness and portrayal of discreteness of the areal contrasts, as revealed in this study. Practically speaking, water management involving these adjacent areas must increasingly draw from the capacity of social-environmental models (see, e.g., Taylor 2005)—as well as participatory social processes and planning—to detect, display, and integrate movements, gradients, and patchiness of key processes (both environmental and broadly social). The tendency of scarcity-based management toward dichotomous spatial designations highlights the need among scientists, engineers, and other groups involved in irrigation (e.g., planners, user groups, and other participatory organizations) to strive to recognize and respond to the often higher degree of connectedness that exists in heavily utilized waterscapes. Greater connectedness requires the scaling-up of processes often inadequately framed at the level of a single geographically contiguous water-users group, or, alternatively, at the level of a homogeneous physiographic or land-use unit. In short, regionally integrated models—scaled beyond the strictly local project units—offer the potential for more fruitful combination with spatially complex and landscape-scale irrigation modeling and management.

NOTE

1. The large projects for water resource development did not disappear completely during the period of neoliberal rule. Numerous large water projects were proposed in Bolivia during 1985–2005. Nonetheless such projects, or the possible plans anyway, did show significant changes relative to the pre-neoliberal period. Private business and extranational ownership became the main frame of reference for this phase of the planning of large-scale water projects. Bechtel's interest in the Misicuni project, for example, demonstrated the shift of possible large-scale water development to private and often international firms.

MARA J. GOLDMAN

10 THE POLITICS OF CONNECTIVITY ACROSS HUMAN-OCCUPIED LANDSCAPES

CORRIDORS NEAR NAIROBI NATIONAL PARK, KENYA

"What exactly is a corridor?" asked Thomas,[1] as we sat in Kitengela town discussing a recent series of politically contentious meetings regarding the construction of a corridor linking Nairobi National Park with the Athi-Kaputiei plains to the south. Thomas is a Kaputiei Maasai and a resident of the greater Athi-Kaputiei-Kitengela area. He thinks a corridor is just another protected area, but he's not exactly sure. He is sure, however, that the community does not want a corridor, seeing it as a threat to their land rights and livelihood practices. Kaiko, my assistant from Tanzania who was with us in the café, listened intently. He knew all about corridors—or at least he thought he did. He participated in meetings in his home village, which resulted in the rejection of a corridor project proposed by an international conservation agency, for much the same reasons discussed by Thomas. But he also worked on wildlife ecology projects—tracking animal movements and noting their migratory pathways, or corridors, in his area. He was intrigued by the way these "Kenyan Maasai" were reacting so strongly to corridors. He understood Kenyan Maasai to be more political and "modern" than they were back home in Tanzania. He was curious to see how they would handle this issue.

We were waiting for our two local collaborators to arrive to go to the meeting that had been called to discuss the corridor. When they arrived, they explained that suspicions were high regarding the corridor and the meeting was expected to be tense; they suggested I stay behind and wait for them in the café. I was, after all, an *mzungu*[2] (a white person), they reminded me, with a smile. This didn't matter to them, they reassured me, but right now I needed to be careful. Not everybody knew me, and my very presence at this meeting—which had already fueled political tension and suspicion of insiders and outsiders alike—could raise questions. They would record the meeting, and we would discuss it when they returned. They returned earlier than expected—Maasai meetings usually take several hours and it had only been two. Kaiko seemed particularly shaken up. It was good I hadn't gone, they explained. The meeting had erupted in violence and with the women in tears. One man was badly beaten and his camera destroyed. He was an "outsider," accused of being interested in *their* land. Several *wazungu* were present and were accused of pushing their interests in wildlife and in land. It was nearly a full year before the national elections in Kenya, but politics had already started to heat up. The meeting—and the corridor issue—was about politics.

One year later (January 2008), the Kenyan election occurred, and political violence erupted all over the country. "Corridors" and other conservation-related politics took a back seat to Politics with a capital P. While watching the BBC website for updates on the conflict, my e-mail box filled with messages from Tanzania: ten messages within the hour, all with the word *corridor* in the title. There seemed to be a movement under way—initiated by a group of concerned conservation scientists (mostly *wazungu*)—to submit data to the Tanzanian government on corridors. The urgency of the situation had been discussed at the annual meeting of the Tanzanian Wildlife Research Institute (TAWIRI) in 2007, where wildlife corridors became a "hot topic."[3] The e-mail was a call for all "local experts" (researchers and conservation practitioners were the only ones on the list) to contribute information to a document about wildlife corridors to be delivered to the government. A draft of the document ("Corridors TZ") outlined the official "statement" as follows: "We support the formal establishment and protection of as many wildlife corridors as is possible in Tanzania to secure national interests (water, power, ecotourism, biodiversity, carbon sinks) in a way that also takes the needs of local communities into account. We recognize that time is very short, in some cases only a matter of months, and we urge the Government of Tanzania to act quickly."

To assist in this endeavor, the document included a breakdown of corridors around the country in a table including the following categories: corridor name; type (see below); degree of urgency (stable, moderate, extreme, or critical), contact (e.g., lead scientist[s]); and references. The e-mail encouraged people to contribute information. A map was attached, made from an earlier hand-drawn sketch by a researcher, of corridors in the Tarangire-Manyara Ecosystem (TME). People were encouraged to create similar maps either using geographic information system (GIS) or by sketching corridors on an attached PowerPoint slide.

The statement was an urgent call for action. Time was short—not only ecologically (e.g., corridors were filling up), but also on the policy front. Tanzanian government officials were meeting in early February to discuss corridors, and the conservation community wanted to make sure they had "scientific" data to draw from. The e-mail thus asked for "experts" to collaborate and provide a summary (with maps) of the state of different corridor hot spots in Tanzania. We had only three weeks to do so. Since my own work is in the TME, an area labeled "critical" on the corridor table, I was drawn into a lengthy e-mail exchange with colleagues (natural and social scientists) over the best course of action. A few days and a dozen e-mails later, we decided to not contribute to the report. Social scientists argued that corridors were complex and political and that it was unwise to provide the government with information without knowing more about their intentions. The information could be used to dispossess local communities of their land. This would not only devastate the communities involved but could jeopardize the relationships we all had with communities in the area—including conservation-related partnerships. I sent an e-mail to the larger group of scientists with an article attached (Goldman 2009), which asked why corridors were being promoted so fervently in Tanzania and outlined the ways they could backfire with communities. I received no response.

The two stories above offer two different perspectives on corridors: confusion, angst, and politics on the side of the community (reflected in the Kitengela case), and certainty, urgency, and collaboration on the side of conservation scientists and practitioners (in the Tanzanian case). In both situations conservationists (scientists, practitioners, and government officials)[4] discuss corridors both as natural components of the ecosystem and as necessary conservation interventions to maintain connected landscapes and viable wildlife populations. In other words, corridors are presented as outside of politics, as necessary and urgent. Nobody in Tanzania or Kenya, outside of confused and skeptical community members and a few social scientists, seems to be questioning the appropriateness—ecologically and socially—of corridors. This stands in direct contrast to other wildlife conservation interventions that have been hotly debated and politically contentious in both countries.[5]

In both Tanzania and Kenya the creation of new wildlife policies and regulations sparked politicized debates locally, nationally, and even internationally. In Tanzania a policy to introduce a new form of community-based wildlife management (WMAs) raised contentious debate, organized community forums, and led to academic critiques (Goldman 2003; Igoe and Brockington 1999). In Kenya an initiative to relegalize hunting and game cropping received similar attention and critique (*Daily Nation* 2007a, 2007b). These two proposals are perhaps explicitly political as they deal with property rights (to land and wildlife), and the distribution of benefits from wildlife conservation. Corridors, on the other hand, are discussed, promoted, and put into policy documents with little notice, and with little high-profile protest. Corridors seem like a simple (nonpolitical) conservation option to conservation agencies and government officials. Corridors are conservation science. Most involved agree that corridors represent a necessary step toward landscape connectivity and thus the improved conservation of wildlife. This is not just the case in Tanzania and Kenya, but globally. In fact corridors represent the most visible expression of the new "landscape conservation" boom (Zimmerer, Galt, and Buck 2004).

In this chapter I ask why corridors are circulating so well, and what the political and ecological implications of their "application" in particular places might be. I use concepts from science and technology studies (STS; boundary objects, standardized packages, and the "bandwagon" effect) to explore the first question—how corridors are being stabilized as (social) facts. Then I discuss the politics implicit in the application of corridors on the ground, in Kitengela.

THE CONNECTIVITY BANDWAGON: CORRIDORS AS TRANSPORTABLE CONSERVATION PACKAGE

> "Connectivity" is a metaphor and an idea that has captured the imagination of conservation biologists around the world. There remains, however, a big gap between the idea of connectivity and pragmatic insights regarding on-the-ground actions that should be taken in the name of connectivity if the goal is long-lasting conservation.
>
> —P. Kareiva, "Introduction: Evaluating and Quantifying the Conservation Dividends of Connectivity"

Corridors offer a structural solution to the complex problem of maintaining functional ecological connectivity. Corridors—as places where animals migrate, and often where native vegetation remains intact—connect otherwise isolated protected areas (e.g., national parks). With habitat fragmentation and destruction recognized as the largest threat to biodiversity globally (Wilcox and Murphy 1985), corridors offer "a practical, on-the-ground strategy for restoring landscape connectivity" (Anderson and Jenkins 2006, 5) and protecting biodiversity. While the concept of ecological connectivity is difficult to measure (including genetic-, population-, and community-level measurements), corridors are easy to visualize.[6] Despite the lack of evidence showing that corridors do in fact *functionally* connect landscapes and wildlife populations, they seem to represent tangible "doable" conservation projects. The "doability" of corridors is obtained through the alignment of different "work" arenas (computer modeling, theorizing, experiments, ecological fieldwork, policy making, funding agencies; Fujimura 1987), but also by the physical construction of demarcated spaces on the ground. Corridors provide strong states with concrete, land-based options for implementing the complex goals of ecological connectivity.

Corridors are quickly becoming the leading conservation intervention in fragmented (and even not so fragmented) landscapes in East Africa and around the world. While scientific consensus on the functional value and cost-effectiveness of corridors may not exist, they are promoted as a prevention measure against the known extinction risks associated with fragmentation (Berger 2004; see also Simberloff et al. 1992; Noss 1987). In places like Kitengela, where habitat fragmentation is occurring quite rapidly, the push for corridors thus takes on urgent tones—to assure that land is set aside for protection before it is too late. Corridors are promoted by scientists (Hilty et al. 2006), international conservation organizations (Bottrill et al. 2006), and national governments (MLHUD 1996; MNRT 1998; Kenya 2007). The title of a recent book, *Applying Nature's Design: Corridors as a Strategy for Biodiversity Conservation* (Anderson and Jenkins 2006), suggests that corridors are not just a good conservation measure, but are in fact the only "natural" choice. In this sense, corridors can be seen as the latest conservation "bandwagon," foreclosing other possibilities.[7]

CORRIDORS AS BOUNDARY OBJECT

According to Star and Griesemer (1989, 393), "boundary objects are objects which are both plastic enough to adapt to local needs and constraints of the several parties employing them, yet robust enough to maintain a common identity across sites." This means they can be translated differently within different social worlds, while maintaining enough common ground across groups to enable communication and work. As such, boundary objects are concepts or "things" that link various actors and agendas in an "arena of mutual concern" (Clarke and Star 2008, 222), enabling "collaboration without consensus" on difficult issues.

As a boundary object, corridors are necessarily ambiguous, with varying definitions across the different disciplines employing the term—conservation biology, landscape ecology, landscape and urban planning, landscape architecture, and wildlife management (Hess and Fischer 2001)—as well as within international conservation organizations,[8] state agencies, and local communities. Corridors come in many shapes and sizes (interpreted and translated differently in different contexts and disciplines).[9] They can be discussed as wildlife migratory pathways, conservation conduits or connecting agents, habitat remnants, barriers (to development), filters (for ecological processes), sources or sinks (for metapopulations), linear fragments, urban greenbelts, conservation extensions (Hess and Fischer 2001; Rouget et al. 2006), and for local communities, "protected areas" and land-acquisition politics. Corridors simultaneously refer to naturally occurring landscape *structures* and a conservation intervention with a particular *function* in mind. Corridors are habitat and conduit, structure and function.

Structurally, one author (Hobbs 1992, 389) suggested, "almost any strip of vegetation could be viewed as a corridor." Yet when defining corridors functionally for wildlife, he argued that "the important component of the corridor is that it allows movement *from* somewhere *to* somewhere else" (389, emphasis in original). While a bit more specific, this definition is broad and ambiguous enough to enable wildlife migratory corridors in particular to be discussed in multiple ways. The Corridors TZ document prepared for presentation to the Tanzanian government outlines five types of wildlife corridors, broadly grouped into two categories: (1) those that connect "two patches of suitable habitat by passing through a matrix of unsuitable habitat," and (2) "an area used by animals to pass from one habitat patch to another." The five specific types of corridors are discussed as follows (2–3): First, corridors can exist merely as "arrows on a map," including poorly documented, but historically known, migratory routes of particular animals, and the "shortest distance between two PAs [protected areas] which animals could travel." So there is potential "functional" connectivity in well-known animal routes, or structural connectivity that may or may not be functional. Nonetheless, by drawing the lines of such corridors on a map, they become, officially, corridors.

The second and third categories of corridors are more troubling: "uncultivated lands between PAs without documentation on animal movement," and "continuous or semi-continuous non-agricultural land between PAs with anecdotal information on movements." Here the structure of a corridor has expanded dramatically to all land that has not yet been cultivated, that happens to lie between protected areas, that wildlife may or may not use. The rationale behind this expanded definition is that in the future the impacts of climate change may lead to wildlife movements to new places, and that corridors should be planned in preparation.

The fourth and fifth definitions listed in the document reflect more of the traditional use of corridors for wildlife migrations: "known animal movement routes between two PAs," and "proposed connectivity of important habitats." In these final definitions animal migratory routes are naturalized as corridors, and corridors are equated with connectivity, which scientists argue is potentially dangerous. According to Weins (2006, 24), functional connectivity is where the "real action is," and it is not necessarily about corridors, but rather about how landscape structure interacts with various biotic and abiotic properties. Weins does acknowledge, however, that structural connectivity is easier to visualize and measure than functional connectivity. We can see corridors in maps and spatial images. In fact the term *corridor* initially came into use in landscape ecology, because of its utility in describing and analyzing landscape structure in aerial photographs and satellite images. A corridor is a linear structure that differs (in pixel display) from the matrix on either side. Yet corridors quickly came to reflect the notion of functional connectivity across landscapes. And pioneering landscape ecologist Richard Forman himself (1991, 81–82) was acutely aware of the value of corridors as a boundary object when he stated that this "single structural object . . . provides a crystal clear objective that landowners, decision-makers and diverse scholars can readily understand and communicate." Landscape ecology contributed to the set of tools that make corridors so attractive and accessible to scientists, policy makers, and conservation practitioners.

CORRIDORS AS STANDARDIZED PACKAGE

According to Fujimura (1992), while boundary objects enable communication, their very flexibility can prevent their stabilization as a fact. Stabilization, she argues, is accomplished through a standardized package, which combines boundary objects with tools, methods, and theory (176). Standardization occurs because of a shared commitment to action (32) and a need to "get the work done" for a common cause. Here, corridors speak to the urgent call to save biodiversity (Terborgh 1991), particularly in Africa (Neumann 2004). Corridors align the connecting power of the boundary object with old and new ecological theories and conservation approaches, and with the power of geospatial tools into an attractive package that is acceptable, translatable, transportable, and fundable across disciplines and communities.[10]

Scientifically, corridors are recognized as a dominant feature of the new landscape-conservation paradigm, but also have a long history in the context of wildlife management (Borner 1985; Harris and Gallagher 1989) and are linked to older equilibrium-based conservation models associated with island biogeography (MacArthur and Wilson 1967). Politically, conservation corridors are often implemented as part of community-based conservation (CBC) but also resonate with the "back to the barriers" conservationist movement opposed to CBC (Hutton, Adams, and Murombedzi 2005), because they strengthen the ecological integrity of national parks and thus the "fortress" model of conservation. In Kenya's new wildlife policy, corridors are listed in the section "Wildlife Conservation and Management *in Protected Areas*" (Kenya 2007, 18, emphasis added), as a way to "strengthen the ecological network of national parks and reserves."

In addition to bridging political positions and theoretical concepts, corridors come with a set of appealing tools—GIS technologies, satellite image analysis, global positioning system (GPS), wildlife-monitoring techniques, and land zoning and/or acquisition procedures. Corridors fit nicely with the spatial, territorial nature of conservation (Harris and Hazen 2006). They are particularly appealing to governments such as Tanzania and Kenya—where the state claims the right to reallocate land for the public good. With the tools to track wildlife movements, the flexibility to call such movements corridors, and the ability to place it all on a map—it is not only "easy" to promote corridors, in the sense of "doability," but incredibly powerful. I will now turn to a discussion of how this was attempted in Kitengela and why it backfired.

TO PROTECT OR DIVIDE THE LAND? CORRIDORS AND LAND POLITICS IN THE ATHI-KAPUTIEI PLAINS

Kitengela is a predominately Maasai community on the southern edge of Nairobi National Park (NNP; see figure 10.1). The park is not fenced on the southern edge, where the Maasai-managed rangelands of the Athi-Kaputiei plains (of which Kitengela is a part) have historically acted as an important dispersal area for NNP wildlife during the wet season. The small park (117 km^2) is a part of the larger Athi-Kaputiei ecosystem, which is twenty-one times larger than the park itself (Reid et al. 2008). Bordering a densely populated urban center, NNP would not have survived this long if not for the hospitable environment for wildlife provided by Maasai living in the area (see figures 10.2 and 10.3). In fact Gichohi (2000) found that many wildlife preferred the grazed and burned pastures of the Athi-Kaputiei plains to the coarser vegetation inside the park. She argued that keeping these pastures open to wildlife was vital to the survival of the park (see figure 10.3). However, in reaction to both Kenyan land policies and high land prices due to the proximity of Nairobi, Maasai in Kitengela have not only privatized their rangelands but have sold large portions of them. With this process has come the erection of fences—clearly delineating the boundaries between land units and keeping unwanted animals (domestic and wild) out. Fences fragment the landscape and deny animals access to key grazing resources (Reid et al. 2008; see figure 10.2).

The acceleration of the subdivision and fencing of land is clearly unfavorable for wildlife as well as for livestock in Kitengela, which need access to vast (heterogeneous) pasture resources to respond to variable climatic conditions. In an attempt to halt and even reverse the process, an innovative program was put in place, in addition to community-based research and action, and people have begun to take fences down and encourage wildlife on their lands. The project was initiated by the local NGO, Friends of Nairobi National Park (FoNNaP) and is now run and financed by the international NGO The Wildlife Foundation (TWF). It protects wildlife pathways and dispersal areas by rewarding people for keeping their land open to wildlife through a lease program. The project began in the locations closest to the park boundary, rewarding participants 300 Kenyan shillings/acre/year (US$3.75) to refrain from poaching, report poaching by others, and

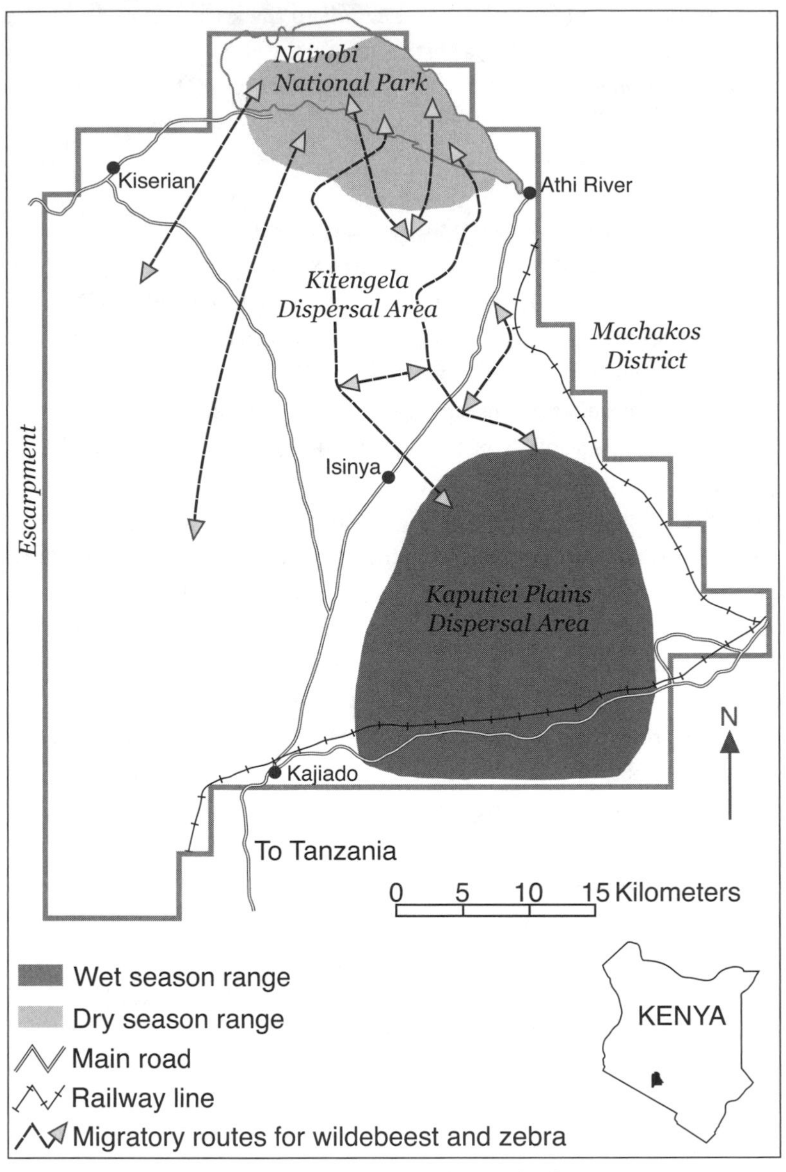

FIGURE 10.1 Map of the greater Athi-Kaputiei-Kitengela plains area. Map adapted from Nkedianye et al. (2009).

FIGURE 10.2 Zebra grazing in Kitengela pasture, in front of a primary school; note fence in the background. Photograph by Mara Goldman, 2007.

FIGURE 10.3 Wildlife and cattle grazing side-by-side in the Athi-Kaputiei plains. Photograph courtesy of ILRI (International Livestock Research Institute, Kenya; photographer: Dave Elsworth).

refrain from fencing or subdividing the land (Nkedianye et al. 2009).[11] The project has been so successful that it expanded from 11 participants covering several hundred acres in 2000 to over 120 participants leasing over 11,000 acres of land in 2007. The project has expanded to other areas within the Athi-Kaputiei plains and currently has a waiting list for new members.

The importance of Athi-Kaputiei plains to NNP was made clear by the park's branding ceremony in December 2006, where the Kenyan Wildlife Service (KWS) promoted a new "image" of the park to the public. The ceremony was held at a primary school near the border of the park (see figure 10.2),[12] and community conservation was the central theme. The ceremony was crowded; community members of all ages were present. Moods were high. There was much talk about successful community conservation and future possibilities to improve livelihoods while supporting wildlife in the area.[13] It seemed that the future of NNP as well as the KWS-Kitengela relationship was a bright one. Less than one month later, everything had changed.

KWS had been working closely with local politicians and the pastoral landowners association (PLA) to come up with a "master plan" for the future management of the area—for the survival of wildlife, livestock, and Maasai. There was overwhelming concern that the park was facing grave threats and that the TWF leasing project was not enough (Mbaria 2006). In fact, in an article in the *East African* (a regional newspaper), just days after the branding ceremony, the East African Wildlife Society (EAWS) criticized the ceremony in particular and KWS in general for an unrealistic and idealistic presentation of the state of the park. The director of EAWS thought the park should be fenced and argued that there was "no more room for theories" if the park was to be saved. He further lamented what he considered "the blind support the wildlife body has given to 'the issue of an imaginary corridor in Kitengela.'" This refers to KWS support of the FoNNaP project, which, while not a "corridor" project per se, was designed to keep land open for wildlife movements across the Athi-Kaputiei plains.[14] There was no agreement within the conservation community on how to "save" NNP. While the FoNNaP project seemed to be succeeding, there were concerns over future funding and the sustainability of the project. Some argued the corridor was already closed, as shown in declining wildlife numbers within the park. And there was an active lobby to degazette the park to make way for the growth of Nairobi and settle poor urban slum dwellers.

In such a climate, a master (land-use) plan was needed if the park was to be saved. The local PLA was working with KWS on the plan. They were particularly concerned with urban growth and increased land sales, and saw an integrated land-use plan as a necessary addition to the lease project to keep land open for both pastoralists and wildlife. Yet KWS had also been working on plans to "compulsorily acquire land in the Kitengela corridor and the Athi-Kaputiei wildlife dispersal area," according to the news article cited above (Mbaria 2006). A draft of their plan was leaked to the press before the community was able to review it, and according to the article (Kinyungu, Kisia, and Mwakio 2006): "The proposed project by the KWS to buy back land and create a wildlife migratory corridor for the park is a massive venture without parallels anywhere in the world. The only other country known to have migratory corridors criss-crossing highways is Canada."

This article sparked the suspicion, anger, and unrest discussed at the opening to this chapter. Rumors and accusations flew in all directions, and several meetings were held throughout Kitengela. Some thought it was just media hype, but there was a draft document with signatures on it. The document was still called the "Master Plan," but was dominated by discussions of corridors and land acquisition, instead of the more integrated planning that people had been discussing for months (KWS, TWF, and KILA 2006). Island biogeography theory was cited as justification for corridors to protect biodiversity. There was no discussion of the positive relationship between wildlife and livestock grazing as shown in previous research. There was no discussion of the integrated management of the landscape promoted by the initial idea of a land-use master plan. There was a map in the document, showing a potential corridor going through land that belonged to individuals. People were hysterical. "They have sold our land," they cried. "We do not want the corridor," they repeated over and over. During this time I asked different Maasai residents what exactly a corridor was. They almost inevitably replied, a bit in doubt, "Well it's a protected area, isn't it?"

For Maasai in Kitengela, a corridor represented land acquisition, and they did not want it or, in many cases, even to discuss it. One young man, when interviewed about a meeting held to discuss the corridor issue, said that the agenda was "to state no to the issue of corridor!" Another described the agenda as "corridor discussion and how to acquire land for animal passage." A third suggested the meeting was held to "discuss the issue of . . . an animal throughway." A corridor in their eyes was land acquisition for wildlife. In all discussions—at the meeting itself and in interviews—people used the English word *corridor* (even when speaking Maa or Swahili). In doing so they were talking about this *thing* that was being planned and mapped that threatened their rights to land and resources; they were not talking about natural migratory pathways used by wildlife.

Maasai in Kitengela know about the pathways of wildlife and know that land-use change and fences are threatening them. They had been working with local NGOs on ways to protect the pathways (by taking down fences). Some were even trained in the tools associated with corridors—GPS and GIS, to map the fences. This enabled a strategic (scientific) selection of households to be involved in the lease program: those located in or near wildlife migratory pathways and dispersal areas. The tools were used to map various resources in the area and show where barriers to movements were, not to suggest new barriers (corridors). According to a community leader and coordinator of the lease project, corridors to Maasai are identified as having a barrier on either side. A corridor is seen as "a specific [wildlife] route that won't change, which doesn't really exist. It brings in the issue of protectionism . . . it will be another park, specifically for wildlife. Because there is a map—a line, at some point there will be a fence, or a wall. It brings fear that there is an identified area that will be protected. And the people begin to wonder if the area outside the corridor, that meant to be 'free of wildlife,' will really be so, if the wildlife will really be restricted to the corridor."

Discussion of dispersal areas, he suggested, does not bring this kind of fear. This is because corridors, even as they connect, also divide. People even feared that they would not be able to access their neighbors on the other side of this "corridor" (pers. comm., Makui Ogeli, 2007) In addition to being scared and angry about the potential of losing their land to a corridor, many people resented that they had not been properly consulted on the project. They had, however, been actively involved in the initial discussions of a master plan, through involvement in regular meetings held by PLA.

Discussions about the master plan were not easy. There were many doubts and much resistance by individuals over the limitations on land sales. Yet there was general agreement that something needed to be done or they (as a community) would lose all their land to outsiders—Kikuyu and Kamba farmers and urban dwellers, upper-class Africans and *wazungu* from Nairobi, and the flower and export industries. The master plan called for integrated, zone-based planning and limitations on land sales to keep urban centers in check and prevent increased land fragmentation. This was quite different than designating and mapping a corridor. The master plan was, as a local community leader explained, about "protecting *their* land" (*eramatare enkop*). The corridor, on the other hand, was about losing their land, having it taken away by the government. This explains the outrage community members had at the news of the pending corridor. Community outrage led local politicians to distance themselves from the corridor, and claim that it was different from the master plan of which they had been a part. The distinction between these two approaches (and the assumed role of *wazungu* in the process) was expressed by a former chief at a politically tense meeting: "I think you have heard what the chairman . . . has said [regarding his role in the master plan]. They made up the Master Plan and are saying this. Let's not fight and let's say that this wildlife are a part of our life. What is burning here is the issue of *wazungu*. They look after wildlife and they collaborated with the government [on the corridor issue]."

Another elder spoke strongly at the meeting linking the corridor directly with KWS: "There is the issue of [the] corridor and we don't like it because we tend to think that KWS wants to take away our lands, and we feel it is not good. We have been living with wildlife and we have had no problems so all the Maasai are against this. That is why we had to plan this meeting." Politics took center stage at this meeting, with the crowd demanding to know who was responsible for the corridor and what they would do about it. The elected county councilor's own re-election was on the line (it was an election year), and he acted strategically. At the first meeting, called just after the news story, he urged people to separate out the corridor from the master plan: "We need to separate the Master Plan and corridor because the issue of the Master Plan had already been discussed. The issue of the corridor has to be addressed properly so that everybody understands much better." At the second meeting—which was called because at the first one too few people attended and little was resolved—he repeated the following sentiment (in near identical form) several times: "Differentiate between corridor and Master Plan because they are different. And we are not going to mix them." As this particular meeting began to erupt in violence, he reiterated his "distance" from corridors: "I am saying that the Master Plan was done as the Council and we marked some borders, [such as] where the town is going to reach. I am standing here in front of you saying that the corridor—I have no idea of what it is."

A few weeks later, tensions began to ease regarding the corridor issue, and local leaders began holding meetings again to discuss the master plan. There had been much discussion, in the tense and political meetings on corridors, about the need for clarification on the issues so that people could contribute to the debate. After hearing more about the focus of the master plan, one elder commented that "there is this problem of people selling the land, but also the land is getting full. This is very good work [the master plan], for stopping town from expanding and for keeping people from selling their land." The elder also reminded everyone that there was a lot of interest in this land by those (NGOs, tour companies) interested in wildlife. There was value, as such, in keeping land open and wildlife around—if they could benefit through ecotourism.

CONCLUSION

When I returned to Kitengela in July 2008, I inquired about the "corridor" issue. I received smiles and chuckles, followed by a reminiscing of how highly politicized the whole discussion had become. People would then tell me that nobody really talked about corridors anymore, but that they were still working on the master plan. The master plan will try, through increased support of the lease project and land-use planning, to protect wildlife migratory pathways. Some were afraid though, that corridors would sneak back into the plans. And their fears were well founded. Conservation work in the area is now being supported by an international NGO that promoted corridors in Tanzania, despite local resistance (Goldman 2009); the Kenyan government added corridors to the national wildlife legislation; and the scientific community is still focused on corridors as the best way to achieve landscape-level conservation. At a wildlife policy review meeting in Nairobi (April 2007), a member of the review committee argued that certain corridors are so important that they will be pushed, if by force, through land acquisition by the government. It seems that corridors are here to stay and that perhaps the Maasai are correct to view them as land-acquisition politics.

Even if corridors may be playing out in Kenya and elsewhere in Africa as land-acquisition politics, that does not diminish their standing as scientific models. Conservation science has always been an inherently political endeavor, as has been well illustrated by political ecologists working in Africa (Neumann 1997, 1998; Schroeder 1999a, 2008). From a political ecology perspective the story I have told here seems like a simple case of conservation politics as usual—the articulation of unequal power dynamics resulting in the application of exclusion-based territorial conservation practice. This, however, tells only part of the story. Corridors do present good options for the Kenyan and Tanzanian states to continue exclusionary territorial-based conservation. Yet in doing so, they are mobilizing large-scale support from international and locally based conservation agencies and scientific communities. Some of that support is political, but some of it comes from scientific theory, methods, and tools.

As a standardized package, corridors have enabled communication and consolidated support across various social worlds: conservation biologists, wildlife ecologists, landscape ecologists, international (and local) conservation NGOs, donor agencies, national and regional government agencies and officials, national park authorities, and local advocates. As all these different groups grapple with complex ecological questions in their "crusade to save biodiversity," corridors represent a "doable" approach to an incredibly complex set of problems. It is doable because the political, financial, social, and scientific pieces are all lined up (Fujimura 1996). Yet in Kitengela, the corridor project failed, at least for now. It was defeated on political grounds. While the very ambiguity of the word *corridor* enabled its use in discussions with community members about wildlife pathways, it also enabled its translation onto local fears, fears based on a history of socially unjust conservation policies. Boundary objects can, and often do, get translated differently within different "social worlds," but they are supposed to maintain enough common ground across groups to enable dialogue. This did happen in Kitengela, until the dialogue turned sour, something directly associated with the historic lack of trust in conservation authorities by Maasai (Wynne 1992).

Understanding corridors as a standardized package uncovers the reasons why they travel so well, and unmasks their image as "natural" objects—in terms of their existence within nature (as wildlife migratory pathways) and as the most obvious (and reasonable) conservation solution. While there is general agreement that ecological connectivity matters (Crooks and Sanjayan 2006), connectivity depends on many factors,[15] and setting up "wildlife throughways" does not necessarily enhance landscape connectivity. It can, on the contrary, further divide the landscape into places of people and places of wildlife. This could be counterproductive if we discover that some wildlife prefer to be in places with people and livestock (see figures 10.2 and 10.3; Reid et al. 2003). It could, in the end, encourage the very fragmentation that corridors are proposed to heal.[16]

NOTES

1. All names used in this chapter are pseudonyms, to protect individual and organization identities.

2. *Mzungu* (pl. *wazungu*) is the Swahili word used to refer to Americans and Europeans, or "white people."

3. I was not present at the conference; this is how others who were present explained to me the way in which corridors had been discussed.

4. These three groups are clearly different in many ways (Walley 2004); my lumping them together here reflects the continuity, to some extent, in the way in which they deal with the corridor issue.

5. While this chapter shows that corridors as a conservation plan have become politically contentious in particular communities, discussions about corridors in general have not taken the same high-stakes political tone that has surrounded the other debates discussed here.

6. See Brosius (2006) for more on visualization techniques for conservation.

7. Fujimura (1988, 261) defines a scientific bandwagon as "when large numbers of people, laboratories, and organizations commit their resources to one approach to a problem because *others* are doing so."

8. Such as the World Conservation Union, the WWF–World Wide Fund for Nature, Conservation International, and in the case of Tanzania and Kenya, African Wildlife Foundation, Frankfort Zoological Society, and the Wildlife Conservation Society.

9. There are, for instance, conservation corridors (Hobbs 1992), landscape corridors (Forman 1991), wildlife corridors (Soulé and Gilpin 1991), riparian corridors (Naimen, Decamps, and Pollock 1993), (vegetation) strip corridors (Forman and Gordon 1981), and continental corridors (Soulé and Terborgh 1999).

10. Here, while drawing on Latour's notion of transportability (1987), it is not as "imutable

mobiles" as much as mutable ones. As Fujimura (1996, 215) argues, "The more portable a tool is, the more reliably it will be reproduced in other situations. Portability refers to the qualities of simplicity and ease of movement and use." The very plasticity of corridors makes their transport easier. See also Callon (1986) on translation.

11. The money has proved particularly valuable in paying school fees and helping families make it through bad years without having to sell their livestock.

12. The school was in Embakasi sublocation. *Embakasi* is a Maa word (the Maasai language) meaning "boundary."

13. I was present at the event and draw from observations as well as informal interviews.

14. Here the fuzziness of the *corridor* word is apparent—referring to wildlife movements and open land connecting a national park to a "dispersal area," which is different than how the word *corridor* is used by KWS, as discussed below.

15. Such factors include (but are not limited to) climate, vegetation quantity and quality, fire regimes, land use, hunting patterns, and community support.

16. For instance, even with the political defeat of corridors in Kitengela, there has been very little discussion of alternative approaches, such as the wildlife-livestock co-use of the landscape, on *ecological* (i.e., scientific) grounds.

Part 3

MARA J. GOLDMAN

CIRCULATION OF ENVIRONMENTAL KNOWLEDGE

NETWORKS, EXPERTISE, AND SCIENCE IN PRACTICE

People's relationship with and knowledge about the environment is strongly shaped by several factors of ecological, social, and political import: local history; the heterogeneity of the physical environment; the heterogeneity of human settlement, institutions, technologies, and social relations; and the spatial arrangements and distance between human and nonhuman agents in particular places. In other words, people-environment relations, and ecological knowledges, are very much shaped by history and place. Yet these relations are also shaped by knowledge-production processes, as well as socioeconomic and political networks, which stretch beyond "the local" and link to faraway places, and other historical moments. As such, local places, ecologies, and knowledge constructions are always only partially local, linked as they are to different natural-social networks, both horizontal and vertical. How then, is expertise—about local process as well as "global" knowledge constructs designed for universal application in different "local" places—determined? Do the same (scientific?) rules of conduct apply when knowledge circulates in nonscientific spaces? Do global (universal) knowledge productions transform as they circulate through and into local ecologies, politics, and culture? These are all important and complex questions that are central to a nuanced understanding of the politics of environmental knowledge; and it is these questions that are the primary focus of this part of the book. Chapters in part 3 address these questions and others, by approaching the production-application-circulation nexus of the politics of knowledge from a circulation perspective, while simultaneously paying close attention to questions regarding the production and application of knowledge.

Chapters in part 1 highlighted the complexities involved in knowledge-production processes among the most privileged of knowledge producers: environmental scientists. While the primary focus of part 1 was on the production aspect of the production-application-circulation nexus, many of the chapters simultaneously addressed questions pertaining to the circulation of scientific knowledge. This is because, as discussed in the introduction, the very production of knowledge is intricately tied up with its circulation—certain ideas become information, and certain forms of information become accepted facts primarily because of the ways in which they successfully circulate through various (social, political, economic, and academic) networks. The specific political, social, and ecological impacts of the acceptance of particular forms of information as fact through their *application* to particular environmental questions was also touched on by some chapters in part 1 (Forsyth, Duval, Campbell). This analysis was further expanded on in part 2, which addressed the specific and highly political implications of the *application* of certain kinds of environmental knowledge on particular peoples and environments. Part 2, while focused on the politics of application, addressed many issues particular to the *circulation* of knowledge—for it is through particular circulation processes that (1) certain knowledge constructs are considered more legitimate, correct, or "true" then others, and (2) knowledge produced in one (situated) location comes to speak to, impact, and otherwise affect knowledge-production processes, livelihood practices, and environments in other (often faraway) places.

In this final part of the book, contributors begin their analyses from the vantage point of circulation. Yet this is just their starting point. As with the two previous parts, all the contributors to this part also analyze, in different ways, the politics associated with the production and application of knowledge, which cannot be separated from questions regarding circulation. In so doing they touch on several overarching themes of the book, and add new angles to particular questions addressed in previous parts. For instance, while scientists, as privileged knowledge producers, were the key focus of most of the chapters in part 1, in part 3 the focus shifts to nonscientific producers of knowledge—some who circulate knowledge back into scientific networks and others who create separate parallel networks. And as chapters in part 2 focused on the often destructive impacts of the application of "foreign" scientific knowledge in local environments, chapters in this part highlight the sometimes promising potential of hybrid knowledge constructions that can occur when foreign scientific knowledge is localized and transformed (Phadke, Rocheleau, Ramisch), as well as the dangers of mistranslation and appropriation along the way (Galt).

Focusing on how knowledge circulates, and what happens to particular knowledge claims and assemblages in the process, exposes the inherent *multiplicity of knowledge production*—within and outside of scientific centers, one of the main themes of this volume. All the chapters in this part illustrate the need to recognize multiple knowledge-production processes, the way they interact with science, and the potential impacts on scientific knowledge and practice as well as on local ecologies and livelihoods. In Ryan Galt's chapter we learn of the transformations in knowledge about pesticide dangers as such knowledge circulates through global produce markets and regulatory systems. By following two specific pesticides as they travel (as "actants") through complex global networks, Galt illustrates how particular scientific forms of knowledge regarding pesticide production, use, and risk came to dominate within U.S. circulation domains. He then uncovers the complex transformations that occur as these pesticides travel beyond U.S. borders and as scientific risk-based knowledge circulates in quite different contexts—to regulate produce imports from Costa Rica back into the United States. Due to weak links in regulatory governance and strong links in global capital prerogatives, scientific knowledge regarding risk is transformed in ways that are potentially hazardous for the people and ecologies involved—both in Costa Rica and in the United States.

The social and ecological consequences of the transport of scientific knowledge around the world is not new and was addressed to some extent in part 2. Yet the consequences need not always be solely negative, and this is clearly illustrated in the chapters by Phadke, Rocheleau, and Ramisch. Roopali Phadke's study shows that when scientific knowledge circulates, even in the form of powerful technological objects such as dams, it encounters and is transformed by nonscientific knowledge, users, and livelihood systems, often in progressive ways. By focusing on social movements in India, Phadke shows how knowledge hybridizes as it circulates between local-global, expert-lay worlds. In the process the very boundaries between epistemologies blur, as does the line between science and politics. She explains how, through this process, local Indian leaders were able to "reclaim" hegemonic technologies, and thus affect the production of new forms of knowledge regarding water management in India. Similarly Dianne Rocheleau's chapter points to the centrality of hybrid knowledge productions to the livelihoods of the people she is working with in Zambrana-Chacuey, the Dominican Republic. She shows how particular kinds of knowledge—ecological, political, social, cultural, economic, and historical—circulate through communities in Zambrana-Chacuey, producing creative and productive synergies, hybrid knowledge formations, and essentially new social-ecological knowledge and practice. In fact Rocheleau proposes that a methodology of "seeing multiple," from situated perspectives within polycentric models is necessary to truly grasp the very complexity of natural-social systems in the local-global sites in which most political ecologists work.

Like Phadke and Rocheleau, Joshua Ramisch is also interested in how knowledge circulates at the interface of scientific and "indigenous" expertise and practice. His chapter explains a programmatic effort by a research team to understand the multiple ways in which agricultural knowledge is performed by the various participants involved in an agroecology project in western Kenya. An aim of the project was to improve the circulation of knowledge between and among local agriculturalists and project scientists, so that more appropriate, resilient, and effective agroecological knowledge about soil fertility could be produced and utilized. Ramisch's approach of viewing the process of knowledge production and circulation as performance highlights the active and creative, if often confrontational, nature involved in building dialogues between different knowledge carriers—a central aspect to any knowledge-circulation process.

Ramisch's discussion of the ways in which Kenyan farmers performed and interpreted agroecological knowledge also highlights another major theme of this volume, *the joint production and recursivity of nature and society*. While the research scientists involved in the project were interested in the specific productivity outcomes of particular crops in ecological terms, local farmers made decisions and produced knowledge that was always simultaneously social and ecological. For instance, crops that were not edible and could not be circulated into the local system of social relations were not adopted as productive by the community, despite yielding high ecological outcomes. Understanding the ways in which nature-society relations were tightly coupled was an important aspect of the project from the perspective of agricultural extension. Similar illustrations are also made in Rocheleau's chapter, where the experiences of the Rural People's Federation of Zambrana-Chacuey are best understood as complex assemblages of plants, people, soils, politics, history, technologies, and economics. While Rocheleau and her colleagues began with a detailed work plan to assess biodiversity in the forests coupled with "social" surveys in communities, they found themselves finding biodiversity in the communities, alongside and wrapped up in social relations. Nature and society mixed in productive and creative ways on people's front yards and gardens—ways that could not be separated into "ecological" or "social" data. And it was in these very "social" spaces that biodiversity was thriving. Recognizing these connections is key, Rocheleau argues, to creating new and more dynamic research, analysis, and action. It is also important in challenging state-based policies (backed by conservation "science") that "forests" can only be protected by being isolated from people (see also Vandergeest and Peluso, chapter 8 in this volume).

The interconnectivity and co-production of nature and society is perhaps most clearly illustrated in Rebecca Lave's study of stream-restoration practice. In her chapter Lave illustrates the production of a new form of knowledge and practice, one designed to reshape streams, which have been disrupted by human activity, back into their "natural" forms. Here the very practice of human manipulation of "nature" is mended through further manipulation of nature into a form described as natural. Lave explains the appeal of such a project, where individuals can feel they are actively taking part in reshaping nature. She also explains the risks involved in the process, something that has not yet received enough attention. Her chapter leaves one with the question of how we know what "natural" is, if streams are being designed to fit a particular format designed by humans. Perhaps even more provocative is Lave's description of the process through which this is occurring—streams are being restored, *everywhere* throughout the United States, according to one particular design—the Rosgen method. Here, Lave highlights a third major theme of this volume—the importance of how *knowledge is packaged and transported*. She explains how it is that Rosgen's method has become the standard protocol for stream restoration despite disapproval and resistance from within the scientific establishment of geomorphology and hydrology. Through strategic marketing, the creation of easy-to-use and standardized protocols, and the creation of large social networks, the Rosgen method travels exceedingly well across disciplines, climate zones, bureaucratic boundaries, and social worlds. Lave illustrates how, through such packaging, the circulation of Rosgen's method succeeds so well, effectively eliminating competition.

The story of the successful packaging of the Rosgen method in Lave's chapter seems, at first glance, quite different than the stories offered by Phadke, Rocheleau, and Ramisch—where dialogue and hybrid knowledge constructions were actively sought and promoted. For in Lave's case, multiplicity is stymied by simplicity and ease of replication. Yet in all the chapters in this part, the successful packaging of information is key to understanding how particular forms of knowledge travel in particular ways and with what effects. For instance, in Phadke's study, a coalition of political activists and local scientists strategically *packaged* a new hybrid water-management model. The new model challenged the idea that one technological model can be essentially transplanted anywhere. Yet the dam-based water model was not eliminated, merely repackaged to fit and travel better locally—with different political and ecological outcomes. Similarly the chapter by Ramisch illustrates a concerted effort by field-based agricultural scientists to repackage soil fertility knowledge in a way that will assure its transport through environments quite different than where many such models originate (recall the discussion in Forsyth's chapter about soil measurements). Ramisch's study, however, nicely illustrates the challenge of trying to fit multiplicity and complexity into neat packages.

In sum, while the chapters in this part vary substantially in their focus and analytical approaches, they all linked in multiple and overlapping ways, and fit in this part for several reasons. They all highlight the various ways that knowledge production and application is impacted by processes of circulation—at the point of production (impacting the knowledge outcome that dominates), at the point of application (where it can be progressively or hazardously transformed into something quite different)—and then again producing new, often hybridized forms of knowledge. In doing so, all the authors draw from different areas within STS to follow objects across networks (Galt, Rocheleau, Phadke), question how expertise is determined (Lave, Rocheleau, Phadke, Ramisch), build dialogues across epistemological domains (Phadke, Ramisch), present knowledge as situated practice (Galt, Phadke, Ramisch, Rocheleau), and talk about technological reclamation (Phadke). Yet they also all are simultaneously grounded in a political ecology framework—one that is based in a local socioecologically complex field site, that is linked to ever-expanding networks of knowledge, power, and practice. As such, all contributions to this part expose the power of networks (social, ecological, political), and the inseparability of natural-social and local-global in understanding politics of environmental knowledge.

11

DIANNE ROCHELEAU

ROOTED NETWORKS, WEBS OF RELATION, AND THE POWER OF SITUATED SCIENCE

BRINGING THE MODELS BACK DOWN TO EARTH IN ZAMBRANA

We all live in emergent ecologies—complex assemblages of plants, animals, people, physical landscape features, and technologies—created through the habit-forming practices of connection in everyday life. We both inhabit and co-create these ecologies of home, often without being able to "see" them clearly. We live in networks of the sort defined by Bruno Latour (2005) as in the assemblages above, yet we are also rooted in specific territories and geographic locations, often several simultaneously and in series. We are both denizens and artisans of the hybrid geographies described by Sarah Whatmore (2002). Human beings are likewise entangled in several related formulations of contemporary nature/culture (Braun and Castree 1998), described variously as meshworks (Escobar 2001, 2004, 2008), rhizomes (Deleuze and Guattari 1987), the network society (Castells 2000), relational places (Massey 1994), complex ecologies (Botkin 1989; Haila and Dyke 2006), and generic models of networks and complexity (Barabasi 2002; Kauffman 2000).

Using selected tools from political ecology, science and technology studies (STS), human geography, ecological science, and complexity theory, we can learn to recognize and to re-imagine these everyday ecologies of home, as seen from the multiple standpoints of complex actors. We also need a prism that reflects the combined light and patterns of "social" and "biotic" life, in a way that helps us to get beyond the nature/culture binaries that suffuse our thinking.

While we inhabit our own everyday ecologies, sometimes we can see the outlines of structure and function more clearly in "the field," that is, someone else's home, workplace, and habitat. The experience and insights of people in the Rural People's Federation of Zambrana-Chacuey in the Dominican Republic played a major role in my own formulation of network metaphors and models applied to social movements, biodiversity, and landscapes. Along with three research colleagues and several Federation members, I documented and analyzed the process and results of the collaboration between this representative people's organization and an NGO as they advocated sustainable farm forestry and social justice. By the end of the first study, I was seeing multiple. My own vision was refocused through the everyday experience, the perspectives, and the data provided by multiple Federation actors, as well as my immersion in the rich, diverse ecologies of their networked lives and landscapes.

In this chapter I make the case for a model of rooted networks, to encompass the complexity of viable, mixed forest and agrarian ecologies. After an overview I summarize several network concepts and models developed in political ecology, STS, geography, and complexity theory and outline an expanded network approach. A return to the field in Zambrana illustrates selected elements of this synthesis and demonstrates the practical origins and applications of rooted networks in political ecology, STS, and conservation ecology.

THE CHALLENGE OF ZAMBRANA

In October 1992 I joined with three colleagues to conduct a four-month study on a farm forestry project in the rolling hills south of Cotuí in the center of the Dominican Republic.[1] The Rural People's Federation of Zambrana-Chacuey (a regional grassroots organization formed during the land struggle of the 1970s and 1980s) and ENDA-Caribe (Environment Development Alternatives Caribbean, a regional branch of an international nongovernmental organization) were collaborating on several joint efforts.[2] The Forest Enterprise Project promoted planting of *Acacia mangium* trees for timber as a lucrative cash crop on smallholder farms (Geilfus 1995). ENDA had negotiated with the National Forest Service, a division of the army (Dirrección General Forestal) to secure permission for legal cutting of this species with special permits from the project. National laws otherwise prohibited the felling of trees, even planted trees on private property. The Federation and ENDA were in the process of constructing a cooperative sawmill with external funding support. The Federation as a whole had embraced the project and supported the formation of a spin-off subsidiary group, the Wood Producers Association, a rising economic and political force within the Federation and the region. Our agenda was to document this case as a model of community-based forestry, and to analyze the interaction of this initiative with gender and class relations in landscapes, livelihoods, and organizations across scales.

We grounded our study in the region, the landscape, the Federation, its members (men and women), their households, and the connections between them. The Federation formed the base for our research on social and ecological dynamics of farm forestry, and was the focus of our systematic, random, and network samples for social and ecological surveys, oral histories, and participant observation in 1992/93,[3] 1996, 1997, 2005, and 2007. Throughout the course of these activities, we encountered braided strands of social and ecological history that linked every feature in this patchwork of farms, forests, gardens, and homesteads to stories of individual lives, families, communities, and social movements.[4]

ZAMBRANA-CHACUEY AS A REGION

Zambrana-Chacuey is a hilly farming region comprising two administrative districts, nestled in the Yamasa Hills near the provincial capital of Cotuí and the Barrick Gold Mine (formerly Rosario Dominicana). In 1992 most of the twelve thousand residents were smallholder subsistence and commercial farmers with one-half to two hectares of land. Land use and cover ranged from pasture and field crops to tree crops, gardens, and forests. Farmers cultivated tobacco, citrus and other fruit trees, shaded cocoa and coffee, *patios* (forest home gardens), and *conucos* (diverse plots of root crops, vegetables, and medicinal crops). Some farmers planted and harvested trees for timber, woodworking, and charcoal. Most households relied on some income from off-farm wage labor (Rocheleau and Ross 1995).

During the 1980s and 1990s Zambrana-Chacuey exemplified simultaneous national trends to strengthen environmental protection and agricultural exports, reconciled under the umbrella of sustainable development. During the Selva Negra (black forest) Anti-Deforestation Campaign, armed troops with helicopters directed enforcement against smallholder farmers, who suffered arrests, fines, and worse for clearing farm plots, making charcoal, and harvesting trees for home use. The state simultaneously encouraged land speculators, ranchers, and agribusiness corporations to acquire and clear more land for agriculture (B. Lynch 1996; Raynolds 1994), an egregious social and ecological contradiction that some authors have overlooked (Diamond 2005). Smallholder farmers increased tobacco and cassava (*yuca*) cash crops in order to survive the decline in coffee and cocoa prices and the suppression of charcoal and woodworking activities.

From 1992 to 2007 farmers relied increasingly on income from off-farm employment, shifting away from tobacco and coffee. Cocoa, coffee, and pineapple prices rose and fell in cycles. A net retreat from coffee was matched by a resurgence of cocoa, based on organic markets and certification. In 2007 pineapple surged in price and in popularity among farmers. During the early 1990s food crops fell in total acreage, production, and diversity, then began to bounce back (upland rice, beans, and root crops) based on higher food prices and market demand. The net result was still a large decrease in food acreage from 1992 to 2007. Timber, in contrast, was a major cash crop by 2007, yet the Association of Agroforestry Producers (APA, formerly Wood Producers Association) grew more timber on fewer farms than in 1993.

THE FEDERATION

Throughout the surveys ran the chronicle of "the Federation" and the undercurrents of resistance, resurgence, and complex relations of power spanning centuries. In 1992 the organization consisted of fifty-nine farmers, housewives, and youth associations from thirty-one communities, with over seven hundred individual members in five hundred households. The Federation directly served over four thousand people and provided broad support to many of the twelve thousand residents in the region. The associations held separate local meetings and sent representatives to the Federation governing assemblies. The organization was rooted in three separate wings of a very broad movement: farm co-operatives; Catholic liberation theology and human rights; and traditional Catholic advocates of "basic needs" (Rocheleau and Ross 1995). Women figured prominently in each, and constituted a fourth, invisible force within the broader peasant movement.

Nurtured in underground grassroots networks and formally founded with the support of the Catholic bishop of La Vega in 1978, the Federation was one of seventeen such regional groups in the larger Confederation Mama Tingo, named for an elder peasant woman leader[5] assassinated during a land redistribution campaign in 1974 (Ricourt 2000). It was part of a wave of land-struggle movements that grew to international prominence in the 1970s, propelled by the convergence of Catholic liberation theology and poor farmers' campaigns for land throughout Latin America. The Federation, like the broader movement that spawned it, adopted the empowerment approach of Paulo Freire (1970) as the prevailing method of training and organization, with a strong focus on encouraging voice and action on the part of those who had long been silenced.

The nonviolent land-struggle movements appealed to long histories in place and the rights of rural people to maintain their lands or to regain lands lost to the U.S.-based sugar corporations, the Trujillo and Balaguer regimes, and their clients. The movement also proclaimed the right and the profound need to create space for displaced and landless people who had migrated from other regions to make new homes and new communities based on a shared sense of purpose, respect, and mutual support (Lernoux 1980). People were not so much claiming ownership as making a statement about the proper use of land, the nature of an agrarian landscape, and their own place in it, through the re-creation and performance of a complex, rooted network, shot through with power, anchored in the soil as well as history and a shared vision of the future.

Most Federation farmers had participated in campaigns for land, free speech, and the right to organize, as well as for schools, clinics, roads, and marketing support for farmers. Men and women used nonviolent civil disobedience, ranging from occupation of underutilized largeholder lands to highway blockades.

They faced armed soldiers and police, jail terms, beatings, and campaigns of intimidation and harassment. Over the years the Federation also served as the main vehicle for popular organization as people in Zambrana struggled through drought, floods, hurricanes, absentee landlords, and boom-bust markets for coffee, cocoa, pineapple, and tobacco.

Over time the Federation emerged as a major actor in the daily life and political development of the region, restructuring social relations as well as the landscapes and ecologies of the region. It acted first through the land struggle and later through the agricultural and sustainable development projects with ENDA, which eventually spawned the Forest Enterprise Project, the Wood Producers Association, and the Federation/Wood Producers Cooperative Sawmill. The people of the Federation also acted individually and collectively, through their everyday farming and forestry practices, to continually remake the rich regional agroforest and the social networks that sustained them.

THE ZAMBRANA STORY IN SEARCH OF A BETTER EXPLANATION

Several contradictions and paradoxes surfaced in our studies of the Federation, the changing composition and pattern of the regional agroforest, and the official maps of forests, deforestation, and reforestation. Among the most striking findings was the multiple nature of the Federation, beyond the formal structure of the Farmers' Associations, Housewives' Associations, and Youth Associations. The Federation was not a mere organization, but rather a specific flexible, dynamic, and self-organized manifestation of much deeper and wider webs of relation, both in the social sense and in terms of actor-network assemblages crossing "natural" and "cultural" lines. Relations of power ran throughout the Federation network, within the membership as well as between the group and other entities (forest service, largeholder farmers, the church, ENDA, the mine, and the new commercial foresters' group [Agroforestry Producer's Association]). Networks, roots, and territories were highly entangled and did not fit within the confines of socially or ecologically focused polygons mapped on two-dimensional Cartesian grids. Multimodal conversations and encounters with a large proportion of the Federation membership about the regional agroforest and the social landscape also provided us with the beginnings of a situated-science perspective. We brought the multiple visions of different actors to the table, based on qualitative and quantitative assessments of the same phenomena from distinct positions in complex networks. To make sense of all this required a model we didn't yet have. The rooted network as a tool offers a way to understand the complexity of the Zambrana story, using existing formulations of networks as a point of departure.

NETWORK MODELS, METAPHORS, AND THEORIES

In general, formal models present networks as existing beyond space and place, above the mess of land, water, blood, and soil. Some social scientists treat network structures as inherently recent phenomena (Castells 2000), contrasting high-technology, postmodern, postindustrial conditions with prior organic, premodern societies. Networks in STS have arisen from social and cultural studies of information and biotechnologies, while much of political ecology has been in the trenches (literally) of rural life. Yet the actor networks postulated by Latour (2005) can allow us to jump scales and to combine humans, plants, animals, machines, and nonliving elements of the planet, from bedrock and hillslopes, to rivers, rain, and sunlight. Political ecology can bring these models "down to earth," to reconcile networks with energy flows, nutrient cycles, and movements of people and other beings in territories and ecosystems.

The convergence of political ecology and STS can bring power into network models of assemblages of people, other living beings, technologies, and artifacts. While STS has focused on the power of technologies and the workings of science within societies, political ecology has focused on relations of power between state and corporate structures and local communities whose livelihoods and cultural integrity are threatened by eviction, invasion, resource theft, and environmental degradation (Peet and Watts 2004; Blaikie 1985). Political ecology has also been about popular resistance to this oppression, as well as organized popular movements to protect their home ecologies, reassert their own worldviews, and reconstruct their own integrated arts and sciences of "production" and "conservation" (Rocheleau 2008; Brosius, Tsing, and Zerner 2005; Escobar 1999, 2008; Peet and Watts 2004; Robbins 2004; Zimmerer 1996), as in the forestry, agroforestry, and ethnobotany work of the Federation.

ANT AS ARTIFACT, SUBJECT TO RECRUITMENT AND REINVENTION

The network, as an enabling metaphor, allows us to reconcile our thinking about cooperation, communities, and local knowledge, with structural explanations of power in national and international structures of economies and politics. Actor-network theory (ANT) offers a way to conceptualize the relationships between humans and the disparate elements that we normally classify as part of "nature" or "culture." It is a conceptual tool to break binaries and explain the power of connections in assemblages of humans and other living beings, technologies, artifacts, and physical features of their surroundings. Actor networks are often represented through a central human actor, augmented and expanded by the number of connections and the weight of the other elements that constitute nodes in the net. The assumption that all connections are positive and can be treated as assets has been the dominant metaphor. Variations of ANT in social science and policy, including social capital analysis (Putnam 2000) and sustainable-development applications (Bebbington 1997) often present all connections as assets. The framing of actor networks as growth engine and robotic augmentation begs the question "Whose network is bigger?" or "What's in *your* network?" (with apologies to a raft of credit card and cell phone commercials on U.S. television).

In contrast, we can transform ANT to fashion complex, polycentric network models that both complicate and clarify our visions of possible futures. We can expand ANT to incorporate the distinct positions and perspectives of multiple groups of people and various species and assemblages of plants and animals, along with artifacts, technologies, and physical elements of their surroundings. It's not just a matter of getting closer, to get the one true story. It's about "getting it" through the eyes of a diversity of actors in distinct positions, in complex actor networks, that are best described as rooted networks and relational webs. As part of a search for viable alternatives to "sustainable development," I propose to recruit the network construct and stretch it, building on selective elements relevant to social and biological science: power and polycentricity, situated knowledge(s), roots and territory, self-organization, and complex constructs that mesh nature and culture.

Networks are ecological and material as well as social, and carry power relations in both the patterns and processes of connection. The combined lenses of ecosystems, networks, and cultural studies can help us to see embedded, uneven, and dynamic relations of power. Explicit models of the type, terms, and degree of connection can incorporate multiple dimensions, including positive, neutral, and negative connections (as seen by a particular actor), strong to weak links, continuous to erratic connections, and dense versus dispersed patterns of connectivity. While many network models focus on hierarchies of degree and pattern of connectivity (Barabasi 2002), the terms of connectivity are a major arbiter of power. They can vary from coerced to voluntary, encompassing relations from slavery to partnership and free association (Rocheleau and Roth 2007).

Place and territory are, at best, underdeveloped in STS and political ecology. To address the entanglement of people in the biotic and physical elements of the material world, and the construction of new ecologies, we need to tie networks to land, to locate them, put them in place(s), though not in simple polygons. Relational spaces and theories of place (Massey 1994) as well as meshworks and territories (Escobar 2001, 2008) tie networks to place, yet we need to further engage the material relationships in socioecological networks linked to multiple territories of extraction, production, circulation, consumption, and transformation.

We can think of "network" and "root" as verbs rather than nouns, to visualize the diverse rooting strategies that connect webs of relation to the surface(s) of the planet, as well as technologies of internal connection within complex entities. Several well-known plants illustrate the varieties of rooting and webbing: taproots in pine trees; the perching of epiphytes ("air plants") in tree canopies; the profusion of new plants produced by "spider plants" outside the pots or the main rooting zone of the parent plants; the soil-building habits of coastal mangroves around their woody stilt roots; and algal mats, which create their own floating worlds from microflora and -fauna, making a seafaring macro-being from microconstituents. Deleuze and Guattari (1987) elaborate specifically on the underground metaphor of rhizomes[6] to describe the entangled realities of connectivity and the complex dynamics of social change.

Community ecology and systems ecology, respectively, model relations among and between species, and flows of energy and materials between living things and their physical surroundings (Botkin 1989), from ecosystems (Odum 1994; Costanza et al. 2001) to ecological networks (Fath 2007). Horizontal flows as well as vertical "roots" tie individual nodes or whole networks to resources in territories of activity, extraction, residence, identity, and influence. We can model terms and pathways of movement of matter, energy, and living beings between nodes in networks and between whole networks to illuminate processes of mobility and circulation as well as extraction, production, consumption, and the terms and types of rooting, and being, in place.

Neural-network models and theories from biological and computational sciences contribute explicit models and robust metaphors to study dynamic self-organization from below as well as the role of already existing structures (Kauffman 2000; Barabasi 2002). Repeated actions create habit-forming practices of connection between neurons in the brain, which create or modify structures, which in turn predispose but do not determine future action. These models also describe dynamic and self-organized phenomena from social movements and organizations to biodiversity in plant and animal communities.

Polycentric governance structures (V. Ostrom 1997; E. Ostrom 2001) provide a point of departure to visualize multiple actors as simultaneous centers of power, influence, and action, rather than single structures, central actors, and simple linear hierarchies. Theories of power and knowledge from feminist poststructural scholarship add two powerful concepts to the mix: situated knowledge and positionality (Haraway 1991; Harding 1986). Each actor (individual or group) has a distinct vision of any given network, based on their position, and their experience of shifting terms and configurations of connection over time.

The resulting artifact, what we might call a poststructural rooted network, incorporates the views from individual nodes (as distinct standpoints or subject positions), to provide a powerful tool for "situated science" in political ecology, STS, and conservation ecology. This networked vision can contribute to critique as well as to the construction of viable "alternative" hybrid sciences that transcend local and global scales, erase nature/culture dichotomies, and join theory and practice. This eclectic tool helps us to "make sense" of complex assemblages of humans, other living beings, and their things, their surroundings, and technologies from distinct subject positions and diverse knowledge perspectives.

As a first step in this process I suggest several specific tasks required to embark on this project: mapping power in networks; mapping rooted networks onto territories; tracing relations of connectivity, autonomy, and sovereignty, as well as mobility, circulation and rootedness; and reconciling complex systems and networks to include assemblages of humans and other beings, their habitats, technologies, and artifacts. Some prerequisites include complicating and expanding our typologies of power; complicating territories beyond fixed polygons; developing a typology of rooting systems and strategies; and integrating hierarchies and self-organization (Rocheleau and Roth 2007). The brief case study below incorporates these various elements through a discussion of women's changing position in the Federation, and the ongoing construction of a complex regional agroforest by multiple actors.

THE WORKINGS OF NETWORKS IN THE FEDERATION

An organizational diagram of social networks in the region readily demonstrates the role of the Federation as a clearinghouse of information and a center of influence in a crowded field of government, church, and civil-society organizations over three decades (figure 11.1). The Rural People's Federation of Zambrana-Chacuey and the region it calls home also embody the kind of multidimensional assemblage described by ANT. It includes the relationship of people to each other (from family and neighbors, to trade and church affiliations, political and social organizations) and incorporates a long list of plant and animal species (wild and domesticated) and physical features of the landscape, ranging from mountains, valleys, rivers, and soils, to springs and groundwater. The network also encompasses technologies, artifacts, and infrastructure: technologies of production, processing, resource management, communication, and transport; infrastructure, such as roads, water collection and distribution systems, residential, commercial, and community buildings, and energy and communication grids; and tools, from plows and tractors to sewing machines, cell phones, and motorcycles. Social technologies and practices also form part of this list of actors in the Federation/campesino network, including practices of organization, education, empowerment, resistance, solidarity, and self-governance.

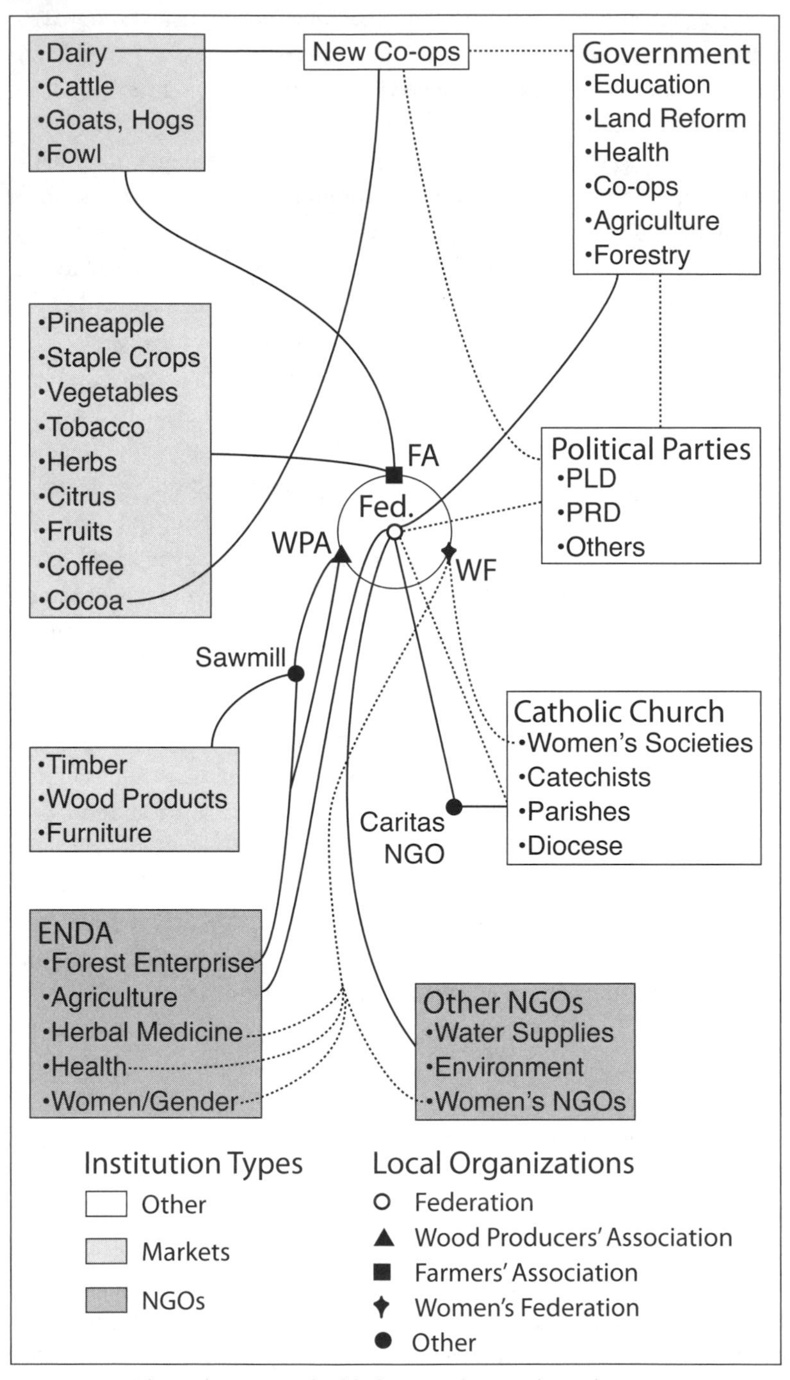

FIGURE 11.1 The Federation embedded in social networks in the region.

The Federation also exemplifies a network-based entity in the more expansive sense of the rhizome metaphors of Deleuze and Guattari (1987), and the characteristics of rooted networks cited above. The organization explicitly addressed the relationships, and the terms of connection, between rural farmers as a group with each other, the land, other living beings, the national political process, and civil society. It tackled the terms of connection to markets for farm products, wood, and agricultural inputs. The Federation sought to shift several relationships toward fair trade, equal exchange, equal rights, and full participation. The members' relationships to each other, national political processes, and markets also hinged on, and impinged upon, the ecologies of their smallholder farms and their connections to land, plants, animals, and the surrounding landscape (figure 11.2).

The politics of the Federation have been explicitly webbed, networked, and rooted, even as they addressed (and sometimes embodied) relations of power. They dealt directly with roots, but included lateral roots in relations of solidarity as well as vertical roots to land, and incorporated roots of different types at multiple scales. The entire Federation was rooted in the twin districts of Zambrana and Chacuey as a regional territory, communities set their roots in local landscapes, and individuals and households drew on roots in small farm properties and specific plots within those. A profusion of tangled roots also crossed each of these scales of social and ecological organization. The plants and animals associated with the households and communities of the membership encompassed both lateral and vertical rooting as they connected to each other, the people of the region, and the soil, water, and landforms. The following examples illustrate selected elements of rooted networks in the experience of the Federation and its members: gender and power in polycentric networks; and reconciling roots, networks, and territories in the regional agroforest.

GENDER AND POWER IN POLYCENTRIC NETWORKS

The Federation consistently used the structure and process of networks (*network* as noun and as verb) to address issues of power and difference. The group had its origins in the politics of resistance against oppressive, unjust, and repressive forces, from a highly militarized national state to hostile agricultural markets and unequal access to land from the local to national level. The founding of the formal organization provided a platform from which to speak truth to power, to enforce the members' own demands, and to resist military and police intimidation through mass mobilization (see Lernoux 1980; Ricourt 2000).

The Federation dealt explicitly with relations of power within the organization itself, in terms of both structure and process. During the 1990s women and men

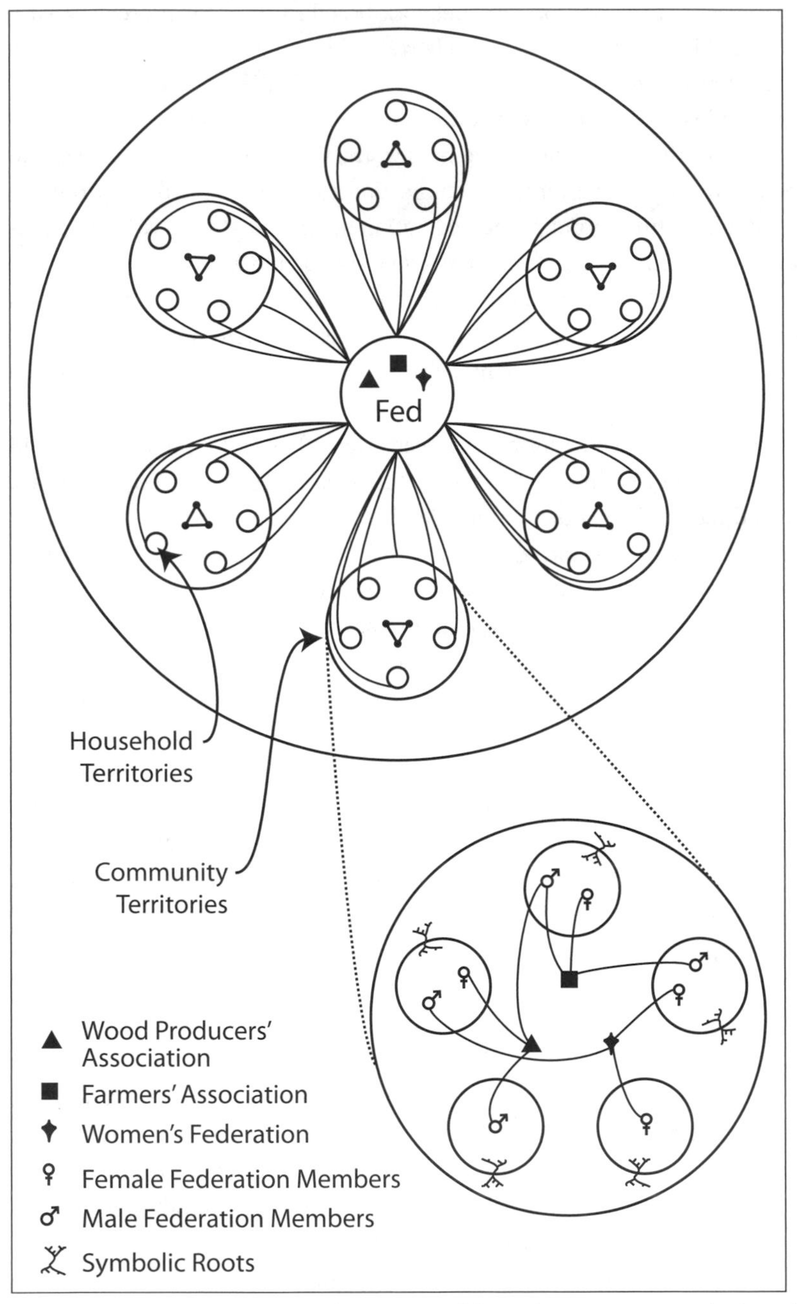

FIGURE 11.2 Rooted networks and territories of the Federation across multiple scales.

participated through the Housewives' Associations (100 percent women), the Farmers' Associations (4 percent women, 96 percent men), and the Wood Producers Association (>95 percent men). Since that time women in the Federation have twice reorganized themselves and renegotiated the terms of their connection and participation, and the Wood Producers have redefined and repositioned themselves as well.

There was a conscious strategy to rely on the diversity of the membership to link with myriad other actors in the organizational landscape (as allies, in solidarity, as clients, or in bargaining mode). An informal division of political labor and social affiliations provided a thick web of connection, communication, and circulation of influence in various church, social movement, government, political-party, and business circles. The membership spanned all three major political parties and more. The hard-won (and ongoing) battle to maintain Federation neutrality with respect to political parties was balanced by individual members with informal connections to ongoing conversations, programs, and government resources linked to party politics. Members likewise bridged various circles within the Catholic church, marketing and commercial networks, and professional and occupational affiliations.

This everyday mobilization of distributed power in polycentric networks had far-reaching consequences. Beyond the land struggle, public services and infrastructure were developed through collective demands (including nonviolent protest and mobilization) to convince state agencies to engage in collaborative efforts with local communities and the regional Federation. Women members of the Federation played a major part within the movement, participating in the active circulation of information and acting to reconstitute places for viable, just, and democratic communities.

The history of women in the regional Federation exemplifies the complexity of the organization, its ideals and contradictions, and its ability to deal with difference and distribute power in networks. From the outset the Federation had a base in women's groups and women's politics of place. Tito Mogollon, one of the founders, noted that the bishop of La Vega commissioned him and three other organizers as human rights promoters in 1974. They approached two women's groups, one in a nearby community threatened with eviction by the Rosario Dominicana (now Barrick) Gold Mine and another in Chacuey. From these efforts emerged the Rural People's Federation of Zambrana-Chacuey as a collective voice and advocacy organization and simultaneously as a center of countervailing power and empowerment.

Women's stories led us beyond the confines of organizations and movements into the realms of sacred space and everyday connections, between humans and other beings, their technologies, artifacts, and physical surroundings. Women's lives in Zambrana-Chacuey led beyond and beneath visible and formal organizations to the web of relationships that predated and gave rise to the Federation. Women did not need permission to join and to govern, or recognition as members and leaders in the Federation, to wield power. Many already had that, in parallel domains of knowledge and authority illegible to more powerful actors (from men at home to government officials). Yet the recognition of women as Federation members and leaders, and their struggles for more equitable structures (1978–2007), enabled a powerful synergy between women and men, and between economic, political, cultural, spiritual, and ecological domains of authority and power.

Our surveys in 1992–93 indicated that 20 percent of the households affiliated with the Federation were connected exclusively through a woman in the Housewives' Association, with no corresponding memberships of men in the household. Women in local associations often brought new initiatives (such as the Forest Enterprise Project) to the attention of the men in their communities and encouraged their participation. The women's groups also maintained continuity of activity, while the Farmers' Association membership and activity tended to rise and fall with cash crops and commercial activity (Rocheleau and Ross 1995). By 1996 some of the women leaders formed a new, explicitly feminist women's board (*junta*) with a focus on economic and political rights for women. Membership surged, and they formed the Women's Federation, as a parallel entity within the original organization. When they encountered resistance to their new status, the Women's Federation broke away to form an independent organization. In 2006 they negotiated with a new slate of Federation officers and re-entered the Federation with greater representation on the board and a stronger role in political and financial decisions. The continuing evolution and revolution of the women's organization within the Federation illustrates the management of power in polycentric networks, from above and below, including power with, power alongside, and power in spite of, rather than the well-worn confrontational models of power over and power against.

The experience of women in the Federation also raises the issue of legibility and the invisible web of relations beneath and beyond the formal organizational structures and recognizable movements. The roots that sustain Federation networks of solidarity and affinity are made and maintained through the continuing performance, affirmation, and creation of positive alternative cultures expressed in values, landscapes, artifacts, rituals, and daily practice that draw their legitimacy from a domain beyond the control (and even the gaze) of recognized, dominant power. This story of women in the Federation is a tale of rhizomes rather than taproots, of a subterranean root mat, a relational web of exchange, extraction, and circulation.

This expanded vision of complex, and sometimes creative, entanglements with power has allowed women in the Federation to imagine and create more just, viable, and humane economies and ecologies, and new ways to be at home within them, while still struggling with unequal and unfair distributions of property, political office, and legal authority. The experience of the evolving, self-organizing women's structures in the Federation also suggests neural networks, habit-forming practices of connection, and struggles over the terms of connection, rather than simple stories of open conflict between diametrically opposed or competing groups.

RECONCILING ROOTS, NETWORKS, AND TERRITORIES IN THE REGIONAL AGROFOREST

The relationships of Federation members to national political process, human rights movements, and markets (from local to global scale) also hinged on, and impinged upon, the ecologies of their smallholder farms and their connections to land, plants, animals, and the surrounding landscape as well as production and resource management technologies. The politics of the Federation explicitly addressed relations of power in the ways that people are connected to land, as well as to other people, multiple species, and a variety of technologies and artifacts. They made a strong appeal to "roots" but not a classic "blood and soil" argument for exclusive rights to a fixed territory by a specific group, based on identity and a long history in place. The Federation incorporated lateral roots in relations of solidarity as well as vertical roots to land. They also combined roots of different types at multiple scales (figure 11.2).

The Federation reconciled networks and territories in daily practice and in history. It linked people "horizontally," between people and other living beings, and vertically, between people and other species with their physical surroundings and, literally, the ground beneath them. The people of Zambrana-Chacuey also brought together notions of fixity and long histories in place, with the experiences of displacement, migration, mobility, multiple complex identities, flexibility, and fluidity. The Federation resolved this paradox by jumping scales, joining people to each other based on co-presence in specific geographic locations and in networks of people linked across separate spaces by shared interests (Women's, Farmers', and Wood Producers Associations) and common values (the Federation and the Confederation Mama Tingo).

The politics of place, power, and changing human ecologies in this context were and are about more than gender, class, racial, ethnic, or anti-imperial struggles over "environment" as a collection of resources in a specific location. Environmental movements as well as rural farmers' land struggles in this region were about the terms of connection between people, and between groups of people, land, other species, artifacts (houses, gardens, tools), and the surrounding physical world. They were also about the terms of connection between local and larger places, both earthly and spiritual. Land was not treated as "real estate," as an exchangeable and interchangeable commodity, but as the ground where body, home, community, and habitat joined in everyday experience as well as in history (Rocheleau 2005). Place was treated not as a container, but as a nexus of relations (Massey 1994), a patterned logic and ethos of contingent connections, rooted in a particular way, anchored in a given space and time.

The bedrock of solidarity among the various groupings within the Federation was and is the shared sense of place, with a common commitment to basic political and human rights for all, as well as land, basic infrastructure, and support services. The material space for community was created through regional and local collective struggles for household plots as private property.

In our field-data collection and subsequent statistical analysis of tree and crop biodiversity, we encountered an invisible, species-rich, regional agroforest,[7] the same patchwork landscape of forest and farms that was being treated by the state as a deforestation crisis zone. A dominant focus on forest as land cover and a selective version of the sciences of conservation and land-use change were being mobilized against the very people who had groomed the biodiverse, culturally rooted agroforest. Their farms, and the surrounding landscapes, including a profusion of forest trees, were almost always mapped into the "deforested" polygons on official maps.

The Forest Enterprise Project, with its eventual focus on a single Australian pulp and timber tree that readily invades cropland, gardens, and riparian forests, was heralded as reforestation. It was actually a successful project for producing smallholder commercial timber, linked to a broader effort to promote agroforestry, medicinal plants, and sustainable agriculture. The expansion of on-farm timber plantations sometimes replaced tobacco or pasture, but it also encroached on the pre-existing diverse mix of native and naturalized trees in the patchwork landscape, and threatened tree diversity in patio gardens, coffee and cocoa stands, and riparian forests.

We made the invisible, species-rich regional agroforest legible to science when we changed the frame of our scientific gaze and the logic of our sampling to see the relational networks of people and plants in place(s). The story of this landscape was very much the story of the Federation and men's and women's politics of place within it, and as such it was embodied in situated knowledge, revealed by multiple land users. This framework provided a countervailing vision to the powerful images of forest and not-forest in neat polygons on standard maps of land use and cover at scales that erased these finely networked human ecologies.

As we proceeded with the sketch maps and surveys of tree and crop species, it became apparent that the patio (homestead) gardens constituted a polka-dot forest. The mainstay of this species-rich agroforest, the patio garden, was largely a women's domain, and equally impressive, the seeds of forest past and forest future were literally wrapped around peoples' homes. The highest biodiversity was found close to—not removed from—the focal point of human habitation. Our surveys also revealed that seeds crossed land-use categories and property lines with impunity, riding on the wind, livestock, or people, or sometimes through purposeful planting by farmers. Our intensive biodiversity surveys in 1996 and 2007 confirmed the existence of a dynamic regional ecology above, below, and beyond the property lines and land-use/cover categories in the maps of resource management and conservation professionals.

The Federation example stands as a formidable challenge to simplistic advocacy for state, common, or private property models as the exclusive precondition for tenure security and strong roots, to enable biodiversity conservation and sustainable resource management. Network models and specifically the notion of rooted networks help to explain the basis and the success of the Federation's approach to roots and territories, mixing a variety of strategies and treating *root* as a verb as well as noun. The land struggle was about more than land, and land was about more than private property. Roots mattered, as well as a place to plant them, but both took many forms that coexisted in complex ecologies.

CONCLUSION

The case study in Zambrana-Chacuey demonstrates the need to develop new models and analyses of rooted networks, relational webs, complex assemblages, and emergent ecologies, reconciled with territories. Self-organization from below is newly legible to formal science through network and complexity theories, and can be modeled along with hierarchical structures. The challenge is to mesh social, ecological, and technological domains in theories and models of rooted networks, relational webs, and self-organized assemblages, all shot through with power, and linked to territories and larger systems. Integrative network models and theories can be powerful tools for thinking and acting in place and across places, to identify instances of viability in actually existing human ecologies and to imagine and foster just and humane alternative futures. The ongoing experience in Zambrana demonstrates promising ways of knowing and being in rooted networks, webs of power, and complex landscapes, past, present, and possible.

NOTES

1. The team consisted of me; Laurie Ross, then a graduate student and now a professor at Clark University; and two Dominican colleagues, Professor Julio Morrobel (then professor of forestry at the Instituto Superior Agricola in Santiago) and Ricardo Hernandez (then a graduate student and local historian and now a professor in Cotuí). We eventually recruited several additional colleagues from the Federation and ENDA to join us in conducting the study. In 1996 and 2007 Professor Luis Malaret, research associate at Marsh Institute, Clark University, joined me to conduct the ecological surveys with forest technical experts from the Federation.

2. The Federation and ENDA (as of 1992–93) sponsored several other projects, including Ethnobotany and Herbal Medicine, Agroforestry for Soil Conservation and Soil Fertility, Small Livestock Production, and Vegetable Gardens, as well as Woodworking, Rattan Furniture, and Metal-Working Workshops.

3. Over the course of four months in 1992–93, we visited and interviewed thirty-one local associations (farmers, housewives, and youth groups) in sixteen communities (out of a total of fifty-nine Federation-affiliated associations in thirty-one communities, each association comprising roughly twelve to thirty people from a farming community in a specific locality).

4. In 1992–93 we combined ethnographic, standard-survey, and feminist methodological approaches, including participant-observation, group interviews, key-informant interviews, life-history interviews, community and organizational histories, detailed sketch mapping, land-use history, land-use simulation board games, and a formal questionnaire and mapping survey (land use, tree species and crops) with a gender-stratified random sample (45) of the more than 700 Federation members in Farmers' Associations and Housewives' Associations, respectively (Rocheleau 1995; Rocheleau and Ross 1995). In 1996 and 2007 we conducted follow-up biodiversity surveys, using a rigorous ecological sampling framework and survey methods in a subsample of the Federation household lands. In 2007 we conducted oral-history, focus-group, and key-informant interviews with Federation members on the history and trajectory of the regional Federation, livelihoods, and landscapes.

5. Florinda Soriana Munoz led and supported peasants in campaigns for land and social justice in nearby Yamasa.

6. Rhizomes are usually horizontal subterranean plant stems, distinguished from true roots in possessing buds, nodes, and usually scalelike leaves.

7. *Agroforestry* refers to the purposeful combination of trees, crops, and animals in managed ecosystems to enhance production as well as conservation, for economic, cultural, and ecological ends. *Agroforest* refers to the resulting socioecological formation as an entity in the landscape.

12

RYAN E. GALT

CIRCULATING SCIENCE, INCOMPLETELY REGULATING COMMODITIES

GOVERNING FROM A DISTANCE IN TRANSNATIONAL AGRO-FOOD NETWORKS

Whereas exposure to pesticides through occupation or accident is basically a local problem . . . the contamination of food must inevitably become a matter of worldwide concern because of the extensive international trade in this commodity.
—P. Hough, The Global Politics of Pesticides

Hence the major effect of the Panopticon: to induce in the inmate a state of conscious and permanent visibility that assures the automatic functioning of power. So to arrange things that the surveillance is permanent in its effects, even if it is discontinuous in its action; that the perfection of power should tend to render its actual exercise unnecessary; . . . in short, that the inmates should be caught up in a power situation of which they are themselves the bearers.
—M. Foucault, Discipline and Punish

Headlines describing various food crises draw attention to our global, industrialized food system. Each day fresh produce arrives from the global South to provision supermarkets in the global North, following the entrenched networks established during the colonial period and becoming subject to agro-food regulations of the territory it enters. In this chapter I show that the science and practice of agro-food regulation in the global North strongly shape agro-food networks and export sectors in the global South. Importantly, this shaping only partially corresponds with the intent of the regulation, because the very place in which regulation is often considered strongest, the United States, has serious weaknesses.

Using the case of the regulation of pesticide residues on food,[1] I focus on flows—material and informational—through specific nodes in the global agro-food network to detail how scientific understandings of pesticide risk, and especially the way these are operationalized in decisions about pesticide-residue monitoring, significantly shape the transnational agro-food system and production at the local level. On the production side of the network, material flows of produce and pesticide residues move from the global South to the global North. On the regulatory side of the network, these foods and toxins flow through ports, with regulators taking samples to Food and Drug Administration (FDA) laboratories. Sampling and testing is shaped by (1) the definitions of which risks matter as determined by the politics and science of risk assessment—consumers' pesticide exposures to residues on food generally carry greater weight than farmworker exposure—and (2) the specific scientific regulatory practices in analytical chemistry—testing for some pesticide residues and not others. Subjecting food and toxins to these risk definitions and scientific practices creates results that travel as imperfect information in agro-food networks to the places of production. In distant locations, these informational flows impact farmers' production practices and beliefs. Through this interchange of material and informational flows in the agro-food network, risk assessments and their always-partial application in agro-food regulation govern from afar and create local policing regimes that ultimately shape economic and ecological activities in places that produce exported food commodities for global North markets. The shaping is highly uneven because of local circumstances and the always-partial nature of enforcement.

Detailing all of the specificities of the pesticide regulatory system, which covers several hundred different agrochemicals, far exceeds the confines of one chapter. Therefore I focus on two pesticides and their use in Costa Rican export vegetable production to illustrate how government at a distance shapes local practice through agro-food networks.

PESTICIDE-RESIDUE REGULATION

Pesticide residues in food generate considerable concern in the general population. Pesticides degrade at different rates depending upon the compound, exposure to sun and other weathering forces, and the ability of organisms to metabolize them. They often degrade to a level below the detection limit of modern equipment,[2] but they remain on food. We do not see pesticide residues, and most of us do not think about them, yet we ingest them with almost every meal.

The health effects of pesticide residues in food remain vigorously debated. Showing the safety or harm of chronic, low-dose pesticide exposure is beyond the limits of toxicology and epidemiology. Regulations pertaining to chemical carcinogens can rarely be based on direct evidence, relying instead upon animal tests and studies of mutations induced by compounds. An inherent problem remains: "the interpretation of such data is fraught with uncertainty and expert disagreements, and the regulatory outcome seems to depend less on science than on the institutions and procedures that are used to resolve the proliferating technical conflicts" (Jasanoff 1987, 203).

Many nation-states attempt to regulate people's exposures to harmful compounds. In the United States, risk decisions about carcinogens were originally decided by a "generic approach," in which exposures were to be minimized. In the 1980s, as a result of a Supreme Court decision and a report from the National Academy of Sciences, risk assessment and risk management focused on the specificities of dose and response, which usurped the generic approach. Industry's agenda strongly shaped this transformation, as their demand for formal risk assessments of individual chemicals aimed to remove the control of risk decisions from agency scientists and bureaucrats who they viewed as having "proregulatory" interests (Jasanoff 1987). A number of authors have critiqued risk assessment and the unexamined assumptions in its decision-making process (J. Brown 1989); my goal here is to trace risk assessment's effects through its implementation in scientific and regulatory practice.

The Environmental Protection Agency (EPA) employs risk assessments to establish the legal limits—called "tolerances" or maximum residue levels (MRLs)—at which pesticides are allowed on certain foods. Current U.S. regulation involves MRLs for 371 different active ingredients of synthetic pesticides (EPA 2007). The FDA monitors residues on foods of plant origin by following these MRLs, a role supplemented by some U.S. states.

Residue regulation is a double-edged sword. On one hand, it helps assure citizens that their food will not poison them, at least not immediately. On the other, it serves as a state-sanctioned legitimation of pesticide residues in food, which many consider an intolerable risk. As Rachel Carson noted, "The luckless consumer pays his taxes but gets his poison regardless" ([1962] 1994, 183).

The pesticide-residue regulation system, and especially its extension across national borders, assumes constant and vigilant enforcement. The possibilities that pesticide residues on foods may exceed MRLs and that foods may contain residues not allowed on that food necessitate constant monitoring for all potential residue combinations if the government were to actually assure consumers that their food complies entirely with U.S. law. In other words, MRLs require and assume an all-seeing and ever-active—panoptic—style of enforcement.[3]

ACTOR-NETWORK THEORY AND POLITICAL ECOLOGY

In the two last decades, actor-network theory (ANT) emerged in science and technology studies (Callon 1986; Latour 1993) and has been adopted by political ecologists (Robbins 2007) and agro-food researchers (Whatmore and Thorne 1997). ANT rejects ontologies based on the separation of nature and society, positing instead that so-called social and natural entities exist in hybrid networks of associations, and that "each of the separate pieces is not independent, but is instead made to be the way it is by virtue of its relationship to all the other parts" (Robbins 2007, 14).

Political ecology has been engaged with the question of pesticide use since the 1980s (Thrupp 1988), and an ANT perspective proves useful for exploring pesticide issues in conventional agro-food systems. Pesticides offer farmers the ability to grow productive monocultures by beating back pest species and creating produce perfect enough to meet stringent aesthetic standards. Yet, as a political-ecological perspective insists, other actors are already involved by the time farmers decide to spray a specific pesticide: pesticide use supports local shops and agrochemical companies, has a long history of state promotion as part of agricultural development, and supports a multibillion-dollar pesticide industry that capitalizes on monocultural agriculture, thereby profiting some of the world's largest corporations. But in turn, as Robbins's use of ANT (2007) demonstrates, unstable monocultures—with their legion of dynamic biological "problems" including insect pests, diseases, and weeds—pose challenges for farmers, and for agrochemical firms, which must create new chemicals to be ahead of the development of resistance. These production networks link together farmers, their families, workers, and farming communities; a variety of insects, fungi, bacteria, and plants; pesticides; state promotion policies; and local and multinational agrochemical firms.

Regulatory networks informed by various scientific endeavors shape these production networks; indeed, they cannot be separated. Toxicologists and epidemiologists attempt to create knowledge about the health consequences of agrochemical exposure. Toxicologists conduct controlled experiments that expose test animals to harmful substances and document health outcomes, including acute toxicity, tumor formation, birth defects, and sometimes other effects. Epidemiologists conduct studies measuring the prevalence of health and disease in human populations, often using study designs that compare two similar populations that differ in their exposure to a certain compound. At the interface between science and policy are the fields of risk assessment, "the characterization of the potential adverse health effects of human exposures to environmental hazards," and risk management, "the process of evaluating alternative regulatory actions and selecting among them" (NRC 1983, 18). Scientists and bureaucrats in the EPA, influenced by chemical-company interests, consumer and environmental-group pressure, and juridical guidelines, use information from these sciences, risk assessment, and risk management to create regulations concerning which pesticides can be used within the United States and which pesticides (at what level) can exist on each food sold within the United States. Once regulations exist, scientists and administrators at the FDA devise sampling protocols and analytical chemistry methods to check imported and domestic produce for agrochemical residues. Analytical chemists play an important role in the scientific structuring of the regulatory networks, because they devise and implement tests that can detect minute traces of chemical components.

These production and regulatory networks intersect with agro-food networks of food commodity and informational flows: certain food commodities are produced in certain places and are subject to the regulations of the place where the food will ultimately be sold. Thus, production, regulation, and trade create a complex transnational network of human actors, institutions, and nonhuman actors (or actants), including farmers and farming communities, hundreds of species at the farm level, agrochemicals and their residues, food commodities, firms, states, scientists, regulators, environmentalists, and consumers.

Rather than being truly "global," which implies a sort of uniformity, these networks are transnational, rooted in specific places (cf. Rocheleau and Roth 2007), and patchily connected by infrastructures allowing material and informational flows. Places in the network include the fields in which crops take root and grow; pesticide manufacturing and formulating plants; export packing facilities; ports; scientific laboratories that formulate pesticides, study pesticide effects, and test foods for residues; legislature buildings where laws are made and passed; supermarkets and other places of produce sales; and houses, restaurants, and other places where people consume these foods and agrochemical residues.

Within these transnational networks of production, regulation, and trade, pesticides have important implications that change according to their position in relation to other actors and actants. A pesticide on a tomato plant in the field protects it from specific pests and can be a boon for the farmer; inhaled by the farmer and workers, it can contribute to poisoning and chronic illnesses like cancer; if detected by testing agencies, it might cause rejection of a shipment and therefore a large economic loss; and as understood by consumers it can create a sense of dread about its negative effects and invoke feelings of distrust toward regulators that allow risky substances on food. I chose two agrochemicals as the actants in the story to illuminate pesticides' various relations to other actors and actants in the network.

TRACING TWO PESTICIDE ACTANTS WITH DIFFERENT RISK PROFILES

Many pesticide "families" exist. Two of the most common are organophosphate (OP) insecticides and ethylenebis-dithiocarbamate (EBDC) fungicides. I chose a representative of each of these two families as the major actants in this chapter because of their very different risk profiles, and the difference in the way their residues interact with the regulatory system. These actants are methamidophos, a systemic OP, and mancozeb, a nonsystemic EBDC.[4] Tracing methamidophos and mancozeb and their regulation through the agro-food network is not meant to provide an overall view of pesticide regulation. Rather, I purposefully selected them as two endpoints of enforcement for risky pesticides: those that are relatively highly monitored, and those that are least monitored.

The German chemist Gerhard Schrader first developed the OP insecticides in the late 1930s (Carson [1962] 1994, 28). Chemists modified them for use in World War II as potent nerve gases (E. Russell 2001). OPs became available to farmers in the United States in 1946 (Shepard 1951, 6), one year after DDT made its debut. In addition to acting as powerful, broad-spectrum insecticides, the OPs are neurotoxins and acutely toxic to humans and other mammals.

The OPs rose to special prominence in agriculture once governments in the 1970s began banning the organochlorine insecticides like DDT, which bioaccumulate in the environment. Relative to the organochlorines, OPs are less persistent, although they present a *much* higher risk of acute poisoning and death to farmworkers (Wright 1990). OPs affect the nervous system by binding to and deactivating the vital enzyme acetylcholinesterase, resulting in uncontrolled, repeated firing at nerve junctions. Poisoning can result in respiratory failure and death. OP poisonings account for around 80 percent of pesticide-related hospital admissions in the United States (Taylor 2001).

Methamidophos is a very acutely toxic OP. Cases of farmworker poisonings often involve methamidophos. For example, in Nicaragua, methamidophos and carbofuran, an acutely toxic carbamate insecticide, caused 77 percent of poisoning cases (McConnell and Hruska 1993, 1559). Methamidophos has been identified as one of the twelve most dangerous pesticides that the health ministers of Central American countries have agreed to ban (Nieto Z. 2001).

As a systemic insecticide, methamidophos works for weeks to kill insects that feed on the plant. This makes it popular with farmers, but problematic for consumers; methamidophos residues have caused a number of consumer poisonings (Wu et al. 2001), including the recent (contested) poisonings in Japan from Chinese dumplings (channelnewsasia.com 2008).

EBDCs—originally developed for the rubber vulcanization process (P. Russell 2005)—were discovered to have fungicidal properties in 1930 at Cornell University (McCallan 1930). EBDCs were first patented in 1934 by Tilsdale and Williams at DuPont, but the first, thiram, did not appear on the market until 1942 (P. Russell 2005). By the 1950s, EBDCs were in use globally (Dich et al. 1997). These were the first fungicides from organic chemistry, classified as such because their chemical configurations included carbon from compounds in fossil fuels.

In contrast to the OPs, EBDCs are not acutely toxic to humans, yet concern focuses on cancer (Dich et al. 1997) and birth defects caused by the compounds and their carcinogenic breakdown product, ethylenethiourea (ETU; Holland et al. 1994). Epidemiological studies show a variety of negative effects of chronic exposure to EBDC fungicides, including thyroid toxicity and tumor generation (Houeto, Bindoula, and Hoffman 1995). A study from the 1970s showed increases in thyroid and liver cancers in regions with higher EBDC exposure as estimated by sales and crop production data (von Meyer 1977). A more recent study showed an increase in relatively rare cancers (thyroid, bone, testis, thymus, and other endocrine glands) in the region in Minnesota where EBDCs are most heavily used (Schreinemachers, Creason, and Garry 1999). Clinical studies of fifty workers with chronic exposure to maneb, an EBDC fungicide, had "significantly increased incidences of various neurologic effects . . . and increases in a variety of other Parkinson-like symptoms (including tremor, ataxia, and bradykinesia)"[5] relative to a nonexposed control group (Ferraz et al. 1988). Toxicological studies on rats have shown EBDCs to be teratogenic, damaging the reproductive organs of males and females, resulting in malformations in offspring and decreased fertility (Houeto, Bindoula, and Hoffman 1995).

Mancozeb, an EBDC based on manganese and zinc, was introduced in 1961, and is likely the most commonly used EBDC (P. Russell 2005). EPA classifies mancozeb and ETU, its breakdown product, as "probable human carcinogens." An assessment by the National Academy of Sciences shows that mancozeb and many other fungicides pose relatively high risks of cancer if allowed to persist on food at their MRLs, with an estimated risk of 3.4 cases per 10,000 individuals in the United States. This is much higher than the EPA's rule of thumb of regulating individual chemicals to produce one cancer case or fewer per one million U.S. citizens (NRC 1987). Under high heat from cooking, EBDCs convert to their breakdown product, ETU, at rates up to 50 percent. In contrast to some knowledge on mancozeb residues, the literature lacks information on home-cooked foods with EBDC residues and their conversion to ETU (Holland et al. 1994).

Specific modes of toxicity influence the effects of these two pesticides on farmworker and consumer health. As Dich et al. (1997, 421) note: "Pesticides with an extremely high acute toxicity may be easily metabolized and eliminated from the body; following long-term low exposure, they may be less toxic and without carcinogenic or mutagenic properties. On the other hand, pesticides with low acute toxicity . . . can accumulate in the body and cause chronic toxicity after long-term exposure even in comparatively low doses."

The first category includes the OP insecticides, as they can be metabolized if the person survives the exposure (although long-term effects often result from these exposures). The second category includes those pesticides that have low acute toxicity and are carcinogenic, like EBDCs, although it is important to note that EBDCs are not known to bioaccumulate.

GOVERNMENT AT A DISTANCE IN AGRO-FOOD NETWORKS

How do these actants, methamidophos and mancozeb, perform on the world stage? The United Nations Food and Agriculture Organization (FAO) Codex Alimentarius has worked to "harmonize" pesticide-residue regulations internationally by setting specific MRLs for crop-pesticide combinations. These are nonbinding standards meant to standardize residue regulations internationally, though many nations continue to use their own risk assessments and regulatory systems (Hough 1998). Despite attempts at harmonization, FAO and EPA MRLs for methamidophos and mancozeb—and most pesticides—remain different (EPA 2007; FAO 2008), and in some cases EPA standards are more lenient.

Not only do MRLs differ between countries, but so do decision-making processes concerning the registration and banning of pesticides. Figures 12.1 and 12.2 show national regulatory status for methamidophos and mancozeb. The maps demonstrate the highly uneven regulations among nation-states. Methamidophos, while registered for a number of crops in the United States, is banned or never registered in a large number of countries, including many in Africa and Southeast Asia. Regulation in Europe, Latin America, and South and East Asia is highly heterogeneous. Mancozeb remains a commonly used fungicide in the United States and is registered in Australia, Canada, India, and many African and European countries. In contrast, Sweden has strongly restricted mancozeb because of its carcinogenic and genotoxic effects, and many African nations have never registered it (figure 12.2).

The global maps of pesticide registrations hint at the complexity of international pesticide regulation and the difficulties faced by various actors in transnational agro-food systems. Thousands of pesticide active ingredients exist in the world, and most have highly heterogeneous regulation across the globe. For farmers and export firms, there is a potentially bewildering array of difference in complying with pesticide-residue regulations of different nations. For national governments, imposing their own regulations on the global flows of food and pesticide residues poses a considerable challenge, as specific production practices in other countries remain hidden from government regulators. If nations succeed in their regulation attempts, these efforts expand the governing of agro-food systems along network filaments that are extensions of power from the nation-state to the specific locales—down to the land user and specific agricultural fields—where pesticide regulations influence farmers' production prac-

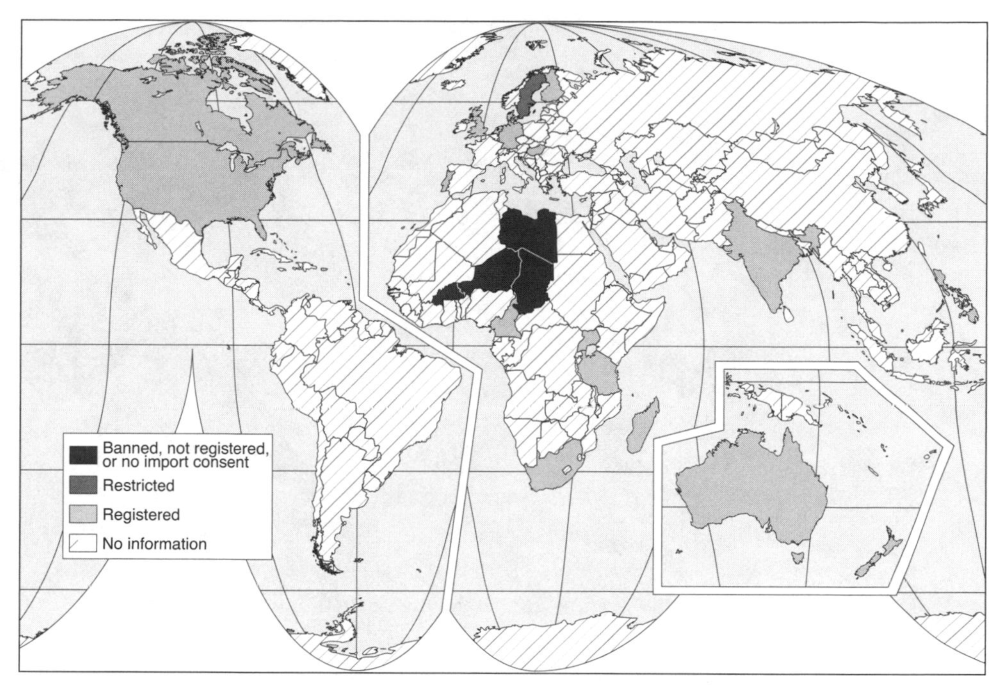

FIGURE 12.1 A global view of national registration statuses of mancozeb. Data from Orme and Kegley (2008).

tices. Latour (1987) calls this type of control "action at a distance," and others have similarly discussed "government at a distance" (Agrawal 2005; Miller and Rose 1990) and "regulation from afar" (Galt 2007).

In the case of pesticide residues, government at a distance involves (1) risk assessment and risk management that set MRLs, (2) sampling of produce to test for residues, and (3) residue testing using the laboratory methods of analytical chemistry to enforce MRLs. MRL enforcement ultimately has important effects on the actions of farmers who are not directly controlled within the territory of the regulatory institution. As Agrawal (2005, 194) notes, government at a distance "overcomes the effects of physical separation by creating regulations known to those located at a distance." Although Agrawal (2005) details the devolution of some control to community forestry councils in Kumaon, India, government at a distance in agro-food networks also becomes enforced by "intimate government" in localities that produce for export. This intimate government often is created by export firms that construct specific understandings of export market regulations and enforce these among the farmers with whom they contract.

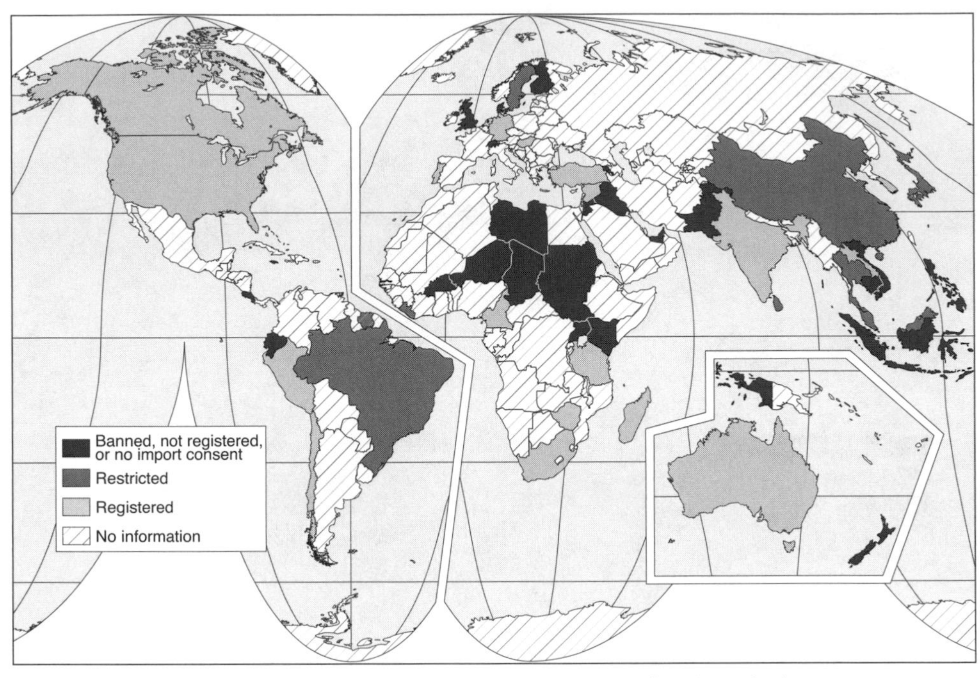

FIGURE 12.2 A global view of national registration statuses of methamidophos. Data from Orme and Kegley (2008).

THE FDA'S UNEVEN ENFORCEMENT OF TOLERANCES

Critiques of U.S. pesticide monitoring of imported food often focus on the low percentages of total imported food that the FDA tests (Wargo 1998; Wright 1990). Less than 1 percent of produce is tested (GAO 1986, 3). While I agree that the level of observation is indeed low, my research suggests that the testing has important but partial effects, as exporters and export farmers have attempted to rationalize pesticide use based on previous residue violations (Galt 2007). In other words, the specific residue tests employed—with their ability to detect some pesticides but not others—have important effects in specific nodes in the agro-food network.

I argue that we need to explore the specific manifestations of government at a distance, that is, the way in which FDA testing occurs, and how this manifests itself in land users' decisions, one of political ecology's historic foci (Blaikie and Brookfield 1987). The specificities of the scientific field of analytical chemistry—its advances, the limitations of its methods, and the resources needed to employ them—all shape the way that government at a distance manifests itself in specific locales of agro-food production tied into export channels.

Methods of analytical chemistry have improved dramatically over the course of the last half century in terms of the number of compounds that can be detected in a single test. For most residue tests, the FDA relies on multiresidue methods (MRMs), which can determine the presence and level of many different pesticide residues simultaneously. The number of residues detected with one MRM has expanded as chemists refine and improve the specific screening modules (figure 12.3). The commonly used MRMs, for example, the Luke method and its subsequent modifications, can determine residues of close to half of the approximately four hundred pesticides for which the EPA has set tolerances (FDA 2004, 3). The MRMs commonly used by the FDA can also determine residues of many pesticides for which the EPA has revoked tolerances, such as DDT and many other organochlorine pesticides. The FDA's *Pesticide Analytical Manual* (FDA 1999, 301-1) instructs its chemists, "Whenever a sample of unknown pesticide treatment history is analyzed, and no residue(s) is targeted, a multiclass MRM should be used to provide the broadest coverage of potential residues."

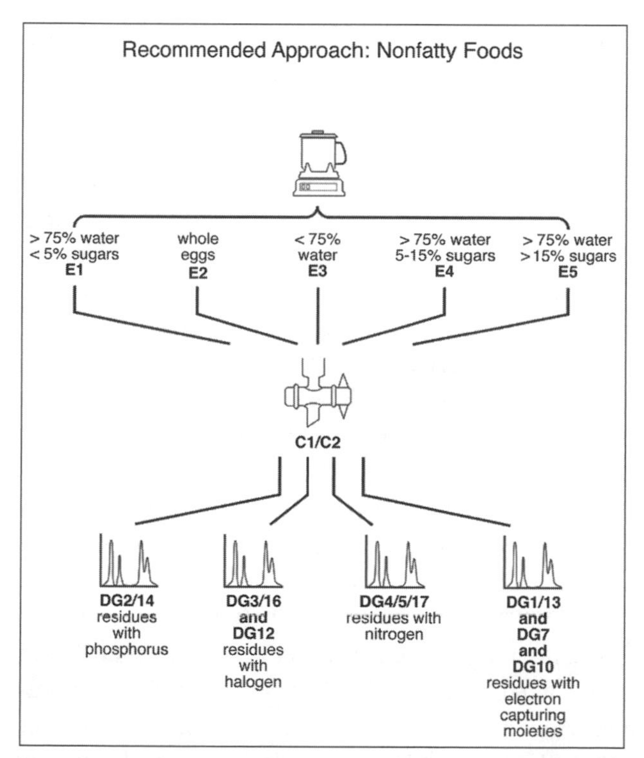

FIGURE 12.3 The residue-testing procedures from the FDA's *Pesticide Analytical Manual*. The various modules (e.g., DG2) detect different groups of pesticides. Reproduced from FDA (1999).

The "ultimate" MRM analyses developed by academic chemists can detect over four hundred pesticides (Stan 2000), but these are extraordinarily complex, and thus expensive, when used regularly in the regulatory setting. Employing these "ultimate," almost panoptic methods is thus out of the question for state agencies generally. The FDA manual (1999, 301-1) blithely notes, "The user may choose as many or as few of these modules as time and resources permit." Time and resources are clearly limited, so actual coverage is far less than provided by the "ultimate" MRMs.

Single-residue methods (SRMs) determine the residue level of a single pesticide or highly related group of pesticides, while selective MRMs determine a small group of chemicals. Both of these are considerably more expensive than MRMs on a per-residue-determined basis and are therefore used much less frequently. The FDA gives little external indication about the circumstances under which its chemists choose to use SRMS. In its manual (1999, v), the FDA notes that "these methods are most often used when the likely residue is known to the chemist and/or when the residue of interest cannot be determined by common MRMs."

These seemingly tedious specificities about residue testing impact our two dangerous actants, methamidophos and mancozeb, and the way they are used and perceived. As an OP, methamidophos belongs to the two hundred pesticides that can be detected by many MRMs the FDA uses. EBDCs, however, including mancozeb, belong to the two hundred that cannot be detected using common MRMs. Instead, SRMs must be used to detect them (Chang et al. 2005).[6] As noted above, EPA considers mancozeb and other EBDCs to be high-risk pesticides vis-à-vis cancer and birth defects and has set MRLs that in theory should be enforced. Additionally, agricultural sectors in many countries rely on mancozeb—it accounts for 30 percent of all fungicides used in Costa Rica (Humbert et al. 2007)—and it appears regularly on produce in developing and industrialized countries when it is specifically tested for (Chang, Chen, and Fang 2005; Hamilton et al. 1997).

Table 12.1 shows FDA tests for methamidophos and mancozeb. While the FDA conducted a handful of tests for mancozeb on imported produce each year in the late 1990s, it no longer tests for mancozeb on imported produce, presumably because of a lack of resources for SRMs. Mancozeb, despite being in heavy use for more almost fifty years, remains a fugitive actant, eluding attempts by analytical chemists and the state to fold it into their project of creating panoptic but affordable MRMs. The state must expend resources to find mancozeb, and chooses not to do so. In contrast, methamidophos remains a pesticide that will be detected by MRMs used by FDA labs.

The characteristics of the various pesticide actants—mancozeb presents a risk

TABLE 12.1. *Number of tests that would detect mancozeb and methamidophos on fresh vegetables imported into the United States, 1998–2003*

	1998	1999	2000	2001	2002	2003
Mancozeb[a]						
All imported vegetables	4 (0)	13 (0)	0	0	0	0
Costa Rican vegetables	0	0	0	0	0	0
Methamidophos						
All imported vegetables	476 (17)	376 (19)	0	46 (16)	66 (12)	39 (15)
Costa Rican vegetables	11 (0)	10 (1)	0	3 (2)	2 (0)	1 (1)

Source: FDA 2008, databases IMVE1998 through IMVE2003.
Note: Numbers refer to number of commodity-country combinations tested, e.g., garlic from Thailand; numbers in parentheses refer to number of commodity-country violations (either in excess of MRL or without an MRL).
[a]Reported as EBDC (identity unknown). FDA laboratories report EBDC fungicide residues as "EBDC (identity unknown)" because all are converted to carbon disulfide for detection (FDA 1999).

of cancer and birth defects, and methamidophos presents a risk of acute poisoning and neurological impairment—have influenced the EPA's creation of MRLs on specific foods. Both their relationship to analytical chemistry (mancozeb remains a fugitive to common MRM tests, while analytical chemists can relatively easily find methamidophos) and the lack of material resources (labor and time to fulfill the panoptic requirements of policing thousands of specific MRLs) shape the actual regulatory apparatus that governs agro-food systems from a distance. How do these uneven relationships shape the control structures of exporters and, ultimately, what farmers do?

UNEVEN ENFORCEMENT AND THE INTIMATE GOVERNING OF COSTA RICAN VEGETABLE PRODUCTION

To understand the influence of pesticide-residue regulation on export farmers, it is essential to consider contract farming because it involves social relationships in which the contractor attempts to directly shape the production process. Exporters with imperfect understandings of U.S. pesticide-residue regulation attempt to fashion the production process through intimate governing. A handful of export firms contract with Costa Rican farmers for export vegetables in northern Cartago and the Ujarrás Valley, Costa Rica. These firms export squash, green beans, chayote (a native cucurbit), and other vegetables to the United States, Canada, and the European Union (figure 12.4).

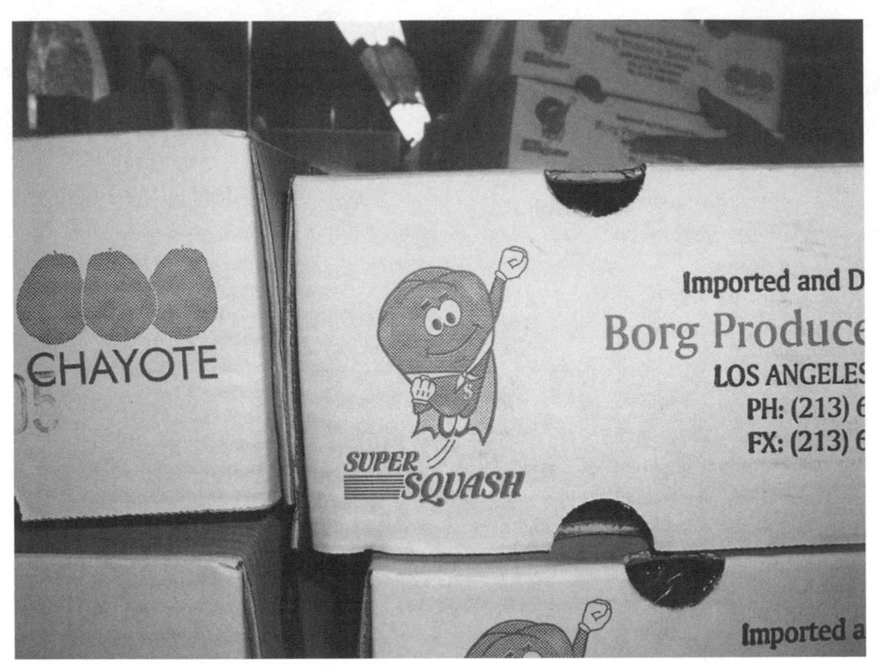

FIGURE 12.4 A box of chayote ready for export to the United States. Author's collection.

Exporters do not have complete understandings of U.S. or FAO pesticide-residue regulations, and they intentionally simplify the message about pesticides to export farmers, emphasizing that farmers cannot use OPs or the organochlorines. The Costa Rican state through its Ministerio de Agricultura y Ganadería (MAG) has also been involved in communicating with farmers and monitoring produce for prohibited pesticides—particularly methamidophos—in the chayote sector. Because methamidophos and other OP residues on chayote and squash have caused violations of U.S. MRLs in the past—resulting in the loss of entire shipments and the revenues they bring—exporters and MAG attempt to exert control over farmers' use of OP insecticides through various policing mechanisms. This intimate governance focuses mostly on shaping farmers' insecticide use, but largely ignores fungicides like mancozeb, as these have not caused violations. As we have seen, they have never caused violations of U.S. residue law in part because common MRMs will not detect them. If, for example, they were detected on vegetables for which they do not have a U.S. tolerance or at too high a level, they would be in violation of U.S. residue regulations.

Data from a survey of 148 Costa Rican vegetable farmers supports the idea that farmers treat fungicides like mancozeb quite differently from OP insecticides like methamidophos. Export squash farmers who use OPs generally comply with requirements for dose and the time required between spraying and harvest (per-harvest interval, or PHI). Their use of OPs generally, and methamidophos specifically, reflect the exporters' and MAG's concern over causing more violations and losing income from rejections (Galt 2007). In this sense, they have become partial, Foucauldian "environmental subjects" created by government from afar (cf. Agrawal 2005). In contrast, export farmers generally violate the PHI for mancozeb, as well as most other fungicides. Many farmers in the area see fungicides as essentially harmless agrochemicals, a situation not helped by their green labels (Galt 2007). Exporters and MAG have done little to change this perception, as they have never experienced direct economic losses due to fungicide residues.

Implicit in farmers' and exporters' understandings of U.S. pesticide-residue regulation is that it is truly panoptic like Bentham's Panopticon. The logic is as follows: if fungicides caused health problems, they would be regulated by the U.S. government, and strict and complete enforcement would guarantee that those fungicides most threatening to human health would be detected and the offender disciplined. This kind of discipline has been the case for OP insecticides, but the lack of negative feedback for fungicide residues serves to reinforce a view of them as safe products.

Let us now return to the discussion of hybrid networks of associations, and the idea that "each of the separate pieces is not independent, but is instead made to be the way it is by virtue of its relationship to all the other parts" (Robbins 2007, 14). Farmers' risk calculations, rather than arising from ignorance of pesticides and acted out through a lack of caution with them—as an apolitical ecology would assert—are strongly shaped by other actors and actants in the agro-food network. Farmers' lack of caution with mancozeb is an artifact strongly influenced by this fungicide's specific chemistry, which is hard for analytical chemistry to find, which makes it expensive to test for, and which means it is not detected and cannot cause rejections even if it violated residue standards. This lack of rejections co-produces and reinforces the idea that mancozeb is "safe" for farmers to use without protections.

These relationships in the agro-food network constitute important *political*-ecological relationships for a number of reasons. First, risk assessment prioritizes certain kinds of knowledge over others—toxicological data is much more valued than epidemiological data (Millstone 1996). This in part results in greater weight being given to consumers' exposures relative to farmworkers' and farmers' exposures, since epidemiological knowledge is vital in revealing the risks of exposure to pesticides at the point of production. Second, enforcement of the pesticide-residue standards is always subjected to prioritization due to limited resources. That certain pesticides are tested for more than others reflects prioritization on the part of FDA officials or scientists, although these decisions remain "black-boxed" since they are not revealed to the public. Third, continued registration of agrochemicals assumes that they can be adequately controlled by state agencies, certainly a questionable, but powerful, assumption.

CONCLUSION

This examination of pesticide-residue regulation and enforcement, traced through the actants of methamidophos and mancozeb and their use in a specific locale, allows for a number of conclusions. First, pesticide-residue regulations act as *de facto* agroenvironmental regulations from afar. Government from a distance can be more powerful than national government; Costa Rican farmers pay considerably more attention to pesticide-residue issues for export than for the national market (Galt 2009). This results from large differences in the enforcement budgets between the countries, and from the subordinate position of Costa Rica in the world economy vis-à-vis the global North. By opting into the world system, Costa Rica must maintain exports and play by the rules of the powerful even to the point that these are prioritized over its own regulations meant to protect its population.

Second, in this era of market-based solutions to environmental problems, this research shows that state regulation *works*, but the current manner in which it works can be improved. The interpretation of pesticide-residue testing as panoptic instills self-regulation by partial environmental subjects through intimate governance of production systems. However, a contradiction arises: although regulators, exporters, and export farmers conceptualize U.S. pesticide-residue regulation as panoptic, it clearly is not. This leads to a serious disconnect where farmers view fungicides like mancozeb as not harmful to human health and exercise considerably less caution with them. Equally problematic is that in continuing with current pesticide registrations and approving new pesticides, the EPA continues to assume the existence of an adequate regulatory apparatus, one that is panoptic when it comes to residues. We should instead see it as demi-optic—half seeing *and* half blind—and recognize that considerable problems result from misconceptualizing it as panoptic. Further studies of the ways in which agencies actually *implement* risk assessments, rather than just the politics of the risk assessment process per se, will allow social scientists to highlight material consequences of these highly contested decisions and to contest foundational assumptions in agrochemical regulation.

Third, the decline in the amount of FDA residue testing between the late 1990s and the early 2000s is disturbing. While the United States in its neoliberal zeal has not revoked MRLs, in terms of pesticide-residue enforcement the United States has become a "weak" state. As McCarthy (2002, 1288–89) notes, "The United States, often portrayed as the gold standard of sovereign state capacity, actually experiences many of the problems and limitations supposedly diagnostic of 'weak' states in controlling its own territory and population." The weakening of the state most recently resulted from tax and program cuts pushed by the Bush administration, which made it no secret that consumer and environmental protections were a low priority relative to profit and war. Significant pushback against the weakening of state regulation is needed on many fronts, especially in agro-food systems.

NOTES

1. Pesticide residues are traces of pesticide that remain on a food product after it has been sprayed in the field or treated in postharvest.

2. Residue detection advanced greatly in the last century. Parts per quadrillion of some substances can be detected.

3. The panopticism discussed here draws on Foucault's discussion (1977) of Bentham's Panopticon quoted at the start of the chapter.

4. Nonsystemic pesticides remain on the plant's surface where they come into contact with pests, while systemic pesticides do this as well as entering plant tissues and moving throughout them.

5. Ataxia is the loss of full control of body movements. Bradykinesia refers to slowness in executing movement.

6. The FDA's manual notes that EBDC analysis requires "special handling of the laboratory sample. EBDCs decompose rapidly as soon as the crop surface is broken and residues contact water, enzymes, and sugars" (FDA 1999, 102–6).

13

ROOPALI PHADKE

RECLAIMING THE TECHNOLOGICAL IMAGINATION

WATER, POWER, AND PLACE IN INDIA

The fear of water scarcity runs deep for farmers in the Chikotra Valley of southern Maharashtra state during the exceedingly hot and dry month of March. Even though his rows of sugarcane crops were already drowning, I watched a farmer in Pimpalgaon village use his diesel pump to pour more river water into his furrows. This same farmer told me that he had been watering his crop for seven days straight. Though he knew that overwatering his fields would not necessarily increase crop productivity, and would likely promote soil water-logging and salinization, from his point of view it was better to water whenever possible rather than wait for the next unpredictable servicing.

This sort of scene is quite common throughout vulnerable agricultural regions of the developing world. Political scientist Robert Wade (1990) has hypothesized that irrigators often apply water at any given opportunity because the technical design of infrastructure does not provide farmers with a sense of water security. Exemplifying Wade's "syndrome of anarchy," the farmer quoted above justified his overwatering to me by arguing that his fields hadn't received irrigation for over forty-five days.

As in many water-scarce areas of India, farmers in the Chikotra Valley have come to depend on dams and canal systems to provide year-round water security. Dams were first built by colonial administrators in this region at the turn of the twentieth century to promote famine protection and drought relief. Yet this model of technological development, which was continued with fervor by postindependence political leaders, paid very little attention to the priorities of local cultivators or their socioecological knowledge and experience with drought protection. The prevailing lack of participatory approaches to water development has sparked intense political controversy for the last three decades in India. Political concerns around issues of water equity, dams, and human displacement have became paradigmatic struggles about balancing the material benefits of industrial development with the ideals of democratic governance and human rights. This sort of activism has become the subject of an entire genre of interdisciplinary social movement research in India, including the work of Amita Baviskar, Dilip D'Souza, Lyla Mehta, David Mosse, and Satyajit Singh.

The academic and journalistic focus on social movements against large dams has paid less attention to the places and spaces where dams are re-emerging from the grassroots. My interest is to highlight this neglected side of water activism. This chapter describes the evolution of a water reform movement in the state of Maharashtra that defies conventional understandings of water politics in India. Social movement leaders associated with the people's science movements (PSMs) in this region have supported the building of new dams and canals as part of an approach toward reimagining development. The PSMs are a distinctive Indian social movement that operates as an umbrella coalition of diverse organizations interested in science and development policy concerns. Begun in the mid-1960s to challenge an ivory-tower approach to science and technology development that failed to address basic human needs, the PSMs have tackled a range of topical fronts including health care, natural resources, and literacy. While there are national-scale organizations and platforms affiliated with the PSMs, most organizations are rooted in local contexts and aim to address specific development problems.

By melding hydrological expertise with local socioecological knowledge about water, PSM leaders in Maharashtra have aimed to reclaim the very technological artifacts, such as large dams, that have become naturalized embodiments of hegemonic knowledge and power. The engineers and environmental activists are materially and cognitively producing new kinds of dams, ones that are productive and restorative, organic and synthetic. Their goals are to make these infrastructure systems stand in for a different set of moral, political, and ecological virtues.

Social movement leaders are helping break down two dominant binaries that are of concern to political ecologists and STS scholars regarding the production and circulation of knowledge. First, movement leaders demonstrate that ecological knowledge is produced in and by social networks working at the intersection of the lay and expert divide. Second, in the local reimagining of technoscientific objects, leaders also demonstrate that knowledge circulates and hybridizes in ways that blur any distinct boundaries between global and local epistemologies.

This chapter explores how this water reform movement is producing and circulating hybridized expertise as part of a broader process of "technological reclamation." I use this term to describe the ways in which technological systems that fail to gain widespread public trust and support are being reconstructed by social movements. To understand how technological reclamation works, I asked, What is being reclaimed, with what instrumentalities, by whom, and for what purposes? People's science efforts in Maharashtra demonstrate that protest politics and alternative design are both vital ingredients for remaking technological systems into models of democratic expertise. NGOs play an important role in brokering both the production and circulation of knowledge toward the reworking of technological projects and objects.

HYBRID KNOWLEDGES AND TECHNOLOGICAL POLITICS

In the conclusion to their 1997 chapter "Nature as Artifice, Nature as Artifact," geographers Michael Watts and James McCarthy argued that science studies, environmental history, and social movement theory were all "critical ingredients of political ecology" (85). While political ecologists have closely engaged with both environmental history and social movement theory, their attention to science studies has been far less systematic. Actor-network theory and feminist approaches to science studies have greatly influenced the work of political ecologists. The scholarship on hybridized expertise and democratic technology development, however, has been largely overlooked. This rich area of research, represented by diverse scholars such as Sheila Jasanoff, Langdon Winner, Alan Irwin, Richard Sclove, and Brian Wynne, has much to contribute to theoretical insights and empirical research in political ecology. Building on this theme in STS scholarship, my concern in this chapter is to theorize how hybrid knowledge gets produced and applied in water resource planning. I want to call attention to how movement leaders circulate knowledge in social networks so that it gains credibility, legitimacy, and political power.

The hybridization of expertise has played an exceedingly important role in how social movements challenge environmental policy and promote alternative technological designs. Political ecologists and science studies scholars have articulated many different approaches toward studying "hybridity." As Sarah Whatmore has described, science studies scholars have discussed hybridity in terms of Callon and Law's "hybrid collectif" and Haraway's "cyborg" (2002, 5). Tim Forsyth describes hybridity as a descriptive term evoked by STS scholars to signal the "processes by which social-natural objects become entwined through social discourse, and henceforth become accepted as objects" (2003, 105). Like the term *co-production*, hybridization is concerned with "the dynamic co-evolution of knowledge and social change" (105).

The political ecology and science studies scholarship on postcolonial technoscience uses the term *hybridity* to refer to knowledge systems that are built on both local experience and expert science. In these instances, the concept of hybridity encompasses the ways in which the global and the local, the traditional and the modern, mix unevenly in discourse and practice (W. Anderson 2002). As Bruno Latour has so elegantly demonstrated, knowledge is produced by local and global actors through the acts of inscribing scientific facts and practices. Anna Tsing (2004) complexifies the routes of hybrid knowledge making by drawing attention to the tensions, the "frictions," between global discourses and local realities. Similarly, Arturo Escobar has described how in the process of place making, social movements enact "global" strategies by producing locality from the "politics of below" (2001, 161).

I situate this study of PSMs in India within the context of a range of theoretical and empirical literature on social movement activism that is concerned with how hybridized knowledges are applied toward producing dynamic rural landscapes (Peluso 1992; Broad and Cavanagh 1993; Wolford 2001). This genre has also addressed the social and environmental impacts of technological projects on rural peoples, including controversies around the construction of highways, oil fields, power plants, and dams. These critiques of technological development have been strongly rooted in both Marxist and poststructuralist theory (Peet and Watts 1996).

Political ecologists and science studies scholars have drawn heavily from Foucauldian theory to demonstrate that technology is a state tool for disciplining society and rendering subjects "legible" (Foucault 1977). Critical studies of "governmentality" have evaluated how knowledge is produced, represented, and contested (Agrawal 2005; Blaikie 1999; Escobar 1995). A range of empirical case studies have informed this perspective, including examinations of state forestry in Indonesia (Peluso and Vandergeest 2001), the Soviet collectivization of agriculture (J. Scott 1998), and the compulsory implementation of "sustainable development" in the Mekong region of Southeast Asia (Goldman 2005).

In the context of water politics and geographies in India, this strand of social movement literature has problematized the appeal to "localness" by drawing attention to the competing narratives for and against development that can live within a single social movement (Rangan 2000). The political ecology scholarship on water resources in India has also explicitly challenged romantic claims that traditional knowledge, and "village republics," can be revived to address postmodern environmental governance challenges (Mosse 2003). As Amita Baviskar has written (2007, 3), the "new public sphere" politics of community-based natural resources development "transects temporal and spatial scales" in ways that blur the boundaries between community and state actors. In his work on groundwater knowledge systems, Trevor Birkenholtz (2008) examines how irrigation technologies have come to exemplify the shifting terrains of power between farmers and state water experts in Rajasthan.

In this chapter, my use of the term "local" knowledge does not attempt to valorize traditional ecological knowledge or argue that place-based, context-driven ways of knowing and seeing the world are a substitute for science. Rather, the term is used to emphasize that knowledge is produced and circulated in networks where nonexperts are agents in knowledge making. I am also interested in symmetrically privileging the nontechnical ways by which knowledge is gained, reproduced, and validated. My emphasis is not on local knowledge as distinct from science, but on the interests, discourses, and narratives that are often absent from conventional state development planning. By evoking the language of hybridity, I aim to move away from the fetishization of indigenity by drawing attention to how new interventions in development practice come about when values accrue to local knowledge (Akhil Gupta 1998; Prakash 1999).

My main concern is to understand how hybrid coalitions and networks simultaneously critique development planning while producing new technological applications that express democratic demands. Dorothy Nelkin (1992) and Langdon Winner (1985) began a wave of influential STS research on the role of social movements in influencing technological decision making. This genre of science studies work has moved beyond its earliest concerns for understanding the political nature of technological controversy toward mobilizing technologies of accountability and transparency in democratic and deliberative governance (Jasanoff 2003; Ezrahi 1990; Irwin 1995; Irwin and Wynne 1996). This is evidenced by the broad range of science studies scholarship on participatory technology assessment. Richard Sclove's work (1995), based in part on the programmatic work of the Loka Institute, has demonstrated the connections between strong democracy and community engagement . Similarly, Alan Irwin's work on citizen science initiatives (1995), ranging from science shops to study circles, has examined the interface of lay knowledge and scientific expertise across a range of European and American technological controversies.

Technological pluralism is particularly difficult to advance in contexts where the infrastructures that enable modernity have become so thoroughly naturalized. Science studies scholars have articulated how infrastructures are rendered stable and can be destabilized through the exercise of political power (Castells 1996; Hughes 1983). As Paul Edwards has argued, mature technological systems, such as roads, sewers, telephones, railroads, and computers, "reside in naturalized background, as ordinary and unremarkable to us as trees, daylight, and dirt" (2003, 185). Yet, as the "connective tissues and the circulatory systems of modernity," these technological systems support social possibilities and create "lawlike" constraints and "systematic vulnerabilities" (185).

Drawing on Edwards's scholarship, I am interested in thinking about how hybridized knowledges are *denaturalizing* infrastructure by embedding it within new material and symbolic contexts. I stir together the rich scholarship coming out of both political ecology and science studies to think about why, how, and when social movement actors build bridges between local knowledge and scientific expertise to reimagine infrastructure. I ground my examination in the work of the PSMs in India. Since the 1960s, the PSMs have been actively challenging the role of technoscience in delivering development. As the opening section of this chapter described, water infrastructure politics are a rich context for seeing people's science at work in western India.

THE PEOPLE'S SCIENCE MOVEMENTS

Questions of epistemological democracy have long dominated Indian academic writings on science and politics. Authors like Shiv Visvanthan, Ashis Nandy, Gyan Prakash, and Akhil Gupta have become important resources for scholars interested in understanding the Indian domain of alternative science and the ways in which science can be redeemed as just, ecological, and moral. I am interested in contributing to this body of Indian scholarship by demonstrating how technoscientific democracy is being asserted by PSMs in India in the water sector.

Among social movements in India, the PSMs are distinguished by their broad membership base and concern for the popularization of science. By the first meeting of the All India People's Science Movement in 1978 in Trivandrum, people's science chapters were already active in the states of Kerala, Maharashtra, Karnataka, West Bengal, and Uttar Pradesh (Jaffry et al. 1983). The most well known chapter of the PSM is the Kerala Sastra Sahitya Parishad (KSSP). Founded in 1957, the KSSP's "science for social revolution" campaigns have reshaped development politics in this southern Indian state. The KSSP is largely credited with initiating participatory development planning as a state mandate in Kerala.

Acting more often as "loose associations," people's science groups operate as decentralized umbrella coalitions unifying the work of disparate, and often contending, activist institutions throughout India (Varma 2001). People's science organizations have dominated the NGO sector, including People Oriented Science and Technology in Madras, the People's Science Institute in Dehra Dun, and the People's Institute for Development and Training in New Delhi.[1] People's science activists, including doctors, scientists, teachers, and students, have been guided by a belief that Indian science and technology development has perpetuated an ivory-tower ethic, where the needs of the masses for safe water, fuel, and health care have been ignored. The "people" addressed by the PSMs have generally been agricultural laborers, small peasants, rural artisans, and urban workers who have been disenfranchised by the postcolonial model of scientific and technological development (Kumar 1984, 1083).

Since the 1980s, the PSMs have been struggling to promote alternative and appropriate models of technological development that represent a hybridization of local and expert knowledge traditions. These efforts often have a distinctive focus on local development concerns and vernaculars. For example, the Dehra Dun–based People's Science Institute is mainly concerned with disaster mitigation, joint forestry management, and participatory watershed development (Chopra and Sen 1991). In another example, the Honeybee Network, sponsored by the Society for Research and Initiatives for Sustainable Technologies and Institutions (SRISTI) in Ahmedabad, promotes the documentation and dissemination of grassroots innovations for biodiversity conservation and sustainable development in the medium of local languages (see Anil Gupta 1997).

In the western state of Maharashtra, people's science efforts have been focused on the reform of state water infrastructure development. Figure 13.1 illustrates the southern region of Kholapur that is the focus of this chapter. In the post-independence period, large-scale irrigation and hydroelectric projects were regarded as India's "modern temples," signifying technological progress and political might (Klingensmith 2007). By the end of the twentieth century, the ecological consequences of these efforts became obvious: salinization, waterlogging, and increased water scarcity. The social costs were equally profound. According to the former secretary of the Indian National Planning Commission, over fifty million people have been displaced by dam projects in India since 1947. In 2000 the World Commission on Dams published its India country report. It concluded that irrigation projects in India have more often reinforced, rather than relieved, existing patterns of social inequity, administrative inefficiency, and environmental degradation (Rangachari et al. 2000).

Accessing water is a chronic concern in rural Maharashtra, where 40 percent of the rural population chronically suffers from drought. Entrenched political interests, including a strong sugarcane-industry lobby, have historically limited the role of citizen participation in the design of water technologies. Since the 1970s, however, a social movement has developed solid technical alternatives for sustainable agriculture development in the region. As an example of people's science organizing, the water reform movement has adopted an approach that melds oppositional politics with reconstructive development efforts. Moving beyond protest politics, these groups have developed innovative methods for implementing public participation in technology design, implementation, and maintenance at the village scale.

The water reform movement has two distinct wings: one constituted by mass farmers' organizations and the other by technical NGOs. The NGOs are based in the urban regions of Pune or Mumbai, and are supported by members and occasional government contracts. The farmers' groups use trade union–style organizing to mobilize their members, who are often connected to local political parties. While their constituencies are distinct, these two groups share an over-

FIGURE 13.1 Map indicating Maharashtra state.

arching commitment toward addressing the social justice struggles faced by both drought-afflicted farmers and dam evictees in rural Maharashtra. Together, the farmers' organizations and the technical NGOs are bridging traditional Marxist organizing with "new social movement" interests in decentralization, democratization, and sustainable development.

Oppositional politics have brought activists into constant conflict with the state government. In their fight over water, the movement has promoted tactics that challenge the state's legitimacy, such as withholding land revenue until water shares are provided. They also use time to their advantage by organizing protests during the construction phase of a dam project when the government is most vulnerable to meeting their demands. Because dams in Maharashtra depend on monsoon rains to fill reservoirs, it is critical that dam construction not be stalled during premonsoon months. Protests scheduled during the months of April and May can effectively delay a dam project for an entire year.

The following case study of the Chikotra Valley Project helps illustrate how this movement is impacting the way the state administers water. While the project remains under construction, as a result of political mobilization it was redesigned to include principles of public participation, equitable water distribution, and ecological restoration. The ultimate fate of the project rests with a reluctant irrigation bureaucracy: one aware of the social unrest around water insecurity in the state yet reluctant to commit to an epistemological revolution that would undermine their authority to govern.

PROMOTING WATER EQUITY IN THE CHIKOTRA VALLEY

The fields alongside the Chikotra River shine a glimmering green during and after the monsoon rains. Lush rice paddies and vegetable and maize crops grow alongside every waterway. Figure 13.2 depicts tall stands of sugarcane along village roads. This same verdant scene quickly withers to shades of brown from October to June. Because of scarce water infrastructure, only 2 percent of land is irrigated in this region; farmers can confidently grow only one crop a year. In addition to a shortage of irrigation infrastructure, water for household use is hard to get. Figures 13.3 and 13.4 illustrate the role village children play in gathering water, often from great distances, for household use.

The Chikotra Valley watershed is marked by the origin of the Chikotra River on the east and the confluence of the Chikotra with the Dudhganga River on the west. In 2001 a medium-size dam was constructed by the irrigation bureaucracy on the Chikotra River. While the Chikotra dam is not one of the largest or most expensive of regional irrigation projects, it has become one of the most controversial, due to the local activist response.

As table 13.1 documents, the original Chikotra project consisted of a sixty-meter-tall dam, with a storage capacity of 1,552 million cubic feet of water. Figure 13.5 depicts this dam site. In addition to the main dam, the project also included two minor storage dams and twenty-five intermittent weirs downstream of the main dam (MKVDC 2001). The project was designed to serve a command area of twenty-seven villages, which all have similar land-tenure patterns and topography, with average landholdings of one to one and a half acres. Because of the limited size of the command area, the government project stipulated that only those farmers whose land was located at a height of less than thirty meters from the riverbed could access water from weir sites. Farmers in the benefit zone were expecting the irrigation agency to grant them individual permits to sink pumps at weir sites and lift water to their fields. Completed in 2002, the main dam inundated 317 hectares of land from five villages. While residential areas in the villages were spared, outlying farmlands were submerged. Residents from these affected

FIGURE 13.2 Sugarcane fields.

villages were not slated to benefit from the irrigation project, not even through the provision of drinking water.

While construction of the weirs began as early as 1994, actual dam work was stalled until 1997. This delay was due to steadfast protests from villagers affected by the project. The conflict between the irrigation agency and the project-affected farmers was finally overcome through the negotiation efforts of one particular NGO, the Shram Shakti Pratistan (SSP). As a technical NGO based in the Chikotra Valley, the SSP has been involved with watershed development experiments in eight villages in this same district since 1993. Responding to the needs of drought-afflicted villagers in the Chikotra Valley, Anandrao Patil, president and founder of the SSP, assumed an intermediary role in the displacement struggle.

FIGURE 13.3 Boys collecting water in the Chikotra Valley.

FIGURE 13.4 Gathering at the well in the Chikotra Valley.

TABLE 13.1 *Features of the Chikotra Project*

Command area	5,630 hectares
Total storage	1,552 million cubic feet
Dam length	860 meters
Dam height	60.2 meters
Land acquisition	317 hectares
Cost	US$33 million (1996)

Source: Data from MKVDC (2001).

Born and raised in the village of Belewadi, located at the center of the Chikotra Valley, Mr. Patil had served as headmaster of the local school for decades. Now in his late seventies, Mr. Patil has garnered strong moral authority in the valley.

In 1996, recognizing the desperate need for water in the Chikotra Valley, Mr. Patil initiated a campaign to build cooperation between beneficiary farmers, the agency, and PAPs (project-affected peoples). After months of village meetings, farmers from the beneficiary villages volunteered to provide over 50 percent of the lands that were required for the PAPs.[2] As a result, at least half of the PAPs were able to cultivate new land close to their villages and receive irrigation benefits. One of the deputy engineers responsible for execution of the Chikotra dam stated that through Mr. Patil's efforts "there was a change of heart among the affected people" after beneficiary farmers gave them land. As a result of this land swap, for the first time in the history of Maharashtra, the *bhoomi puja* (the inauguration ceremony) for the dam project was conducted by PAPs themselves. By 2002 the irrigation agency reported that 95 percent of PAPs had been assigned new lands within the valley.

While general resolution of the PAP conflict meant that the Chikotra project could proceed, the SSP took up a new battle with the agency over the distributive benefits of the irrigation project. The NGO cited three main objections against the design of the Chikotra project. First, the project provided water to only twenty-seven out of a total of fifty-two villages in the basin. In addition, village residents who were landless or employed outside of the agriculture sector were not provided any domestic water security. Second, the agency plan stipulated that they could afford to allow only those farmers whose lands were located at a height lower than thirty meters from the river to have access to water stored behind weirs. While the agency has argued that the cost of lifting water to a height above thirty meters is simply uneconomical, it is also the case that the steepest lands in the valley are cultivated by the most powerless and marginal farmers.

FIGURE 13.5 Construction of the Chikotra dam.

The designation of the command area for the Chikotra project has resulted in disenfranchising farmers within a beneficiary village. For example, the village of Belewadi Hubbalgi has 350 families and 500 hectares of land. The topography of the village is such that only 10 percent of village land falls under the official command area of the Chikotra project. Third, the SSP objected to the agency rule that irrigation permits could be granted only to individual farmers as opposed to water users associations (WUAs). This decision meant that only those farmers who could afford to purchase or rent pumps could access water.

Protesting dams has become very commonplace in India. In the Chikotra case, activists went beyond the mode of critique toward designing a viable alternative plan for the government irrigation project. Before a new distribution plan could be offered, the SSP required data about how much water could be stored in the valley, which methods would be used to store it, and the minimum assured level of water that each family could be provided. This meant investigating soil and water conservation measures, as well as potential changes to local cropping patterns toward greater water efficiency. The SSP cast a wide call for help to design an alternative plan. University professors, college students, and engineers came to their assistance. Faculty and students from both Pune University and Shivaji University in Kolhapur conducted soil and hydrological experiments in collaboration with local farmers. The most important source of technical assistance, however, came from engineers and economists associated with the NGO Society for Promoting Participative Ecosystem Management (SOPPECOM) in Pune.

SOPPECOM activists, agronomists, and engineers have accumulated twenty years of experience designing technical projects to serve principles of social equity and ecological regeneration. They have been involved with some important community-based research projects, including the Baliraja dam, Pani Panchayats, and the Uchangi project (R. Phadke 2003). SOPPECOM projects promote the equitable distribution of water, the integrated use of surface and groundwater, and farmer participation in irrigation design and management (SOPPECOM 2002). SOPPECOM's model of community-based research includes collecting oral histories and conducting community mapping to gauge local water needs. Under a banner of "Some Water for All," SOPPECOM helped the SSP design an alternative basin development plan that would provide equal access to water for all families from the fifty-two villages, including project affected areas, in the Chikotra Valley.

Based on their work with local farmers, SOPPECOM's work in the Chikotra Valley was guided toward scaling up models of equitable water development that have shown success at the microwatershed level. One of the challenges involved in river basin development is overcoming typical upstream/downstream user conflicts. Sociological studies of irrigation development have long documented that upstream users generally appropriate, at times illegally, an inequitable share of water, leaving tailenders with little or nothing in the balance (Horowitz 1991; Wade 1988). Solving this problem requires stringent local governance, as well as decentralization of both water storage and distribution networks.

Decentralization is the overarching design feature of the alternative Chikotra Valley plan. This approach integrates microwatershed projects with large-scale lift irrigation. The annual rainfall, cultivable area, and potential for water conservation differs from village to village in the basin. SOPPECOM engineers suggested an action research program to construct farm ponds, small percolation dams, and soil-conservation projects in every village in the basin. Through these construction efforts, water tables and soil-moisture levels would increase throughout the year, limiting the need for exogenous water sources. Exogenous irrigation water, however, would be available to supplement local supplies, bringing better balance between water-rich villages and those with limited opportunities for rainwater harvesting (SOPPECOM 2002).

The alternative plan premises water distribution on a notion of intervillage and intravillage equity. The aim is to have the surface and groundwater that is stored within the Chikotra basin be equitably shared among all fifty-two villages on a per capita volumetric basis. Under these distribution principles, the command area of the Chikotra irrigation project can be extended from an original 15,000 to 24,000 hectares for two seasons of crops. SOPPECOM engineers estimate that each family can receive approximately three thousand cubic meters of irrigation per year, regardless of their physical location in the valley or whether they hold land (Datye et al. 1997). WUAs play a very important role in the alternative plan. The alternative proposal suggests that permits be granted to registered water users cooperatives, instead of individual farmers. WUAs are being established in every village in the valley. Finally, in the people's science spirit, the alternative plan establishes the need for an action research program to continue involving local residents in the process of data gathering and planning, particularly for siting irrigation outlets.

In 2000 the alternative proposal was submitted to the agency, as well as local elected officials. Having the support of leading Maharashtra water activists, such as Vilasrao Salunke, the proposal was reviewed at the highest political levels. In January 2002 Maharashtra Chief Minister Vilasrao Deshmukh pledged his support for a redesigned Chikotra project. The irrigation agency was directed by the chief minister to conduct a feasibility study regarding project implementation. The agency reported back to the chief minister that there was "theoretically" enough water to provide every village in the basin with minimum access.

While the chief minister signaled his support, agency officials, particularly the minister of irrigation, have been vehemently against the alternative proposal. Foremost, agency officials have argued that the cost of providing lift-irrigation facilities to every village in the Chikotra Valley will be financially unbearable for the agency. Irrigation officials have also cautioned politicians against setting a powerful precedent in the state by supporting a widespread notion of water equity. In response to the financial claims made against the alternative Chikotra project, the NGOs have sought financing from an external multilateral or bilateral source, such as the World Bank or the Danish International Development Agency. Any such funding would likely be channeled through the Irrigation Department and would require their support of the project.

While the Irrigation Department has fought against setting a precedent for water equity in Maharashtra, it is the prospect of such a precedent that has been motivating NGOs. For both the engineers and activists involved, the Chikotra project has the potential to demonstrate that water can be equitably distributed and used in an integrated manner at the scale of a river basin. From their original goal of helping the farmers of Chikotra Valley get access to water, this struggle has now come to symbolize the need for broad and sweeping changes in water development throughout Maharashtra.

The implementation of the alternative Chikotra project now hinges on successful financial support from national and international agencies. Since the total project is not yet complete, it remains uncertain whether the government will meet its promise to deliver greater water equity. The irrigation bureaucracy's reluctance to fund a "democratically just" water project demonstrates its continued political ability to resist epistemological challenges from below. While it may be premature to celebrate the Chikotra case as an example of the democratization of water infrastructure planning, it is nonetheless possible to draw some lessons from this case, and the broader PSM, about the future of dam development in India. The most important lesson may be that social movements are becoming increasingly important agents for challenging conventional models of expertise and creating new networks that circulate local knowledge in ways that speak to power. While this movement is deeply rooted in the local context, it resonates with shifts in the international development community arguing for greater transparency, participation, and devolution of decision making.

The case also demonstrates the importance of analyzing the knowledge politics at work in social movement reform campaigns. The Chikotra case offers some useful insights into the role technical NGOs play as boundary organizations inhabiting the space between social movements and government agencies. David Guston (2001) defines boundary organizations as those groups who straddle scientific and political institutions. Yet, rather than blurring boundaries, these organizations police the boundaries between science and politics so that they may project authority by showing a "responsive face to either audience" (405). Guston argues that while the politicization of science, and the scientization of politics, are slippery slopes of reason, the boundary organization "does not slide down either slope because it is tethered to both, suspended by the coproduction of mutual interests" (405).

In this water case, NGO experts play this boundary policing role when they translate political goals into technological projects. While NGO leaders are closely affiliated with social movement organizations and consider themselves activists, as development experts they are also epistemologically tied to their government peers. Several of the NGO leaders involved in the Chikotra case were retired government officials who had once worked alongside current agency administrators. These individuals represent local knowledge when they help negotiate with government officials, particularly when they describe local priorities and categorize the ways in which local people work with and understand natural resources. As translators of hybridized knowledge, the NGO leaders also play a powerful role by placing technological alternatives into the language and norms of agency protocols. These strategic moves are enabled by the fact that NGO experts can rely on, and draw from, their social positions to gain authority as translators and hybridizers of knowledge.

In the context of boundary organization work, this case study raises interesting questions about the circulation of knowledge within social movement networks that is germane to other resource struggles. In particular, can technical NGOs better support hybridized knowledge by becoming part of the official development process? Would their involvement as arbiters of local knowledge help limit the need for communities to protest government efforts? If they served as government consultants, would that undermine their epistemic flexibility and ability to politically represent local communities?

TECHNOLOGICAL RECLAMATION

Reflecting back on the earlier discussion of science studies scholarship on democratic technological development, one of my aims in telling these stories has been to decenter dams as artifacts embedded in naturalized infrastructures. This decentering move enables us to see dams not as natural solutions to drought, but as sites of dynamic and contingent expertise-building exercises. Dams built in the Nehruvian era in socialist, postcolonial India symbolized a sweeping utilitarian logic. In a Latourian sense, dams have been the perfect example of the "immutable mobile." While this mindset continues to pervade large-scale water development in other regions of India, a new imaginary is emerging around dam development that denaturalizes and reconfigures these infrastructures.

In addition to demonstrating how knowledge is being hybridized in Indian development practice, the Maharashtra water reform movement helps articulate how hybridization can be a part of a broader process of "technological reclamation." The word "reclamation" evokes a range of meanings connected to concepts of regeneration, rehabilitation, and restoration. Irrigation textbooks define reclamation as the act of restoring land to productive use. Using the example of water reform in India, I want to demonstrate how social movements move beyond simply producing hybridized knowledge toward applying that hybrid knowledge in ways that unsettle political authority. These "reclamation" acts involve material, symbolic, and discursive reconstitutions.

In the material dimension, the water reform movement is reclaiming dams so that they serve a different set of political and economic goals. With a Marxist intent to redistribute the gains of development, the redesigned dam projects aim to bring more water to greater numbers of formerly disenfranchised people. The deconstruction and redesign of technologies by social movements involves re-embedding the technological system or artifact into the local social fabric. In the Chikotra case, the technological imaginary was reconstituted through acts of participatory action research. As a result, new actors and old experts found new roles to play in the development process. This has allowed naturalized categories to be dismantled and recast as containing social elements. An important example of this is the changing definition of PAPs. Once largely perceived by Indian administrators as the victims of development, these groups of people were transformed into project beneficiaries in the case study.

The symbolic and discursive dimensions of technological reclamation are equally important. Dams have immense rhetorical power in India as embodiments of state authority. When social movements reclaim dams as organic technologies, firmly connected to local people and place, they come to stand for new configurations of nature, culture, and power. In the case study examined, the redesigned dam project symbolizes far more than water storage. It signals a return of dignity and respect to those who have tilled the land for generations.

In both the material and symbolic dimensions, these experiences with water reform are very important for understanding the processes that enable new knowledge to emerge and circulate in the processes of environmental governance. Mistrust of science and technology has motivated many social movements around the world to challenge development efforts related to energy, biotechnology, and land use. In such contexts, a theory of "technological reclamation" is not intended to guide policy, though it has great policy relevance. Rather, it can be a powerful conceptual tool for thinking about how social movements deploy material and discursive strategies. In addition to asking what is being reclaimed, it is important to ask, with what instrumentalities, by whom, and for what purposes?

This case study suggests that success at "technological reclamation" can be measured by different gauges, including the generation of new data, the increased agency of disenfranchised actors, and ultimately government reform. This case also demonstrates that protest politics, knowledge brokering, and alternative technical design are all important ingredients for remaking technological systems into models of democratic expertise. Social movements produce various social goods by engaging in these processes. In this example of people's science organizing, it has meant more acceptable designs, better knowledge, more political buy-in—and eventually and hopefully more water for the most vulnerable members of society who suffer from chronic water insecurity.

NOTES

1. For a more thorough analysis of the evolution of PSMs in India, see R. Phadke (2003).

2. These farmers were paid for their land by the district collector, who then changed title over to displaced farmers. The rest of the farmers were to be either provided monetary compensation or given similar land belonging to the government.

REBECCA LAVE

14 CIRCULATING KNOWLEDGE, CONSTRUCTING EXPERTISE

Proponents of ecological restoration claim that we can transcend the defensive stance of traditional environmentalism and repair the damage already done (Jordan 2000). The tremendous appeal of this image of humans as a positive contributor to environmental health has rapidly made restoration a driving force of the environmental movement and an institutionalized commitment at all levels of American government. Stream restoration, in particular, attempts to restore the natural functions of riparian systems, typically by undoing anthropogenic harm. Restorationists try to bring back both ecological functions, such as habitat for salmon or other endangered species, and physical functions, such as ability to transport the flows of water and sediment received from the surrounding watershed. Stream restoration is thus a multidisciplinary project bringing together aquatic ecologists, botanists, engineers, and fluvial geomorphologists (who study the ways that flows of water and sediment shape riparian landscapes).

Because it is implicated in a range of critical issues such as protection of endangered species, flood prevention, drinking-water quality, and outdoor recreation, stream restoration receives tremendous amounts of public attention and funding. Motivated by hope and new regulations, organizations ranging from local advocacy groups to the federal government spend more than a billion dollars and countless hours of time each year attempting to restore streams to a more natural condition (Bernhardt et al. 2005). Unfortunately, the demand for stream restoration projects far outstrips knowledge of how to implement them; streams are highly complex systems, and there is a good deal of uncertainty about restorationists' ability to produce the positive effects for which they strive. There is thus great demand for the creation of a field of stream restoration science: agencies have funds for such projects, communities are pushing for them, developers must complete them to mitigate project impacts, and many people want to be trained to implement them. Despite this, the nascent field of stream restoration has yet to gain a strong foothold in the academy; there was until recently no way to get even a certificate in stream restoration at an American university, much less a degree.[1]

Dave Rosgen, a consultant in Colorado, stepped into this breach by developing a relatively simple system for classifying and reconstructing stream channels.

Rosgen uses classic 1950s and 1960s work in fluvial geomorphology as the basis of a purportedly universally applicable alphanumeric classification system that divides channels into seven categories (A–G, each with multiple numbered subdivisions; see figure 14.1) based on a relatively brief evaluation of their physical form (Rosgen 1994). According to Rosgen (1996), practitioners can use this classification system as a basis for designing stable natural channels. These claims of universal applicability and stability are deeply appealing for resource agencies. Since the mid-1990s, Rosgen's work has been adopted by key federal agencies including the Environmental Protection Agency, the U.S. Fish and Wildlife Service, the National Resources Conservation Service, and the U.S. Forest Service, as well as the natural resource departments of more than a dozen states. Many of these agencies now require use of Rosgen's approach, and it is thus central to the expanding market in stream restoration services.

Rosgen's success raises questions about the construction of expertise and the production and application of scientific knowledge more generally. Here I use the circulation of Rosgen's knowledge claims as an entrance point for analysis of the interlinked character of knowledge production, circulation, and application, drawing on data from interviews, participant observation, surveys, and document analysis.[2] I argue that the foundation of Rosgen's success is his ability to simplify complex knowledge into standardized and easily disseminated forms that serve the needs of the growing restoration market and the state, and thus to build an epistemic community of consultants and agency staff who actively support and promote his work. By creating and promoting a standardized package of restoration techniques, Rosgen has come to effectively control the stream restoration field and market despite the vehement opposition of university- and agency-based scientists.[3]

This emphasis on the role of political-economic forces runs counter to what is perhaps the most common theoretical tradition that political ecologists turn to when attempting to pull insights from STS into their field: actor-network theory (ANT). Promulgated by Latour, Callon, Law, and others, ANT is one of the most powerful explanatory frameworks in science studies. The attempt to combine critical studies of society/nature relations and ANT has produced a rich body of work with provocative epistemological and ontological implications (Whatmore 2002, 2006; Thrift, Harrison, and Pile 2004; Lorimer 2007; and Robbins 2007, among many others). Yet ANT's basis in individualistic, phenomenological theory leads it to ignore political-economic relations. I thus attempt to demonstrate here that political ecologists wishing to draw insights from STS into their analyses while preserving their political commitments might be better served by strands of STS that address structural power and inequality.[4]

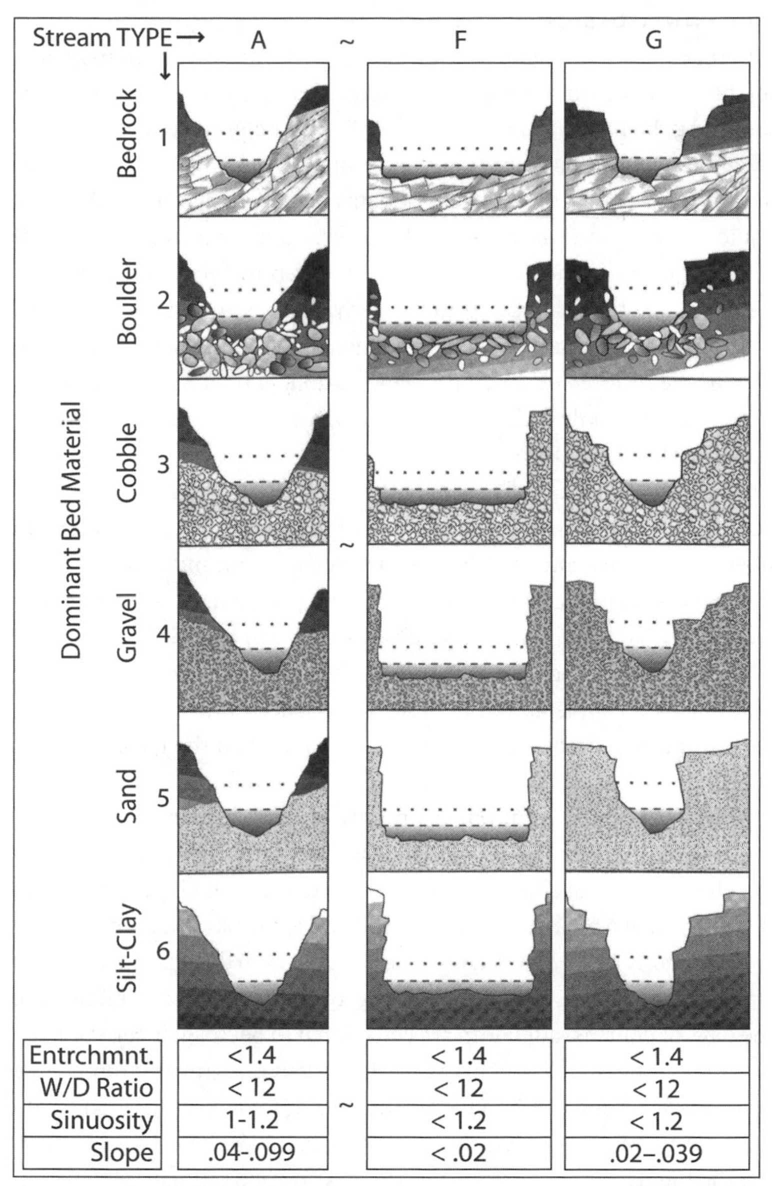

FIGURE 14.1 Excerpt of Rosgen Classification System. Reprinted by permission of Wildand Hydrology.

NATURAL CHANNEL DESIGN

Rosgen's approach, Natural Channel Design (NCD), has three main components. Each is notable for both simplifying the complexities of fluvial geomorphology and standardizing their application in practice. The first component of Rosgen's work is the alphanumeric classification system mentioned above, which allows practitioners to evaluate the condition of a stream channel using a prescribed and relatively quick technique (Rosgen 1994). The second component of the NCD approach is a set of guidelines that specify a forty-step restoration design process. The steps range from assessing the underlying causes of the issues to be corrected, through consideration of passive management alternatives, developing a design for the restored channel, implementation, and monitoring (Rosgen 2007). Rosgen's are the only codified standards of restoration practice available in the United States.

The last component is a set of structures that can be used to implement restoration designs by stabilizing a particular part of the channel to prevent it from moving sideways or downcutting. These structures perform many of the same functions of more traditional hydraulic engineering structures, such as those constructed by the U.S. Army Corps of Engineers (USACE) in their flood-control projects. Rosgen's structures, however, are constructed from locally available boulders and logs with roots and branches still attached, and thus have both a much more natural appearance and more habitat value than the typical USACE project.

While generally excluded from university curricula because of academic opposition to his work, Rosgen has been extremely successful at disseminating his NCD approach via a series of short courses. Rosgen teaches roughly a dozen such courses each year, all of which are heavily attended by agency staff and private consultants despite a fairly steep cost ($1,500–3,000 per person depending on length of course, not including transportation, lodging, or meals). After completing the twenty-nine days of courses, attendees are considered to be Rosgen-certified. Attendance at the short courses is spurred by the fact that a growing number of agencies have made training in NCD a requirement for consultants bidding on restoration projects. More than ten thousand students have been through Rosgen's courses in the last twenty years. While there is no good estimate of the number of restoration practitioners in the United States, this is surely a high proportion of them (see Lave 2008 for a rough estimate).

THE ROSGEN WARS

In response to this success, a vocal university- and agency-scientist opposition has been denouncing Rosgen as a charlatan in print and at conferences and short courses in what have come to be known as the Rosgen Wars. For his opponents, it would be bad enough if Rosgen training were only treated as necessary, but it is unacceptable that it is increasingly treated as sufficient. Students who have completed Ph.D.'s in fluvial geomorphology or hydraulic engineering are being turned away from restoration jobs because they are considered unqualified. Professors and full-time consultants with decades of experience cannot bid on projects because they have not studied their own subject as taught by Rosgen. The NCD approach is now central to the lucrative stream restoration market, and is increasingly seen as a more legitimate basis for restoration practice than academically produced science and training.

Every Rosgen critic I interviewed addressed this issue. A geomorphology professor described how he first learned of Rosgen when one of his students "applied for a job and was told he wasn't a geomorphologist because he wasn't Rosgen-trained." And a Montana consultant with nineteen years of experience spoke for many when he described how he has simply abandoned public-sector work because there are so many agencies that will "write into the RFP that you're going to do it this [Rosgen's] way. I've stopped trying to fight that battle and I don't respond to those RFPs anymore. I got burned out trying to educate them."

Rosgen's critics, a disjointed collection of university- and agency-based scientists and consultants with advanced degrees, are attempting to restore their traditional monopoly of scientific legitimacy and their access to the restoration market by undermining Rosgen's. While they put forward literally dozens of common criticisms of Rosgen's work (Lave 2008, 2009), their primary critiques fall into three main areas: criticisms of his design approach, his practice as a scientist, and the scientific content of his work.

For many of Rosgen's critics, the most powerful condemnation of his design approach stems from its effects on the ground. Every critic I interviewed alluded to the high potential for project failure because of the "cookbook" nature of Rosgen's approach. They argue that by breaking his classification system down into a step-by-step, recipe-like process, Rosgen ignores both the complexity and the specificity of stream channels in favor of universalizing and simplistic techniques that can be used anywhere by people without the experience to understand when such techniques do or do not fit. While moderate members of the opposition acknowledge that many non-Rosgen approach projects fail as well, the general consensus seems to be that Rosgen's students are more prone to design failures.

Critics see these design failures as problematic for three main reasons: they waste public funds, they may in some cases cause irrevocable ecological harm, and widely publicized failures may undermine the entire project of stream restoration by eroding public support. As described by a senior scientist at the U.S.

Bureau of Land Management, the issue is that "*huge* sums of federal dollars [are being] wasted on inappropriate 'stream restoration.' . . . When you consider that each of the people Rosgen trains applies his method to several hundred streams over their career, the potential environmental degradation is stupendous! These are things that can't be undone! So there's a huge amount of money that's been wasted, but there's also tremendous amounts of damage to stream systems, irreparable sorts of damage. . . . It's awful to watch people waste so much money and think they're doing good when they're doing harm."

Rosgen's critics in academia and the research branches of federal agencies care about restoration practice, but restoration *science* is their life's work, and many are outraged by what they see as Rosgen's departures from accepted scientific practice. Some Rosgen critics point out that he has little formal training in fluvial geomorphology, the subject he teaches, and argue that his recently granted Ph.D. is illegitimate because it required no course work (which is not uncommon in English universities) and was granted under the supervision of Dr. Richard Hey, who is effectively Rosgen's employee.[5]

More important for his critics than credentials, however, is the fact that Rosgen does not comply with what they see as the norms of scientific practice: he does not publish in peer-reviewed journals and, most critically, he does not make the data sets behind his knowledge claims and sediment transport models available for review. To Rosgen's critics, this is not simply a failure to pay respect to sacred cows; by refusing to provide access to the data supporting his work or to participate in the peer-review system, Rosgen has in their eyes opted out of the project of science.

Even without access to the data supporting Rosgen's knowledge claims, there are some that critics see as clearly wrong in light of the current consensus in the field of geomorphology. The most common criticism is that contrary to Rosgen's claims, you cannot use the current form of a channel to diagnose the processes that shaped it: what a channel reach looks like now does not tell you why it looks that way, and it certainly doesn't tell you where the channel is going next. A second major critique of the scientific content of Rosgen's approach is that it is based on the concept of bankful discharge, the insight from Wolman and Miller (1960) that the discharge that controls channel form is not the rare big flood, but the more regular flow that occurs somewhere in the ballpark of 1.2 to 2 years. In a later paper (Wolman and Gerson 1978), Wolman himself argued that the bankful discharge concept does not apply in arid and semiarid environments. Others have argued that bankful discharge is meaningless in systems that have no floodplains (Doyle et al. 2007). To Rosgen's critics, his reliance on the bankful discharge concept is a fatal flaw in his approach.

A last key sticking point is Rosgen's claims about channel stability. There is a widespread perception that both Rosgen and his students favor Stability: channels that are not going to move no matter what flows hit them. As a staff geologist with the California Department of Fish and Game described it, "There's a basic misunderstanding in the idea that here's where you put your meanders, here's where you put your bank revetments [and there they remain]. Meanders meander, but you're developing static control measures. People forget that the channel is dynamic. . . . Rosgen folks are talking about *stability* stability: they're saying it's not going to change." This is in direct contradiction of the current scientific consensus, which emphasizes rivers as dynamic systems. These days, channel migration is a goal.

ROSGEN STRIKES BACK

Rosgen and his supporters have pushed back against this barrage of criticism in conference presentations, short courses, and rebuttals to journal articles. Rosgen's supporters use a number of strategies, but the most important are realigning the axis of conflict, underlining the importance of practical experience, and asserting critics' ignorance of Rosgen's approach.

First, to play down the view of the Rosgen Wars as the entire weight of the scientific establishment versus Dave the Cowboy, Rosgen redefines the axis of conflict as NCD versus traditional hydraulic engineering. This strategy holds great power in the larger stream restoration community, as many consultants, agency staff, and grassroots volunteers initially became involved in stream restoration in outraged response to a USACE flood-control project.

Second, Rosgen and his supporters continually underline the importance of practical experience and applicable solutions. They argue that because critics do not actually implement projects or develop applied tools, they have nothing useful to offer. A senior U.S. Forest Service staff member and Rosgen Wars moderate put it this way: "Academics lament that people don't come to them for restoration advice. Maybe they understand rivers, but they haven't been able to communicate that well to people. And many don't have practical experience. It's easy to be critical of failed projects, but they can't say, in most cases, 'This is what I've done that worked better.'" During the level I course I attended, Rosgen said: "There's a rule you guys: if you criticize someone else's restoration project, you'd better be able to explain what went wrong and give them some advice about how to fix it. . . . You can't come along and just be the critic, you've got to say what they should have done." Among practitioners, this kind of argument has serious traction.

Finally, Rosgen and his supporters have had great success in pointing out critics' ignorance. None of Rosgen's major critics have attended his short courses, the primary way in which his NCD approach is taught. Thus some of what critics perceive to be direct hits against Rosgen's approach miss the target entirely. For example, critics claim that Rosgen's approach focuses on form at the expense of process, but in fact he incorporates hydraulics and sediment transport into his design standards and devotes the entire nine days of his level III course to the study of these processes. A second example is critics' objections to Rosgen's reliance on the bankful discharge concept. Because Rosgen does not determine bankful discharge from field indicators, the technique on which critics' objections focus, their criticisms are irrelevant to the actual practice of NCD (see Lave 2008, 137–41, for an extended discussion of this point). The obvious inapplicability of some of the most common critiques of Rosgen's work throws into doubt other issues that critics raise.

THE IMPOSSIBILITY OF CLOSURE

So far the conflict over NCD has been impossible to resolve. The arguments described above have been in heavy rotation since the mid-1990s without either side managing to convince the other of the error of its ways. This is striking given that the debate over the NCD approach seems to center on an empirical issue: does it work or doesn't it? Why has it proved so difficult to answer this seemingly simple question?

To date, there has been no broad survey comparing outcomes of stream restoration projects. The complexity of riparian systems and the high level of uncertainty of restoration science make producing such conclusive data a considerable challenge. At the most basic level, it is not clear how to establish equivalency between projects in different watersheds and climates, or between project designers with highly disparate levels of formal education and practical experience. Further, developing common criteria for evaluating project outcomes is difficult given that for Rosgen and his supporters the goal is a channel that does not move, while for his opponents the goal is a dynamic channel adjusting to changes in inputs of water and sediment.[6] If not a broad comparative study, how about good case studies to provide indicative, if not definitive, data? Here, too, we come up short. While critics have conducted a few excellent case studies of failed restoration projects (most notably Kondolf, Smeltzer, and Railsback 2001; and Smith and Prestegaard 2005), these studies have at best limited relevance to the debate over Rosgen's work: the projects' designers did not follow anything close to the complete NCD approach.

Rosgen's supporters, for their part, have not conducted the type of detailed published case studies necessary to meet academic standards of evidence. Instead, they refute their critics with anecdotal evidence about NCD projects, typically claiming success rates of 80 percent and above.[7] Given that Rosgen's supporters are just as committed to healing riparian systems as their critics, their claims for project success cannot be dismissed out of hand. Indeed, some of them are very convincing. For example, Buck Engineering, a North Carolina firm highly experienced in implementing NCD projects, offers clients warranties on their projects that cover the costs of repairing or replacing any structure that moves for five years after project completion. Although there have been several severe hurricane seasons in North Carolina since the warranty program was established, it has been a *net source of revenue* for the company: once in place, the Rosgen structures Buck Engineering installs rarely move. Still the informal presentation of these claims (as well as the geomorphically short time frame over which many projects have been monitored) limits their plausibility with Rosgen's critics.

CONSTRUCTING SCIENTIFIC AUTHORITY

STS scholars have put a great deal of time and ink into demonstrating the erosion of scientific authority in Western culture, but that authority is not diminished to the point of total ineffectiveness, as dissident scientists have found to their cost (Delborne 2008). Why then have Rosgen's opponents, who include many of the most respected academic and agency researchers on stream restoration, been unable to use their societally granted monopoly on the definition and production of science to exclude Rosgen? How has Rosgen been able to uphold his claims to scientific status and his centrality to the restoration market despite his lack of training and institutional affiliation?

I argue below that Rosgen has built his scientific authority and market share by focusing on a particular kind of circulation of knowledge that is heavily implicated in, and shaped by, the production and application of that knowledge. He has packaged his new knowledge claims, particularly his classification system and standards of practice, in simplified, easily applicable forms, and established an epistemic community of practitioners and regulators trained to use them through his short courses. This community is linked by (1) Rosgen's classification system as a common language for communication, (2) his design approach as a set of shared methods and standards of practice, and perhaps most importantly, (3) his short-course series, which has functioned as the primary training venue for stream restoration professionals since the mid-1990s. By crafting a knowledge product and a means of circulating it that are tailored to the needs of the restoration market and regulatory field, Rosgen has become the dominant force in stream restoration.

OPERATIONALIZING AUTHORITY: THE NCD EPISTEMIC COMMUNITY

The NCD epistemic community consists of project funders, regulators, managers, and designers at NGOs and state and federal agencies; consultants; and a handful of academics. I highlight in this section a few of the key players in that community.

One of Rosgen's earliest and most important endorsements came from Luna Leopold, who is widely regarded as the father of stream restoration in America. Leopold provided Rosgen's primary training in fluvial geomorphology, as well as giving Rosgen enormous legitimacy by supporting his work, co-teaching his initial short courses, and writing the foreword to his first textbook. Because of Leopold's eminence in the field, he was also able to serve as a buffer between Rosgen and advocates of traditional hydraulic engineering until he established a track record.[8] Without Leopold's support, both supporters and opponents agree that Rosgen's work would likely never have achieved its current prominence.

A second key source of support for Rosgen has been the U.S. Forest Service (USFS). Rosgen spent nearly two decades as a hydrologist with the USFS, and it was during these years that he began to develop his extraordinary stream database, as well as the classification system that serves as the basis of his work. Despite the fact that the USFS effectively fired him in 1985,[9] the agency became his first stronghold of institutional support after he started working as a consultant. The USFS ordered copies of his first textbook, *Applied River Morphology*, for *every* hydrologist on their staff, the only time the agency has ever made such a purchase (interview with John Potyondy, USFS, April 29, 2004). And over the years the USFS has sent literally thousands of their employees to Rosgen's trainings. The agency has also funded an enormous number of restoration projects across the United States, and in many cases USFS request for proposals require Rosgen's approach.

A third key source of support is the U.S. Environmental Protection Agency (EPA), a crucial early promoter of Rosgen's work, which sponsored special Rosgen courses for their employees as well as local and state officials. In 2006, the EPA began to play an even more critical role in Rosgen's success by commissioning him to develop a tool for assessing channel stability and sediment supply—the *Watershed Assessment of River Stability and Sediment Supply* (Rosgen 2006)—which is now the agency's recommended approach for developing sediment TMDLs (total maximum daily load) *nationwide*. This explodes Rosgen's reach from just those streams considered for bank stabilization or reconfiguration to all flowing bodies of water in the United States.

SIMPLIFICATION AND THE ROLE OF STANDARDIZED PACKAGES

In her classic paper "Crafting Science: Standardized Packages, Boundary Objects, and 'Translation'" (1992), Joan Fujimura elaborates on the role of standardized packages in science. As conceptualized by Fujimura, standardized packages are combinations of theories and standardized methods that "serve as *interfaces* between multiples social worlds . . . [facilitating] the *flow* of resources (concepts, skills, materials, techniques, instruments) among multiple lines of work" (170). Crafting a set of techniques and ideas into a standardized package enables them to be picked up and widely used, both facilitating cooperative work and increasing their influence. As Fujimura puts it, by presenting a coherent set of specified and easily applicable methods, standardized packages allow people "to get work done . . . , to construct and solve 'doable' problems" (177).[10] I argue in the sections that follow that Rosgen's scientific and economic success is a direct result of his ability to craft his work into a standardized, and thus easily circulatable, package

The Rosgen Classification System as Communication Tool

Even the staunchest Rosgen critics grant his classification system's utility as a communication aid (cf. Roper et al. 2008). Stream restoration requires coordinated efforts among a number of different disciplines, most notably biologists, ecologists, engineers, and geomorphologists. These biology- and physics-based disciplines have very different training, causing confusion over even basic things, such as which bank of the river is left and which right![11] Communication is thus a significant hurdle: if differently trained experts cannot agree on a description of the current character of the system they wish to modify, how can they design its future form? Rosgen's alphanumeric classification system fills this gap, creating a simple shared terminology that allows practitioners to quickly grasp the key morphological characteristics of a stream system. As a senior staff member with the California Department of Parks and Recreation described it, "Hydrologists speak very well to hydrologists, engineers speak very well to engineers. What I think Dave did was transcend those . . . very real vocabulary barriers and provide a way for people to actually have a common understanding and a common language."

Beyond *disciplinary* boundaries, the Rosgen classification system is also intended to transcend *geographic* boundaries, providing a universally applicable framework for describing streams. As a professor and staunch Rosgen supporter said, "It allows communication, because people know what you're talking about when you're talking about a C4 no matter where you're at." By providing a lingua franca, Rosgen's classification system has become a central support of stream restoration practice in the United States, a key component of the standardized package that is NCD, and thus a factor linking practitioners and regulators all over the country into a pro-Rosgen framework.

Rosgen as Provider of Shared Methods and Standards of Practice
Rosgen has developed the only set of purportedly universally applicable methods for channel reconstruction projects, in large part because the current emphasis on complexity in geomorphology makes the project of developing a rival set of standards of practice appear ludicrous to academics. Because of liability issues, the American engineering community has started the process of developing standards of practice that transcend Rosgen (see Shields et al. 2003, 2009; Slate et al. 2007), but given the required rigor of new standards of practice for licensed engineers, that process is unlikely to bear fruit for many years. For the foreseeable future Rosgen's design guidelines provide the only solid footing of specified, agreed-upon practice for restoration practitioners. His forty-step method is thus a key part of the standardized package he presents, and a critical element of his success.

Even though critics argue that the cookbook nature of Rosgen's design guidelines enables poorly trained practitioners to wreak havoc on streams, practitioners are not the only NCD constituency; the standardized form of his design approach is just as critical to the functioning of the *regulatory* community. Regulators at the local, state, and federal levels have been confronted over the last fifteen years with the Sisyphean task of choosing consultants, managing contracts, issuing permits, and writing legislation to produce successful stream restoration projects. How is someone with little or no relevant training to evaluate the differences and decide between a proposal prepared by an academically trained fluvial geomorphologist and an NCD practitioner? And on what basis can they justify that decision to their superiors and the public at large?

Content seems the obvious ground for distinguishing between the approaches, but because Rosgen and his opponents agree on the basics of hydrology and fluvial geomorphology up through the 1960s, there is sufficient overlap between the two camps that it requires a substantial knowledge base to see the differences, much less to understand why they matter. Several of the biology-trained restoration practitioners I interviewed said that they could not differentiate between the material presented in Rosgen's initial short course and that presented in the initial course of a series developed by Rosgen critics. Experience could be another possibility for justifying the decision to select one camp of consultants, but there are very experienced practitioners in both. What remains, then, is the bureaucrat's safe haven of justifiability: the application of accepted standards. The only set of standards, spelled out step-by-step, against which it is possible to check a channel reconfiguration design are Rosgen's. Thus the standard of practice supplied by Rosgen's "cookbook" approach is critical to its utility for the new field of stream restoration.

Disciplinary Reproduction and Training

An appealing standardized package is not, in itself, enough to ensure success; prospective users must learn the package exists and be trained to use it. Perhaps the most important element of Rosgen's success has thus been his role as the primary trainer of restoration practitioners in the United States. Stream restoration has had such a sudden rise to environmental and economic prominence that there is a serious disconnect between the jobs available and the pool of qualified practitioners. The National River Restoration Science Synthesis project found that the restoration market in the United States is approximately $1 billion per year, and that amount is growing steadily (Bernhardt et al. 2005).[12] Universities have been slow to respond with programs focused on restoration science or practice. Rosgen has used his series of short courses to fill the gap, creating a system for circulating knowledge of the NCD approach and in the process creating a community of practitioners sharing the bond of an intensely formative educational experience.

There are a number of notable differences between short-course series and the course work required for a professional qualification such as a master's degree. First, the amount of material that can be covered in a short course, even the five- to ten-day short courses that Rosgen teaches, is substantially smaller than the material in a university course. Thus, the range of information presented in short courses is necessarily much narrower, although that narrow focus means that the material covered is far more precisely targeted. Second, because short courses compress a great deal of content into a very short time period, students have little time to reflect on the material presented, ask questions, and attempt to put it to use. Trying to absorb the content of a short course has more in common with attempting to drink from a fire hose than the more measured cup of university courses. Thus students often report that they come out of a short course thinking they generally understood the content presented, and then encounter substantial problems when they try to apply what they thought they had learned. These students can't sign up for office hours or ask questions in lecture the following week; their only recourse is going to additional short courses. Finally, short courses create a total-immersion experience, far from family, work, or other instructors that might question the premises being taught. They thus have the capacity to create intense bonds among students, and even faithlike conversion experiences.

All three of these characteristics are common to most short courses, but there are also aspects distinctive to Rosgen's courses that help build his students into an epistemic community. First, Rosgen's courses are distinguished by an intense focus on teamwork. Students are divided into teams that work together in the field, and then spend long evenings working up their data. Ties among teammates are even more firmly cemented by the communal presentation and critique of their data, an educational hazing experience that is the source of bonding even among those who took Rosgen's courses at different times.

Second, there are notable differences between the materials distributed in Rosgen's courses and those in other short courses. In Rosgen's courses, distributed materials underline the legitimacy of the information presented: a bound, tabbed course notebook with background material and detailed instructions for the field exercises and, most importantly, a copy of one of Rosgen's textbooks. This is a far cry from the ad hoc three-hole-punched handouts typical of courses organized by academics and agency researchers. There are also take-home gifts in Rosgen courses—mugs, tape measures, and signed certificates of completion—which allow students to advertise the fact that they've been through the courses and are members of the NCD community.

Perhaps the most telling difference between Rosgen's courses and his opponents' is their treatment of the relative uncertainty of restoration science and practice. In a mail survey I conducted of students from the level I Rosgen and a competing short course taught primarily by his academic critics, I asked, "Based on the information presented in the course, how predictable or unpredictable do you think restoration work is?" Responses were on a scale from 1 to 10, where 1 was totally unpredictable and 10 was totally predictable. The mean level of certainty posited by Rosgen level I course respondents was 6.5 out of 10, almost 2 points higher than the 4.8 posited by academic course respondents (figure 14.2). Even more telling, while 83 percent of Rosgen-course respondents believed that restoration practice was more predictable than not, only 43 percent of academic short-course students did. At a very basic level, the Rosgen short courses encourage participants to view stream restoration as a doable project, a key component of the success of NCD as a standardized package.

Rosgen's short courses have been highly successful in building an epistemic community of people employing and promoting his NCD approach. Consulting firms and regulatory agencies throughout the United States are now heavily staffed with people who associate stream restoration with NCD. This has serious consequences for restoration practitioners with conventional academic training. Even if a request for proposals does not specify that consultants must use NCD, if that is the only method with which the client agency is familiar, consultants proposing to use any other approach are fighting an uphill battle. Steve Gough, a restoration consultant in the Midwest, described a project that had been in litigation for more than a year because ascending levels of the client's supervisors, all of whom had been to Rosgen's level I course, objected to the fact that Gough did

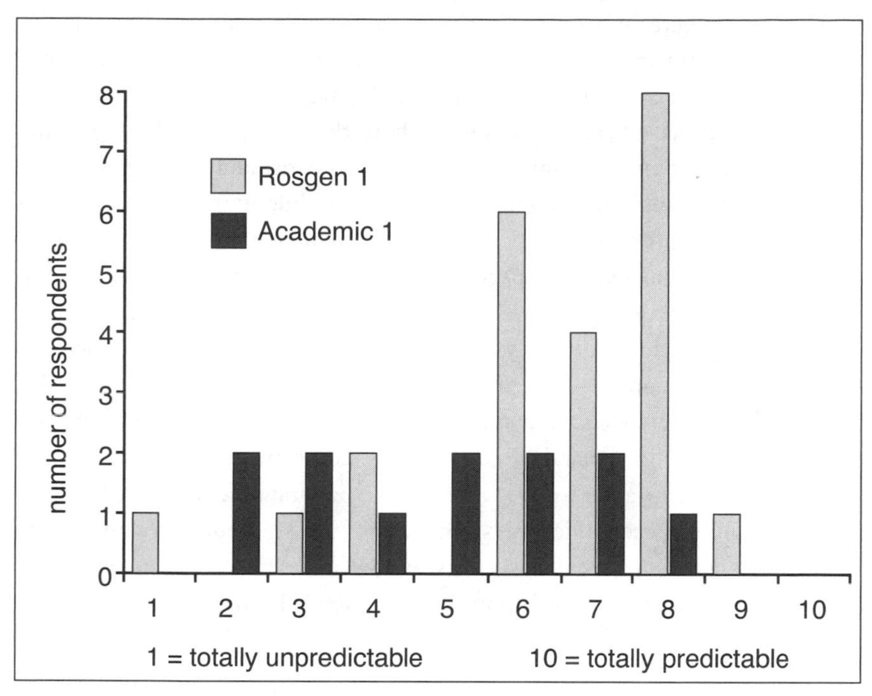

FIGURE 14.2 Level of certainty of restoration work according to short-course participants.

not use NCD: "The case has been settled twice, but each time someone showed up and said, 'Wait a minute, this guy didn't use the Rosgen system? This can't be right!' It's reached the level of absurdity. Another guy comes in the room and says, 'What? You didn't use the Rosgen system!' And everybody goes, 'Oh shit, did you have to say that? Now we have to open it up again'" (interview with Steve Gough, Little River Consulting, June 19, 2006).

Clearly, Rosgen's ability to get his knowledge claims into circulation has been critical to the near ubiquity of his approach in U.S. stream restoration.

CONCLUSION

By developing a so far unrivaled system for stream restoration education, as well as a common language and standards of practice for his students to draw on, Rosgen provides a powerful and coherent framework for the stream restoration field. In creating and circulating this standardized package of simplified tools, he has built a broadly distributed and powerful epistemic community of regulators, funders, designers, and even a few academics who back his claims to scientific status and insist upon the legitimacy, indeed the superiority, of his approach. Rosgen thus transcends the traditional role of consultant as applier of science developed elsewhere, and functions instead as an active producer of knowledge, an arbiter of legitimacy, and a powerful rival to established scientific institutions.

It is critical to note that Rosgen's success in structuring the stream restoration field would have been impossible without the demands of the expanding restoration market and the support of the state, as consultants and agency staff both use NCD and insist that others use it as well. Highlighting the importance of political-economic forces in Rosgen's success also demonstrates the profoundly interconnected character of the production, circulation, and application of science. Rosgen produced his knowledge claims with circulation and application in mind, crafting his classification system and standards of practice into efficiently teachable and easily applicable forms. Similarly, Rosgen circulates his knowledge claims—via short courses and a standardized package of techniques—in ways designed to meet the demand of consultants and agency employees for easily implementable restoration tools and accessible restoration education. Ignoring the role of political-economic forces in the Rosgen Wars would have missed the most critical factors in Rosgen's success, as well as the interconnections among the production, circulation, and application of knowledge in his NCD approach.

NOTES

1. There are a few environmental management programs in the United States that grant master's degrees with concentrations in stream restoration. These programs are based on Rosgen's work, and it seems no accident that none of them is at an R1 (top-tier research) university.

2. The research on which this paper is based included more than sixty interviews with consultants, agency staff, NGO staff, and academics involved in stream restoration, including Dave Rosgen and most of his major critics; participant observation at short courses taught by Rosgen and his critics, and at the conferences where many battles in the Rosgen Wars are fought; analysis of the way the fight has been carried out in print in journals, textbooks, and proceedings papers; mail surveys of short-course participants to learn more about how they select courses to attend and what they get out of them; and geomorphological case studies of completed restoration projects.

3. For some of the more influential journal articles critical of Rosgen's work, see Gillilan (1996); Juracek and Fitzpatrick (2003); Kondolf (1995); Kondolf, Smeltzer, and Railsback (2001); Miller and Ritter (1996); Roper et al. (2008); Simon et al. (2007); Smith and Prestegaard (2005). For conference presentations, see Ashmore (1999); Doyle, Miller, and Harbor (1999); Simon et al. (2005); Shields and Copeland (2006); Kondolf (2007).

4. Several comments on an earlier version of this chapter suggested ANT as a relevant analytical framework. And to some extent the hierarchical and agonistic dynamics characteristic of ANT do seem to describe Rosgen's struggle: this is clearly a case of one person attempting to enroll many others in supporting his truth claims. However, it is not accurate to describe the Rosgen Wars as a struggle between competing networks: the opposition to Rosgen's work

is too loose and disorganized to qualify as a network. More importantly, ANT does not acknowledge the impacts that political-economic relations can have on scientific production and circulation, and so would ignore the central role that markets and the state have played in Rosgen's success.

5. Hey teaches the Rosgen level 1(e) short course, Fluvial Geomorphology for Engineers, twice a year. It is important to note that Rosgen could not have received his Ph.D. without meeting the standards of two outside examiners; the decision was not Hey's alone. Thus this criticism may be less damning than it appears.

6. While the two camps have very different understandings of what constitutes success, there is some overlap in their definitions of failure: cases where the newly designed channel and the structures installed to hold it in place wash out completely as the stream carves a new channel for itself.

7. The lowest claim I have seen comes from Mondry, Melia, and Haupt (2006), who report a stability rate of 74 percent for Rosgen structures installed by the North Carolina Ecosystem Enhancement Project in restoration projects two to five years old.

8. A case study on one of Rosgen's early projects in the 1992 National Research Council report *Restoration of Aquatic Ecosystems* (474) describes Leopold serving as arbiter between the USACE and Rosgen on the design of the Blanco River project: "The project almost failed to materialize when COE [the USACE] subjected the unique design to expert review and was told by its reviewers that the new system would not contain flood flows. The project design was then sent for review to Professor Luna Leopold at the University of California, Berkeley, Department of Geology and Geophysics; Leopold praised the project and expressed confidence that it would work. On the basis of his recommendation, COE withdrew its reservations, and the project was allowed to proceed."

9. Rosgen angered his superiors and was given a mandatory transfer they knew he would refuse, in effect firing him.

10. As J. Scott (1998) has demonstrated, simplification and standardization are not simply tools of science, but also critical to the function of the modern state. This is particularly notable given the powerful role the state has played in the Rosgen Wars.

11. Geomorphologists reference right and left off the flow of water and sediment downstream; fisheries biologists reference off upstream movement of anadromous fish.

12. According to a report by the Environmental Law Institute (ELI 2006), the federal government alone spent more than $2.9 billion on compensatory mitigation projects required by Clean Water Act permits in 2006, although the report does not make clear how much of this was spent on streams. Most likely, the majority of the funds were for wetlands restoration, but even if only one-fourth of the $2.9 billion were spent on streams, this would suggest a large rise in spending on restoration in the United States, since most restoration projects are financed by private developers and state agencies.

15

JOSHUA J. RAMISCH

EXPERIMENTS AS "PERFORMANCES"

INTERPRETING FARMERS' SOIL FERTILITY MANAGEMENT PRACTICES IN WESTERN KENYA

This chapter discusses the portrayal of "local knowledge" as it interacts with and challenges the knowledge of outsiders, in this case researchers like me. I propose that the standard explanations of farmers' "experiments" with new agricultural practices or varieties are limited fundamentally by a failure to understand the ways in which agricultural practice is actually an embodied "performance" of farmers' knowledge interacting with social and environmental contexts over time. Case studies from a development project's experience of farmers' experiments with cereal-legume rotations illustrate how farmers and researchers construct soil fertility management knowledge. By discussing the presentation and circulation of knowledge at the interface between "outsiders" (development change agents or researchers) and members of "local" communities, this chapter contributes to the burgeoning literature on farmers' "experiments" and whether it is useful to think of such experiments—or the "local knowledge" of rural people—as embodying "knowledge systems" distinct from (or analogous to) "scientific" knowledge. It also illustrates the complexities of "performing" and communicating knowledge at the interface between different development actors.[1]

The examples come from the Folk Ecology Initiative (FEI), a community-based learning and development project that worked to broaden the repertoire of soil fertility management and adaptation strategies available to smallholders in western Kenya.[2] It consciously used an adaptive learning process of dialogue between farmers' local ecological knowledge ("folk ecology") with outside knowledge systems to develop a shared, "dynamic expertise" of soil fertility management (Ramisch et al. 2006). The FEI also tested whether community-based learning and farmer-led experimentation could reduce the epistemological and communicative distance between local communities and scientists (Sikana 1993).

FARMERS' KNOWLEDGE AND AGRICULTURAL EXTENSION

The FEI's approach contrasts significantly with the mainstream, "transfer of technology" model of agricultural extension. In such a model, agricultural innovations (new crop varieties or husbandry practices) originate from scientific research activities, such as replicated experiments within the controlled conditions of research stations or laboratories. National agricultural services then use extension agents who are trained to communicate the scientific knowledge to farmers in ways that will increase the likelihood that technologies will be adopted. In such a model, farmers passively receive innovations, and neither extension agents nor scientists are likely to engage with farmers' existing knowledge except to identify "gaps" that need improving. Farmers are also not seen as the generators of new knowledge or technologies: their contribution is limited to voicing their problems to scientists through extension agents, so that new technologies can be developed to address them.

The FEI also contrasts with more participatory approaches to extension, such as the "farmer field schools" that gained popularity in the late 1990s. These approaches critiqued conventional extension's failure to improve smallholder farming systems as a failure to adapt scientific recommendations to those systems' complex and locally variable conditions. Farmer field schools attempt to build farmers' own understanding of science using structured curricula and weekly meetings between farmers and extension agents on village-based "demonstration" farms (Dilts and Hate 1996). Nonetheless, these "schools" still operate with a "transfer of technology" model, from scientist (or extension agent) "teachers" to farmer "pupils." Although the teachers undoubtedly learn from their pupils' experiences in these field schools, that learning is not explicitly used to develop new technologies and is more likely to be useful only for improving the teaching methods or curriculum of future field schools.

The basic assumption of the FEI was that new and useful soil fertility management practices originate from farmers and scientists alike. Our intention was to make the farmer field school strategy more meaningfully interactive, bringing together the knowledges of farmers and scientists through dialogue and colearning. The FEI model did not automatically presuppose vast differences in epistemology between farmers and other populations (Millar 1993); neither did it assume important synergies between different knowledge sets (Sumberg, Okali, and Reece 2003). However, the FEI did believe that identifying and understanding differences (and similarities) where they exist must constitute a starting point for any collaborative venture, especially given the many ways in which concepts of soil fertility might be embedded within more holistic concerns about crop performance, climate, pest or weed ecology, and markets.

UNDERSTANDING "KNOWLEDGE" AS PERFORMANCE

The discourse relating to "performance" typically relates more to its (re)presentational dimensions (e.g., the performative discourse of "gender" in the scholarship of Judith Butler [1990]) or to its outcomes (e.g., crop performance in its agronomic sense, integrating a given crop's yield in response to husbandry or inherent resistance to pests). Although both these aspects are important for the coming discussions, the sense I want to emphasize for this paper is the *process* of performance, along the lines introduced by Paul Richards (1989, 1993).

The sense that Richards invokes is of a smallholder farmer acting as a skilled musician attempting to play a complicated piece. Each witnessed performance therefore demonstrates something about the farmer's own ability to farm, the resources and knowledge available, and the unfolding challenges (environmental, sociocultural, economic) faced in that particular season's context. Richards attacked the idea that rural Africans use a simple list of rules and decisions as part of a prior body of "indigenous technical knowledge" by presenting a more nuanced, embodied image of knowledge put into action, deployed in contingent response to unfolding events over a growing season. "The crop mix [that researchers might observe] . . . is not a design but a *result*, a completed performance. What transpired in that performance and why can only be interpreted by reconstructing the sequence of events in time. Each mixture is an historical record of what happened to a specific farmer on a specific piece of land in a specific year, not an attempt to implement a general theory of interspecies ecological complementarity" (Richards 1989, 40, emphasis added).

This model of performance has been attacked in turn as overly populist or unhelpfully holistic, and the musical performance metaphor has often been understood simplistically as a synonym for "improvisation" (intuitive, uncodified, tacit knowledge, etc.). However, Richards (1993) anticipates these critiques and defends the notion of performance as the outcome of actions by "skilled [agricultural] performers." They may plan or indeed improvise their behaviors—and in the agricultural realm must do so by "hitching a ride" on inherently unpredictable natural processes—but do so within the domain of their own evolving knowledge and abilities.

Others have adapted the idea of farming as performance: Batterbury (1996), for example, argued that farmers' performances in the semiarid Sahel are guided by plans—not so much musical here, the analogy he deploys is of building a house in stages, with multiple objectives of material comfort, social prestige, and so forth. A farm is never finished; neither are the skills needed to farm ever completely mastered. It is a lifework that may or may not reflect the "evidence of learning," of even temporary mastery over natural forces, or an improved ability to interpret natural conditions to one's advantage (Orr, Mwale, and Saiti 2002). Sara Berry (1993) also depicts knowledge as a "work in progress," partial and imperfect for any given actor, but simultaneously the legacy of the dynamic accumulation of that actor's lived experience and learning from interacting with external forces and agents.

Combining these notions of "performance,"[3] I draw the following lessons:

1. In all cases there is a score or text (either explicit or implicit) that represents an *ideal* which we attempt to the best of our ability to interpret through our own actions.
2. Whatever the performer's own abilities, a performance is itself *purposeful*, a set of actions guided by the underlying ideal or our ideas about how to achieve it (e.g., how best to plant crops or deploy livelihood options in a given season). Within the FEI, the term "local logics" (Misiko 2007) emerged as a way of explaining the different motivations guiding performances.
3. A performance is a contingent experience lived in real time, whether we start out with a proactive or reactive agenda (or some combination thereof). It is *unrepeatable and irreproducible*, fundamentally partial and imperfect by virtue of being grounded in a given situational context.

The implications of these ideas is that to understand smallholders' agricultural practices, we need to apprehend not just the "ideal" forms of agricultural production or local practices but also their dynamic and embodied nature. If agricultural performances are indeed dynamic (and they do embody an actor's skill, knowledge, and resources in striving for an "ideal" outcome), the fact that they occur in real time becomes fundamental: each season, each crop therefore represents a *new* performance with new knowledge and conditions and not merely a repetition of the "same" activity.

Finally, since the performance of local agricultural practices invites the interest of outsiders such as researchers or agricultural extension agents, let us consider a further lesson:

4. The performance itself must be *interpreted* to give it meaning and so that learning can occur.

Interpretation by the performer her/himself (against the "ideal" or against the caliber of previous performances) certainly informs future performances and decisions and may also be summarized or explained to outsiders not versed in the local context. Such explicit explanation notwithstanding (and certainly more so in its absence), outsiders will also make their own interpretations of the performers' activities, decisions, and outcomes against metrics of their own, based either on the "ideal" of other agricultural contexts (e.g., other local practices, the "on-station" results of researcher-managed trials) or a synthesis of diverse, "local" sites guided by theory. To crudely simplify, local agricultural performance builds and explains itself inductively, while outsiders' interpretations of it will more likely derive from deductive logic.

The goal of the FEI was to bring together local and outsiders' knowledge and experience to create a "dynamic expertise" of soil fertility management, which was relevant to the site(s) of action and jointly held by farmers and researchers (Ramisch et al. 2006). Acknowledging the performative nature of both agricultural practice and its interpretation meant that the FEI engaged in intensive, ongoing, and profoundly iterative activities to put farmers and researchers in constant dialogue about soil fertility topics. The outcomes of many farmers' performances, over multiple sites and multiple seasons, allowed the FEI participants to discern the "local logics" that had shaped experimentation. The following sections discuss some of the challenges in actually implementing this vision of collaborative learning, illustrating the performance nature of farmers' experiments and agricultural practices.

SETTING THE STAGE

SITE DESCRIPTION

Western Kenya is an area historically neglected by central government authorities, where rural populations contend with poor infrastructure, poor market access, high rates of HIV/AIDS infection, and widespread, semipermanent out-migration of youth (predominantly young men; Crowley and Carter 2000). Both land and labor shortages interact with biophysical challenges (such as soil fertility decline) to limit agriculture.

The FEI involved four ethnically distinct communities chosen along an agro-ecological and population-density gradient. Ethnolinguistically the three sites in Vihiga and Busia districts are predominantly speakers of Luyia dialects (a Bantu language with many subgroupings), while the population in Teso district is predominantly Teso-speaking (a Nilotic language). All four sites had previous contact with either international or local NGOs working on soil fertility management. Farmers participated in the FEI as members of groups organized at the village or community level. These typically included 10–25 households, either as part of existing self-help or women's groups, or as previously active self-help groups that were reconstituted after the FEI was established.

SOIL FERTILITY AND LEGUMES IN WESTERN KENYA

Poor soil fertility occurs within a context of many other interconnected challenges. Besides the widespread nitrogen (N) and phosphorus (P) deficiencies reported for western Kenyan soils (Jama et al. 2000), agrarian populations face crop pests and diseases, devastating weeds such as *Striga hermonthica*, climatic variability, and marketing problems (Misiko 2007). The FEI was one initiative among many where scientists from the Tropical Soil Biology and Fertility

(TSBF) Institute collaborated with western Kenyan farmers to research, discuss, learn, and adapt soil fertility management technologies to the local ecological conditions.

This chapter concentrates on the FEI's work with cereal-legume rotations for improving soil fertility. These activities were based on the logic that testing a wide range of legumes would allow farmers to select the most appropriate legume(s) for various soil, climate, and cropping conditions. The attraction of legumes rests on their multiple potential uses (for food and fodder, biomass incorporation, and nitrogen fixation). These multiple aspects imply reduced costs for investment in soil fertility, soil structure improvement and erosion control, income generation through marketing and seed production, and the benefits of crop rotation (as compared to further, continuous cereal cultivation), such as breaking crop pest and disease cycles (Misiko et al. 2008). Western Kenyan farming systems already make use of legumes such as common bean (*Phaseolus vulgaris*), typically intercropped with maize or other cereals. Other legumes that researchers promoted in the region included soybean (*Glycine max*) and the inedible cover crop mucuna (*Mucuna pruriens*).

RESEARCHERS' KNOWLEDGE AND THE CEREAL-LEGUME ROTATION "DEMONSTRATIONS"

The cereal-legume rotation component of FEI's activities began with collectively managed "demonstration" plots in each participating community, initiated in the first, "long rains" season (March–July) of 2003. Almost immediately thereafter individual households began their own "experiments" inspired by or adapted from these collectively managed sites.

The collectively managed demonstration plots were located on host farms in each community. Groups ensured that host farmers were popular, well-integrated with their communities, with easily accessed farms representative of local soil types and history of cultivation (e.g., cultivated continually for 20–50 years). The demonstration plots in these farms were all classified as "infertile" during the participatory soil characterization phase and were specifically selected to "see if the new technology worked."

This demonstration trial was designed to illustrate two new technologies that TSBF researchers felt would be useful for improving soil fertility in western Kenya: (1) the potential of legumes (soybean, mucuna, and a local choice such as groundnut or yellow grams) for improving crop yields when grown in rotation with staple cereal crops, and (2) the optimal use of mineral fertilizer (providing both N and P) in legume-cereal rotations. While the "demonstration" approach was considered ideal for showcasing known legume technologies to farmers, since these technologies' actual performance under different farm-level circumstances was unknown, an experimental element was unavoidable. For a variety of reasons, however, the demonstration trial plots did not include randomization of treatments or internal replications as would have been the case in a purely researcher-managed experiment.

The main reason for not including replicate versions of the same treatments was an effort to maintain what farmers and researchers alike considered a "clean" and "easy-to-follow" design. Plots were not randomized for a similar reason, to ensure that all the plots of a legume under different fertilizer treatments could be easily seen and compared side-by-side, rather than at scattered and unpredictable locations within the field. This was not just the research team's version of a "simplified" or "farmer-friendly" watered-down science, but a layout that was discussed with the participating communities to correspond with what appeared to be the experimenting style typical of the study areas (e.g., unreplicated trials of new treatments immediately beside existing, known practices). In the words of the farmers, "too many plots" (or multiple plots essentially "showing the same treatments") were "confusing" at best and "wasteful" of scarce inputs and land at worst. Both farmers and researchers agreed that if these were concepts that had already been "tested" (and therefore understood to be "proved") in decades of previous research, it was better to use the little land that each group dedicated to "demonstration" for "as many technologies as possible."

Since the initial, collectively managed experiments provided the first occasion for most of the farmers to see or work with the new legumes or husbandry practices, it is worth reflecting on how these sites of experimentation "performed" or expressed researchers' knowledge and expectations in particular. As will be seen below, the collective experiments often provided a "script" of ideas, practices, or raw materials (e.g., new species) for later, individually conducted experiments. Thus, the performance of researchers' knowledge in the collectively managed sites is noteworthy both (1) as a set of *explicit* statements about how best to adapt or indigenize scientific practices to make them more legible and malleable for the local participants in the FEI, and thus more useful for generating and sustaining the emerging "dynamic expertise" on soil fertility management; and (2) as an *implicit* assumption, on the part of farmers and researchers alike, that the emerging "new" knowledge would be best seen at (and spread from) the collective sites, even if much of their design and content would reflect or showcase "local" practices and priorities.

This last point reflects the dominance of the "mother-baby" trial model for collaborative farmer-scientist experimentation (Snapp 1999). In its simplest form, the model uses two kinds of learning sites within a given local context to collect data: the central "mother trial," where technologies are tested under researcher-managed conditions (e.g., optimized crop husbandry, such as carefully measured spacing of planting within and between rows, basal fertilizer applications, regular weeding, preventative pest control, etc.) and a series of satellite "baby trial" plots of the same technology tested on farmers' own farms, under their own management conditions.

The mother-baby model originated in Malawi for participatory plant breeding (the name was coined by one of the farmers), where data collected on crop performance under the mother trial's researcher-controlled conditions could be compared with the results of the same technologies on adjacent farms under farmer-managed conditions. As such, it can admit a range of learning objectives, from testing potentially complex questions on the mother trial, to gaining experience with the new technology on the baby trials. However, there is still the expectation that "new" learning spreads mostly from the collective to the individual experimentation activities. Similarly, there is little opportunity for feedback from the individual experimental offspring "babies" back to the collective "mother" except to validate or contradict collective findings.

The mother-baby model persists throughout participatory agricultural research, including situations like the FEI's, where the mother trial was jointly designed and controlled by researchers and farmers and not a source of "pure" agronomic data collection. The FEI adopted the mother-baby terminology with some trepidation, since it seemed to imply a top-down (or at least vaguely condescending) attitude toward farmer experimentation, which clashed with the FEI's stated objective of using the learning embodied by individual experimentation to feed back into choices made on the mother trial sites. Unlike the research team, however, participating farmers had no issue with the terminology, and even noted that "babies grow up and have children of their own" (although no one suggested that children can also teach their mothers . . .).

INTERPRETING FARMERS' EXPERIMENTS

TERMINOLOGY: EXPERIMENTATION AS "TRYING"

The local (Kiswahili) term for "experiments" is *majaribio*, literally "things that are tried." The collectively managed "demonstrations" were referred to typically as *majaribio ya kuelimisha* (things that are tried for purposes of teaching), a special case of the general category of "things that are tried." The contrast between this Kiswahili terminology (which was jointly agreed between farmers and researchers) and the English word "demonstration" is important. In English we were merely *showing* farmers a range of previously validated technologies (often referred to within TSBF project documents as a "basket of options") with the objective that farmers would choose the most appealing ones. The Kiswahili term belies a deeper level of trying (*kujaribu*) and therefore contingent experimentation: these technologies had yet to prove themselves in the local context, and these demonstrations were for many farmers the first opportunity to see these cereal-legume rotations in operation.

Reinforcing our research team's own language of "validated technologies" or "best-bet options," farmers told us, "We are counting on you to bring us new, good things (*vitu vipya vizuri*) that you have seen elsewhere and that can make a difference for us here." This assumption that all the treatments on display in the demonstrations—including control treatments—were "good things" inherently worth trying led many participants into troubling efforts at rationalizing the outcomes. Many participants (or passersby) who saw the miserable performance of maize grown in the unfertilized control subplots of the cereal-legume demonstrations therefore initially reported disappointment with the project: "When I saw those stems without any cobs, looking so dry, I thought: if [you, the] researchers can't grow maize here—ayi!—what can we do. We were counting on the project for help but even me I can grow better looking maize than that" (female farmer, Ebusiloli).

Finally, the differing attitudes toward the role of the demonstration plots can be seen in light of the poor state of the farms that were selected to host them. The research team had hoped to showcase the technologies on sites that represented the (low) soil fertility norm in each community. The participating communities, however, had deliberately selected their absolutely worst fields (e.g., most depleted or *Striga*-affected). This hidden agenda of "testing" the technologies was only learned by the researchers when it became obvious that some of the demonstrated rotations were producing far below the typical potential for their agroecological zones. This discovery was indeed one of the first expressions of the local logics guiding farmers' experimental practice. Their logic—which was widespread but initially covert—was that if potential solutions could be arrived at here, on the absolutely worst sites, then they would have the answer to problems of agricultural productivity on other, less difficult plots.

PARTICIPATORY MONITORING AND EVALUATION OF THE "DEMONSTRATIONS"

Visits to the demonstrations were regularly held by both farmers and researchers to assess the ongoing performance of the experiments. Farmer groups would meet at least weekly (if not more often during weeding seasons) at the plots to carry out independent participatory monitoring and evaluations (PM&E), that is, to record observations free from researchers' influence. Monthly dialogues were also organized between the research teams and each farmer group, which frequently included other interested farmers from outside the groups. Such latter dialogue was largely unstructured, but typically included discussing the concepts behind the demonstrations, and whether, how, or why those concepts could explain the differences in plot performances as they appeared.

The farmer-generated concept of crop "performance" (*kufanya*, both "to do" or "to make" in Kiswahili) illustrates the process-based nature of knowledge about soil fertility and crop health acting over time (and in comparison to other contexts). According to farmer notes and discussions, "good" performance for maize (*mhindi uliofanya vizuri*, "maize that is doing well") was reflected in the growth of the plants (e.g., leaves that were dark green, long, and wide; tall and sturdy stalks; an impressive rate and uniformity of crop growth in a plot), the number and size of cobs and grains (e.g., multiple, large cobs on every plant; large, pure-colored grains, growing in many full lines on every cob), and the weight or quantity of maize at harvest.

The use of these local logics for evaluating "good" results at different points in the growing season proved much more robust than relying solely on the researchers' criteria (Ramisch et al. 2006). This is certainly what has been observed elsewhere with more standard mother-baby trials (Snapp 1999). While the TSBF scientists expected the demonstrations would show tangible benefits of mineral P fertilizer (in particular) on legume productivity, the PM&E activities revealed that farmers were not always so impressed by these benefits. Comments would point to the continuing and visible importance of constraints other than soil fertility (e.g., crop diseases like maize streak virus, erratic rainfall, and *Striga* infestation), which were also beyond the control of either farmers or researchers.

However, the process nature of how experiments' performances were being evaluated can be seen in the shift in group behavior over the life of the FEI. By 2006 in most sites, the farmer groups were combining their routine visits to the collective demonstration sites with visits to the experiments that individuals were conducting on their own farms. By the end of the FEI project in 2008, the collective demonstrations still retained a certain symbolic value (discussed below) but the center of attention had very clearly shifted toward assessing the viability and utility of the technologies on offer in a broader range of "real-life" contexts over multiple seasons. These individual experimental sites also had the virtue of showcasing the performance of the most favored technologies, whereas the demonstration sites now only served the role of testing and promoting additional new and unfamiliar technologies.

INDIVIDUAL EXPERIMENTS AND LOCAL LOGICS

Participant observation and in-depth interviews were conducted in 2005 among forty households, selected because they were testing for themselves some of the options related to the demonstrations.[4] This investigation of the *majaribio* being initiated spontaneously by households or individuals showed three broad categories of activities: (1) "validation" of demonstration findings, (2) streamlining the use of the technology, and (3) finding new uses (cf. Ramisch et al. 2006).

Validation Experiments

While the mother-baby model assumes that household experimentation would strive to achieve similar results to those observed in the demonstration, in practice few farmers were so impressed by the advantages of legumes that they could justify trying a cereal-legume rotation (see comments about new uses and cultural factors below).[5] Where such validation experiments did occur, we often heard expressions like "I'm trying my luck," which implied beliefs attributing the "good" performance observed on the collective plot to multiple factors that might not necessarily be reproducible on another site or in another season.

Such *majaribio* also typically attempted to replicate the crop husbandry of the demonstration as exactly as possible, including elements that the research team did not initially consider part of the technology being "demonstrated." Describing the cereal phase of his *majaribio*, where maize was intercropped with common beans exactly as he'd seen it on the collective site, a male farmer in Matayos reported: "I have learned that you can get very good yields by line planting beans instead of broadcasting them [the way we normally do]."

"Streamlining" Experiments

Much more common were efforts by nearly all the studied farmers (37/40 or 92.5 percent) to "streamline" or modify the rotation scheme in various ways. The most common was adapting the temporal logic of legumes' soil fertility effects on cereals (crop rotation) into a spatial logic (intercropping). This was particularly evident with the intercropping of maize into fields of soybean. Other adaptations included varying the cropping densities (usually to increase them) and applying low rates or no mineral fertilizer on legumes.

Each of these adaptations revealed very strong, local logics behind agricultural practices. The first has to do with the centrality of maize in the local farming systems. While conceding that legume rotations did boost cereal yields, farmers also stated that they were unwilling to sacrifice even one season of maize production by growing a sole crop of legumes and would therefore try to get the legumes to "do their work" (*kufanya kazi yao*) through intercropping. Soybean became the intercrop of choice in individual experiments essentially because (unlike the vigorously bushy mucuna vines) the plants did not appear to compete with maize in the demonstration plots. Farmers also reported that they had observed that soybean did not seem to suffer in the shade of mature maize plants in the demonstrations, even though this was not an aspect that the research team had thought was being shown and there were actually very few plots within the demonstrations where soybean and maize were planted alongside each other.

It is worth noting that while farmers in all sites generally felt that mucuna was the legume that provided the greatest boost to cereal yields, very few were actually planting it in individual trials in 2005. Farmers would politely praise mucuna's power when evaluating the demonstration plots at the end of the season, but its seeds were conspicuously abandoned after the harvest of those same plots, even while every last soybean seed was gleaned from the collective sites. Unlike soybean, mucuna is too aggressive a vine to intercrop, it is inedible, and the only market for its seeds would have been to researchers or other project farmers. Thus, while a farmer who planted a mucuna rotation would (hypothetically) regain the maize harvest foregone during the "legume" season by the increase in the following season, this was unacceptable culturally. As one female farmer stated, "We Ateso people do not farm something that cannot be eaten." Farmers would explain that growing crops that everyone else in the community also grows does not just provide food for the household but guarantees the means of reciprocation, thereby ensuring one's place in a social system. Even if one's maize performs badly, it is better than growing a field of only mucuna, which is a crop that no relative or friend would ask for during hard times. As a result, the research team was almost ready to dismiss the further use of mucuna in the FEI until we caught wind of an underground resurgence of interest in the plant later in the project (discussed below).

Another important local logic that was revealed by farmers' "streamlining" experiments relates to the decisions made to forego inorganic fertilizer applications on the legume crops. The new legumes, having been "demonstrated" as playing a role in "improving soil fertility" were therefore conccived of as being a form of *mbolea* ("manure" or "fertilizer"). The implication was that the legume should provide benefits in its own right regardless of how much "better" it might have grown with the further addition of N or P inputs as in the demonstrations. As a male farmer in Chakol told us, "A fertilizer should not [itself] be fertilized, unless it confers tremendous benefits."

Experiments for New Functions

The third motivation for *majaribio* was to find new functions for legumes, beyond any role they might have been seen to play in improving soil fertility. By far the most common new use (25/40 or 62.5 percent) was to smother weeds on unproductive plots. Many farmers visiting the demonstration plots were struck by the reduction in *Striga* incidence, especially on plots previously planted with mucuna and soybean. One female farmer's "experiment," typical of these *majaribio*, intercropped soybean and maize, during which she marked clearly where each crop was in the first rainy season and interchanged their positions in the following season. With *Striga* a major problem, three-quarters (30/40) of the farmers interviewed were planning to use some form of targeted plantings of soybean (often intercropped with maize).

The evolving nature of technology decision making can also be seen in the resurgence over time of farmer interest in the inedible cover crop mucuna. While farmers like the Teso woman quoted earlier were initially and openly skeptical about including it in future experiments, a message began to gain momentum within local knowledge networks that mucuna could be planted "to exterminate *Striga*." This message was usually masked from researchers, who had not thought to make *Striga* management a learning objective in the demonstrations or the FEI research. However, farmers were making observations of the performance of *Striga* on the demonstration plots before, during, and after the legume rotations and continued to do so in their individual *majaribio*. Other research has shown that as legumes build a soil's fertility, they may cause the seeds of the parasitic *Striga* weed to germinate without the means to survive (i.e., "suicidal" germination increases; Misiko et al. 2008).

Finally, many of the new functions being tested for new legume species were inspired by (and compared against) the traditional uses of common beans. Because of its perceived ability to intercrop with staple cereals, many farmers tested planting soybeans in the same hole as maize, which is the current practice with common bean. Soybean leaves and other residues were also tested for novel uses, such as burning and using the ashes to produce traditional salt (10/40, 25 percent), as livestock fodder (10/40, 25 percent), or as an edible green vegetable (3/40, 7.5 percent).

SYMBOLISM: "EXPERIMENTING FOR RECOGNITION"

A fourth motivation for *majaribio* was to reaffirm social ties with the research team, or as a male farmer in Butula put it, "to show a good example and vision." These *majaribio* showcased experimenters' farms as good examples. For example, all five cases of households planting the mucuna-maize rotation (as "verification" experiments) took pains to "do [the rotation] as it was on the demonstration," including applying mineral fertilizer and planting the "experiment" on land that was not particularly degraded. Whatever its limitations as an inedible cover crop, mucuna was a researcher's idea, and showing "loyalty" to researchers (including "showcasing" it on land where it seemed guaranteed to perform well) was a way of getting social recognition. Within the project, "good" experiments were jointly identified by farmers and researchers and frequently visited as learning fields. Visitors (especially from outside Kenya) were often taken to such farms by project staff to illustrate the "success" of the learning process. However, such experiments were few and usually not sustained for many seasons. This implies that the motive for such experiments may have been mainly to show solidarity with the research process, and in turn enhance the social standing of the farmers in question (Ramisch et al. 2006).

Finally, the symbolic value of experimentation—which could be summarized in the visibility of new crop varieties, the novel crop husbandry, the collection of data, and the regular visits by the farmers' groups or outsiders—played an important role in establishing group identity. By 2008 farmers in Chakol were even referring to their collective experiments as a "church." When taking their lessons about soil fertility management to a different site four kilometers away, the local facilitator stated: "This was part of our scaling out mission, we decided to preach the same gospel. We liked the trial as designed by TSBF, because it showed clear lessons. So we replicated it here for ourselves too, even though many of us already knew the key lessons and some are not practicing them."

Retaining a central venue for collective experimentation, therefore, represented both a symbolic devotion to the FEI project (and its methods) and a belief that having a dedicated site for "demonstrating" new technologies was important for group solidarity. Interestingly, members continued to "preach" certain "lessons" (legume species or husbandry practices) they themselves did not use, perhaps reflecting a belief that these options might still be potentially useful in the new community. Whether future "churches" would also replicate the original experimentation format and technologies, however, remained an open question.

IMPLICATIONS OF EXPERIMENTS AS "PERFORMANCES"

The collaboration between researchers and farmers in the FEI reveals important nuances relating to soil fertility management knowledge and how we (as farmers or researchers) test what we think we know. If agricultural practice is indeed a "performance" in the sense outlined above, it means that knowledge is tested and recreated to some extent every season. The FEI team never believed that the "demonstration" sites were going to showcase cereal-legume rotations as scientific facts, at least not in the stereotypic sense implied in some mother-baby trials where the collective plot serves only to locally incarnate some universal truth. However, it is fair to say that the FEI team—again, researchers and farmers alike—was surprised by the extent to which the mere "testing" of new legume varieties in the various sites would highlight the diverse local logics and motivations for managing soil or indeed for engaging in farming.

The challenge for future participatory research of this nature lies in fully grasping the unrepeatable nature of agricultural production in a given spatial and socioeconomic context. As we learned, each season is essentially an opportunity to "try" (*kujaribu*) to combine existing knowledge and resources in the face of that particular season's challenges. In the FEI, new legumes were being "tried" and evaluated not only for their possible effects on soil fertility (the explicit purpose of the project) but also against their compatibility with a maize-centered farming system, their complementarity with existing legumes such as common beans, and for viable new uses such as weed suppression.

The local logics and past experiences guiding the assessment of how well the new legumes "did" (*kufanya*) could be broadly characterized by farmers in conversation as "ideal" soil management behaviors or crop performances, but could be much better grasped (implicitly) through the interrogation of multiple, individual experimental efforts observed over a range of seasons. The process of deciding which elements of a technology "work" and are worth maintaining must therefore take place over considerable time, with repeated stages of "trying," each of which is testing distinct objectives. All of which suggests that farmers' experimentation is a much richer terrain for investigation than commonly accepted, even if there is also no easy shortcut to understanding local agricultural practice and the knowledge and attitudes that underpin it.

The FEI ambitiously strove to generate "dynamic expertise" from dialogue between actors but soon had to recognize the challenges inherent to such a goal. Through the course of activities such as the discussions and learning about the cereal-legume rotations, it became obvious that agricultural knowledge and practice were themselves inherently dynamic and continuously evolving for any given actor. Furthermore, as researchers and farmers interacted, it became clearer too that the best role for "science" was in interpreting and explaining previous "performances" and outcomes (such as *Striga* suppression) rather than promising future benefits such as improved soil fertility. For both farmers and researchers, knowledge and experience were ever-shifting, built on the lessons of past seasons but not necessarily in equilibrium with or able to anticipate how to adapt to the next set of social and environmental conditions.

NOTES

1. As shown throughout this volume, the literature on local environmental knowledge is vast and growing. The project built on several strands of this literature, seeking to move beyond understanding the "scientific merit" of local soil fertility management practices (e.g., Oberthür et al. 2004; Ettema 1994; Sikana 1993) and their relevance within broader sociocultural contexts (e.g., Winklerprins and Barrera-Bassols 2004; Amanor 1994).

2. The project *Strengthening "Folk Ecology": Community-Based Interactive Farmer Learning Processes and Their Application to Soil Fertility* was implemented from 2001 to 2008 by the Tropical Soil Biology and Fertility Institute (TSBF-CIAT) with local governmental and nongovernmental partners and funding from the International Development Research Centre.

3. I could also acknowledge other performing arts or sports where knowledge and technical skill are embodied as legible actions. Scholars might even consider how we as authors or presenters have succeeded or failed in expressing ourselves clearly, intelligently, or provocatively, given time constraints and lack of rehearsal.

4. This chapter discusses only the cereal-legume *majaribio*, but farmers were experimenting with many technologies that had been demonstrated by the FEI.

5. Only ten farmers (25 percent) tried a soybean-maize rotation and only five (12.5 percent) a mucuna-maize rotation.

MATTHEW D. TURNER

CONCLUSION

This volume began with a vignette explaining how an unlikely place—*A barren stretch of ground in the Sahelian region of West Africa*—can attract a great deal of attention, reflecting different knowledges and interests regarding its past, present, and future use. The vignette illustrates how complex and intertwined the politics of knowing, managing, transforming, and conserving nature can be. We used the vignette to illustrate the limitations of understanding environmental problems as (1) an issue related to the production of environmental knowledge (i.e., how the remote-sensing scientist characterizes the land unit in question); (2) the application of environmental knowledge (i.e., how a local conservation agency applies a land classification to the area to push a management agenda); or (3) the circulation of environmental knowledge (i.e., how particular claims of "degradation" of such arid land units become common knowledge). We argued that these three spheres (production, application, and circulation) are not easily separated, and are indistinct at best. Yes, environmental knowledge is produced, circulated, and applied in conservation and development programs. But circumscribing our analyses around these activities to understand them ignores the fact that they are highly intertwined and obscures a politics that emerges from their very entanglement.

Simply put, environmental knowledge is locally produced and deeply contextualized. It is not only "produced" *with* an intended application but often *through* application. The patterns of circulation often determine what sort of knowledge production is supported (and thus, ultimately, *produced*). Moreover, circulation influences the understandings, methods, and interpretations at other sites of production or application. While we may seek to explain the politics that surround the activities of production, application, or circulation of environmental knowledges, our explanations must transcend these conceptual categories. So, while contributions to this volume are grouped into different parts, contributors used these "foci" as starting vantage points to raise particular sets of questions, not as boundaries for their analysis. The different parts, therefore, reflect the starting point of the chapters but not their explanatory scope.

All contributors presented theoretically informed empirical analyses to illuminate an environmental politics that is obscured unless one fully engages with the entangled nexus of production, application, and circulation. Different contributors adopt quite different positions and provide contrasting empirical explorations of theory-laden concepts such as networks (Galt, Lave, Rocheleau,

Taylor), nonhuman agency (Ingram, Duvall, Taylor), hybridity (Fujimura, Phadke, Vandergeest and Peluso), boundary objects and metaphor (Goldman, Nadasdy, Rocheleau, Vandergeest and Peluso), standardized technomethodological packages (Fujimura, Forsyth, Goldman, Lave, Phadke, Zimmerer), scale in knowledge and governance (Campbell, Duvall, Fujimura, Zimmerer), packaging/translation of knowledge for circulation (Duvall, Galt, Goldman, Lave), complexity and scientific uncertainty (Campbell, Duvall, Forsyth, Fujimura, Taylor), science and state interests (Galt, Phadke, Vandergeest and Peluso, Zimmerer), and boundary work around knowledge systems (Campbell, Duvall, Fujimura, Galt, Ingram, Lave, Nadasdy, Ramisch). By drawing on different theoretical and analytical tools and discarding false boundaries between the production, application, and circulation of environmental knowledge, each chapter provides not only a fuller account but reveals political processes that would remain hidden otherwise.

For example, the production of environmental scientific knowledge is shown to be strongly influenced by application intentions and interests. This may be most clearly shown in Lisa Campbell's chapter, where scientific debates about sea turtle ecology (e.g., sustainable use, recruitment, mobility) among conservation biologists are strongly shaped by divergent understandings of the human role in conservation. In his chapter on the regulation and monitoring of pesticide residues of internationally traded vegetables, Ryan Galt shows how environmental science is enrolled in the service of the market with significant implications for the production of scientific knowledge (regarding the extent of contamination and its health effects). Our knowledge of food toxicity is strongly shaped by the rate and type of sampling permissible within underfunded regulatory agencies, which must not hold up the flow of perishable products to fill U.S. dinner plates. Moreover, the establishment of residue standards depends heavily on value-laden calculations of risk based not solely on biochemistry and human physiology but also on assumptions about food provenance, consumption, and production practices.

Such influences are not just evident among the "applied sciences." As Mrill Ingram demonstrates, our views of microbes as organisms to be controlled (stemming from the health sciences) has historically led us to ignore the complex relationships between humans and microbes. This effectively leads to a hollowing-out of microbial agency by ignoring their responsiveness and subjectivity in relations with humans as well as their own complex ecologies (separate from relations to humans). Similarly, Paul Nadasdy demonstrates how the management imperative of wildlife ecology has contributed to the discursive reference to wildlife populations as if they were agronomic crops to be *harvested*. In contrast to First Nation peoples, such language both influences and reflects how managers and scientists think about wild animals (e.g., populations to be managed at equilibrium) while reducing their subjectivity and agency as organisms. The political implications of "management" is further explored in Peter Vandergeest and Nancy Peluso's chapter, which traces the co-production of science and social orders in the development of *political forests* in Southeast Asia. Southeast Asian states invested and developed forest management strategies during the same historical period when they were involved in counterinsurgency actions in heavily forested areas. State-sponsored scientific practice and military operations transformed *jungles* into separate, well-managed (and controlled) territories of agricultural land and scientifically managed forest. This reterritorialization of agriculture's relation to forest was not solely the result of military interests but was supported by international conservation interests that viewed agriculturalists in forests as counter to preservationist goals. This chapter thus highlights the often hidden links between the production, translation, circulation, and application of scientific knowledge—in a way that shapes the production of certain kinds of "natures" as well as livelihoods.

While highlighting the links between state and other institutional interests on the application of scientific practice, contributors to this volume do not present this relationship as linear and determinant. The web of influence and power relations is much more diverse and multidimensional (Taylor). For example, much work on colonial environmental science in Africa has tended to reduce scientific knowledge claims as simply stemming from European biases against African smallholders, coupled with the political interests of the colonial state. Chris Duvall's excavation of the scientific debates about the ontological status of ferricrete exposes a much wider debate within the colonial scientific community (see also Campbell's chapter), a community that was not isolated from colonialist interests, but which was also influenced by discipline- and experience-bound spatial and temporal framings. These divergent framings very much led to significant debate about the human role in ferricrete formation—with different consequences for understanding human-landscape interactions. Different understandings of ferricrete were then applied in particular ways by colonialists to further particular political interests in West Africa.

Duvall's chapter highlights the complexity involved in the challenge of understanding natural processes that occur at time scales greater than we can observe them. This is of course one of the challenges of environmental science in general, which necessarily seeks to understand complex interactions among multiscaled systems, not controlled in laboratories, but operating within and across organisms, landscapes, and ecosystems. Coming to grips with such complexity is a daunting task for which scientists often rely on models, analogies, or metaphorical logics to make sense of complex systems, and nonobservable phenomenon. This is most evident in new fields of study or in situations where prior scientific research is sparse. Joan Fujimura traces the development of "systems biology," showing how scientists have approached the complex biochemical interactions operating at and across subcellular (e.g., ribosomes, mitochondria), cellular, organ, system (digestive, nervous, etc.), and organism levels. Biologists have oscillated between reductionism and holism, with "systems biologists" treating the living organism as machine to find a middle ground between these two unsatisfactory poles. The use of the machine analogy to make sense of the complexity of ecological community, atmosphere, or ecosystem is very common within the environmental sciences. Treatments of these biological/ecological systems as machine analogues have wider political implications when one considers the ontological differences between mechanics, biology, and ecology. The "command-and-control" view of microbial relations described by Ingram could reflect in part the treatment of the human body as a machine.

Another complication of the machine, or other metaphors, models, and analogies used to understand environmental complexity, is that they assume some degree of universal application. A model is, by its very nature, a tool of simplification and generalization. Models and metaphors not only attempt to make sense of complexity *in place*, they are also often used as a universal generalizing tool of analysis. Yet once one recognizes that environmental knowledge is necessarily local with significant limitations to its relevance in other contexts, the politics of substituting environmental knowledge produced in place with circulating models, metaphors, and knowledge from quite different places is revealed. Tim Forsyth's chapter shows how scientists' and managers' reliance on a tool, the *universal* soil loss equation, both filled a gap of scientific understanding of soil erosion in northern Thailand and discounted alternative more local understandings of erosion. This is despite the fact that this model was developed to estimate erosion potential under quite different environmental conditions. The widespread circulation of concepts, understandings, models, and analytic tools developed in one context for application in quite different contexts reflects the economic and political power differentials between sites of the production and application of "scientific knowledge." This process also highlights the nature of the scientific enterprise itself, which seeks to generalize (even for that which is not generalizable) through the translation and packaging of knowledge claims (e.g., into models and through metaphors) for wide circulation.

In the realm of conservation science, concepts that are abstracted sufficiently from local context, that seem universally applicable and relevant, are most likely to gain widespread influence. Rebecca Lave's chapter outlining the popularity of the Rosgen "natural channel design" assessment method for stream channel restoration is an illustration of how the influence of scientific ideas in conservation is very much shaped by their transportability. Despite widespread questioning and dismissal by fluvial geomorphologists, the Rosgen method has proven to be much more influential within the stream restoration field than academic understanding, because it provides standard assessment guidelines *useful* across a wide range of situations. The popularity of the Rosgen method is thus not only about "science" but about the applicability, marketing, and usability of science in practice—it provides a model, a tool, that can be easily taught and therefore transported and applied widely. Often the influence of a concept is enhanced when different actors in conservation science can simultaneously hold slightly different understandings of it (e.g., boundary object). As illustrated by Mara Goldman's analysis of conservation corridors, boundary objects that simultaneously invoke ecological theory (connectivity and gene flow) and can be viewed as physical objects (paths) on the ground are particularly powerful, given that they can be "real" to a range of actors involved in conservation (scientist, manager, policy maker, public). These chapters, along with that written by Karl S. Zimmerer, demonstrate that the influence of conservation ideas in circulation is further reinforced if they are bundled with a standardized package of technology and methods that, through their use, reinforce the concept itself. The standardized packages of economic-environmental models of irrigation, spatial analysis of land use and remote sensing, and channel categorization based on widely available data have helped reinforce the circulatory success of the watershed planning, ecological corridor, and natural stream channel approaches, respectively.

The widespread circulation of scientific models and ideas has significant implications for people that live and work in landscapes where ideas *touch down* and are applied. The effects are familiar to political ecologists. Maasai pastoralists in Tanzania are excluded from paths constructed by conservationists as corridors, within zones where wildlife and livestock have moved. Bolivian *cholo* peasants and *indios* find their upland fields in areas classified as "water surplus" from which water is to be harvested to meet the "water-scarce" lowlands. First Nation peoples' relationships and understandings of wildlife are disregarded, reducing their effective engagement with, and access to, wildlife. But the politics do not simply lie within the realm of "application" where fully formed scientific knowledge claims are used strategically by resource users, conservationists, and government officials to make claims on local resources. The politics of conservation and development described in this volume involves multiple political arenas animated by competing interests and knowledge claims, which while not necessarily focused on the distribution of resource control in Tanzania, Bolivia, or the Yukon, may strongly shape it. Political ecologists seeking to understand the politics of resource access in places around the world are prone to fail unless they are able to follow the extralocal networks that in many ways shape the conceptual templates through which local conflicts must be waged. In short, researchers and participants need to understand the "heterogeneous constructions" that underlie environment politics (Taylor)—and occur across networks that often stretch far from the "sites" of application.

While all contributions to this volume acknowledge the importance of the circulation of environmental knowledge as it relates to environmental politics, the chapters in the third part focus specifically on how we should conceptualize knowledge circulation and the inherent contacts among multiple environmental knowledges that occur along the way. Actor-network theory has inspired a plethora of network speak that seeks to break the dualism of the human and nonhuman, and "follow" knowledge as it is produced, circulates, and is applied. As argued by Peter Taylor, early ANT work is a poor fit to political ecology given its playfully poor recognition of differences between the human and nonhuman and its limited treatment of social power. These limitations produce a flat relational ontology where actants interact across space but seemingly not within and between places (social and ecological context). Dianne Rocheleau's chapter, using casework from Zambrana in the Dominican Republic, outlines an alternative conceptualization of "rooted networks"—networks that are tied to social and biophysical geographies and power-laden in complex ways (i.e., not only power over). These are not the networks invoked by references to our networked world, networked elites, or Twitter accounts. These are the multiple social and material networks connecting people, places, and ecologies arising from everyday livelihood activities and political struggles.

The chapters by Tim Forsyth and Paul Nadasdy present cases showing how nonscientific, place-based understandings of the environment are likely to be more nuanced, historically informed, multifactorial, and meaning-rich than those imposed from above by short-term, superficial scientific engagements with these places. Both argue in different ways for taking nonprofessional lay knowledge seriously, not by placing it on a pedestal as if it were fixed in stone, but through cross-cultural engagement that does not seek to understand through full translation (reduction) into a single epistemology or cosmology but through dialogue at points of common framings. As such, they reflect Rocheleau's call to create networks of situated knowledges—views from different places along complex networks. This would require a comfort level with the incongruencies and inconsistencies that are inherent in the processes of knowledge exchange, dialogue, and hybridization. A major barrier is the power-laden relationship between the "scientist" and everyday citizen. The chapter by Josh Ramisch on participatory soil fertility research describes the epistemological barriers to such exchanges and a model for joint research that can help facilitate more participatory dialogues. These free-flowing engagements are not predictable—if local citizens gain decision-making authority, the interventions that they identify as "appropriate" may deviate from what environmentalists or appropriate technologists would predict or support. This is demonstrated vividly in Roopali Phadke's chapter, which shows a different side of the water politics in India where the water reform movement in the state of Maharashtra has sought not to eliminate dams but to reconfigure their relationship with irrigation networks. In so doing, activists have utilized an icon of disembodied Western technology and state power (the dam) for their own purposes by changing its use as informed by their own socioenvironmental understandings of irrigated agriculture.

When taken as a whole, this volume thus presents a conceptualization of environmental politics that is broader than traditional treatments by STS scholars (e.g., among scientists and labs for resources and influence; within and between private and public institutions over science policy; in public involvement in science-laden political debate), political ecologists (e.g., struggles over access to resources, the use and manipulation of fully formed scientific claims for political ends), and other social scientists (e.g., environmental movements, environmental justice, social construction of nature). As such, the chapters add to a growing body of political ecology and STS scholarship that is moving beyond their historic foci of the application and production/circulation of environmental knowledges, respectively. How the *barren stretch of ground in the Sahelian region of West Africa* eventually is managed is shaped not solely by political struggles over its use but also by scientific understandings of other *barren stretches* elsewhere. These (mis)understandings fill the void produced by the ignorance of lay knowledges of this place by Western managers. It is in fact only through their *management* that many similar stretches of land around the world come to be truly known in any way by Western science. In these ways, landscapes are simultaneously materially produced and partly learned.

Environmental politics are tied to the co-production of science, social order, *and* material landscapes. Much has been written about the production of hybrid, cyborg, or neoliberal natures. The rich empirical work provided in the preceding chapters goes beyond simply documenting the multiple meanings attached to nature's objects by tracing how knowledge politics are inscribed in landscapes. Socionatures are ontological. The knowledge politics that surrounds conservation and development practice shape the placement of earthworks in India, the United States, and West Africa; shifting agricultural patterns with the effective resettlement of people in Southeast Asia, Bolivia, and Tanzania; the condition of soils in Thailand and West Africa; the public health of people in the United States and Costa Rica; the viability of wildlife populations in the world's oceans, the Yukon, and Tanzania. As Tim Forsyth admonishes us, understanding and acknowledging the knowledge politics that run through the tripartite categories of production, application, and circulation is crucial not only to identify who wins and loses with environmental change and conservation but also to be more successful in *knowing* nature by acknowledging the positionality and normative basis for all of our environmental work. This, it seems, is a necessary starting point for creating more responsible interactions within the socionatures we inhabit.

REFERENCES

Abreu-Grobois, A., and P. Plotkin. 2007. Global assessment for *Lepidochelys olivacea* for the IUCN Red List. IUCN/SSC. Marine Turtle Specialist Group. http://www.iucn-mtsg.org/red_list/lo/RLA_Lepidochelys_olivacea_revised.pdf.

Adams, W., and D. Hulme. 2001. Conservation and community: Changing narratives, policies and practices in African conservation. In *African Wildlife and Livelihoods: The Promise and Performance of Community Conservation*, edited by D. Hulme and M. Murphree, 9–23. Oxford: James Curry.

Agrawal, A. 1995. Dismantling the divide between indigenous and scientific knowledge. *Development and Change* 26 (3): 413–39.

———. 2005. *Environmentality: Technologies of Government and the Making of Subjects*. Durham, NC: Duke University Press.

Agrawal, A., and J. Ribot. 1999. Accountability in decentralization: A framework with South Asian and West African cases. *Journal of Developing Areas* 33:473–502.

AKCRSC (Aishihik-Kluane Caribou Recovery Steering Committee). 1996a. Minutes from the meeting of the Aishihik-Kluane Caribou Recovery Steering Committee. April 18. Prepared by Rob Moore. Burwash Landing, YT.

———. 1996b. Minutes from the meeting of the Aishihik-Kluane Caribou Recovery Steering Committee. February 6. Prepared by Paul Nadasdy. Haines Junction, YT.

Akera, A. 2007. Constructing a representation for an ecology of knowledge: Methodological advances in the integration of knowledge and its various contexts. *Social Studies of Science* 37 (3): 413–41.

Alatout, S. 2006. Towards a bio-territorial conception of power: Territory, population, and environmental narratives in Palestine and Israel. *Political Geography* 25 (6): 601–21.

Alcorn, J. 1989. Process as resource: The traditional agricultural ideology of Bora and Huastec resource management and its implications for research. *Advances in Economic Botany* 7:63–77.

Aleva, G. J. J. 1994. *Laterites: Concepts, Geology, Morphology and Chemistry*. Wageningen, Netherlands: International Soil Reference and Information Center.

Alexander, L. T., and J. G. Cady. 1962. *Genesis and Hardening of Laterite in Soils*. Washington, DC: USDA Soil Conservation Service.

Ali, Ismail bin Hah. 1966. A critical review of Malayan silviculture in the light of changing demand and form of timber utilization. *Malayan Forester* 29:228–38.

Alm, J., J. Swartz, G. Lilja, A. Scheynius, and G. Pershagen. 1999. Atopy in children of families with an anthroposophic lifestyle. *Lancet* 353:1485–88.

Alon, U. 2003. Biological networks: The tinkerer as an engineer. *Science* 301:1866–67.

Amanor, K. S. 1994. *The New Frontier: Farmers' Response to Land Degradation; A West African Study*. London: Zed Books.

Anderson, A. B., and C. N. Jenkins. 2006. *Applying Nature's Design: Corridors as a Strategy for Biodiversity Conservation*. New York: Columbia University Press.

Anderson, W. 2002. Introduction: Postcolonial technoscience. *Social Studies of Science* 32 (5–6): 643–58.

André, V., G. Pestaña, and G. Rossi. 2003. Foreign representations and local realities: Agropastoralism and environmental issues in the Fouta Djalon table lands, Republic of Guinea. *Mountain Research and Development* 23 (2): 149–55.

Arnold, D. 2006. *The Tropics and the Traveling Gaze*. Seattle: University of Washington Press.

Asch, M. 1989. Wildlife: Defining the animals the Dene hunt and the settlement of Aboriginal rights claims. *Canadian Public Policy* 15 (2): 205–19.

Ashmore, P. 1999. What would we do without Rosgen? Rational regime relations and natural channels. Paper read at 2nd International Conference on Natural Systems Design, Niagara Falls, Ontario.

Assies, W. 2003. David versus Goliath in Cochabamba: Water rights, neoliberalism, and the revival of social protest in Bolivia. *Latin American Perspectives* 130 (30): 14–36.

Aubréville, A. 1947. Erosion et bovalisation en Afrique noire française. *Agronomie Tropicale* 2:339–57.

———. 1949. *Climats, forêts, et désertification de l'Afrique tropicale*. Paris: Société d'Éditions Géographiques, Maritimes, et Coloniales.

Aufrère, M. L. 1932. La signification de la laterite dans l'évolution climatique de la Guinée. *Bulletin de l'Association de Géographes Français* 60:95–97.

Aymonim, G. 1983. André Aubréville (1897–1982). *Bulletin de la Société Botanique de France* 130 (3): 257–61.

Bakker, K. J. 2000. Privatizing water, producing scarcity: The Yorkshire drought of 1995. *Economic Geography* 76 (1): 4–27.

———. 2003. Archipelagos and networks: Urbanization and water privatization in the South. *Geographical Journal* 169 (4): 328–41.

Bakker, K. J., and G. Bridge. 2006. Material worlds? Resource geographies and the "matter of nature." *Progress in Human Geography* 30 (1): 5–27.

Barabasi, A. L. 2002. *Linked: The New Science of Networks*. Cambridge, MA: Perseus Publishing.

Barad, K. 2003. Posthumanist performativity: Toward an understanding of how matter comes to matter. *Signs* 28 (3). http://www.journals.uchicago.edu/Signs/journal/issues/v28n3/032806.text.html (accessed April 2007).

Bassett, T. J., and D. Crummey, eds. 2003. *African Savannas: Global Narratives and Local Knowledge of Environmental Change*. Oxford: James Currey.

Bassett, T. J., and K. B. Zuéli. 2000. Environmental discourses and the Ivorian savanna. *Annals of the Association of American Geographers* 90 (1): 67–95.

Batterbury, S. P. J. 1996. Planners or performers? Reflections on indigenous dryland farming in northern Burkina Faso. *Agriculture and Human Values* 13 (2): 12–22.

Baviskar, A. 2007. *Waterscapes: The Cultural Politics of a Natural Resource*. New Delhi: Permanent Black.

Bebbington, A. 1997. Social capital and rural intensification: Local organizations and islands of sustainability in the rural Andes. *Geographical Journal* 163:189–91.

Beck, U. 1992. *Risk Society: Towards a New Modernity*. London: Sage.

Benjaminsen, T. A., and G. Berge. 2004. Myths of Timbuktu: From African El Dorado to desertification. *International Journal of Political Economy* 34 (1): 31–59.
Berger, J. 2004. The last mile: How to sustain long-distance migration in mammals. *Conservation Biology* 18 (2): 320–31.
Bernhardt, E. S., M. A. Palmer, J. D. Allan, G. Alexander, K. Barnas, S. Brooks, J. Carr, et al. 2005. Synthesizing U.S. river restoration efforts. *Science* 308:636–37.
Berry, S. 1993. *No Condition Is Permanent: The Social Dynamics of Agrarian Change in Sub-Saharan Africa*. Madison: University of Wisconsin Press.
Bertalanffy, L. von. 1933. *Modern Theories of Development*. Translated by J. H. Woodger. London: Oxford University Press.
———. 1952. *Problems of Life*. London: Watts and Co.
Bhaskar, R. 1979. *The Possibility of Naturalism*. Brighton, UK: Harvester.
Biagioli, M., ed. 1999. *The Science Studies Reader*. New York: Routledge.
Bik, E. M., P. B. Eckburg, S. R. Gill, K. E. Nelson, E. A. Purdom, F. Francois, G. Perez-Perez, M. J. Blaser, and D. A. Relman. 2006. Molecular analysis of the bacterial microbiota in the human stomach. *Proceedings of the National Academy of Sciences USA* 103 (3): 732–37.
Birkenholtz, T. 2008. Contesting expertise: The politics of environmental knowledge in northern Indian groundwater practices. *Geoforum* 39:466–82.
———. 2009. Irrigated landscapes, produced scarcity, and adaptive social institutions in Rajasthan, India. *Annals of the Association of American Geographers* 99 (1): 118–37.
Bitterman J. 2008. France milks cheese for all it's worth. CNN.com/Europe, May 2. http://www.cnn.com/2008/WORLD/europe/05/02/french.cheese/index.html.
Bjorndal, K. A., and A. B. Bolten. 2003. From ghosts to key species: Restoring sea turtle populations to fulfill their ecological roles. *Marine Turtle Newsletter* 100:16–21.
Bjorndal, K. A., J. A. Wetherall, A. B. Bolten, and J. A. Mortimer. 1999. Twenty-six years of green turtle nesting at Tortuguero, Costa Rica: An encouraging trend. *Conservation Biology* 13:126–34.
Black, M. 1962. *Models and Metaphors: Studies in Language and Philosophy*. Ithaca, NY: Cornell University Press.
Blaikie, P. 1985. *The Political Economy of Soil Erosion in Developing Countries*. London: Longman.
———. 1999. A review of political ecology: Issues, epistemology and analytical narratives. *Zeitschrift für Wirtschaftsgeographie* 43 (3–4): 131–47.
Blaikie, P., and H. Brookfield, eds. 1987. *Land Degradation and Society*. London: Methuen.
Blaser, M. 2005. Pathogenicity and symbiosis: Human gastric colonization by *Helicobacter pylori* as model system of amphibiosis. Paper presented at "Ending the War Metaphor: The Future Agenda for Unraveling the Host-Microbe Relationship." National Academy of Sciences, Washington, DC, March 16–17. http://www.iom.edu/Object.File/Master/25/911/Blaser%20War%20Metaphor%20Talk%20REV.pdf.
Blockstein, D. E. 2002. How to lose your political virginity while keeping your scientific credibility. *BioScience* 52:91–96.
Bolen, E. G., and W. Robinson. 2003. *Wildlife Ecology and Management*. 5th ed. Upper Saddle River, NJ: Prentice Hall.
Boonkird, S. A., E. C. M. Fernandes, and P. K. R. Nair. 1984. Forest villages: An agroforestry

approach to rehabilitating forest land by shifting cultivation in Thailand. *Agroforestry Systems* 2:87–102.
Borner, M. 1985. The increasing isolation of Tarangire National Park. *Oryx* 19 (2): 91–96.
Botkin, D. 1989. *Discordant Harmonies: A New Ecology for the Twenty-first Century*. Oxford: Oxford University Press.
Bottrill, M., K. Didler, J. Baumgartner, C. Boyd, C. Loucks, J. Oglethorpe, D. Wilkle, and D. Williams. 2006. *Selecting Conservation Targets for Landscape-Scale Priority Setting: A Comparative Assessment of Selection Processes Used by Five Conservation NGOs for a Landscape in Samburu, Kenya*. Washington, DC: World Wildlife Fund.
Boulding, K. 1956. General systems theory: The skeleton of science. *Management Science* 2:197–208.
Bowen, B. W., W. S. Grant, Z. Hillis-Starr, D. J. Shaver, K. A. Bjorndal, A. B. Bolten, and A. L. Bass. 2007. Mixed-stock analysis reveals the migrations of juvenile hawksbill turtles (*Eretmochelys imbricata*) in the Caribbean Sea. *Molecular Ecology* 16 (1): 49–60.
Bowen, B. W., and S. A. Karl. 2007. Population genetics and phylogeography of sea turtles. *Molecular Ecology* 16:4886–4907.
Bower, B. 2004. One-celled socialites. *Science News*, November 20, 330–32.
Braun, B. 2002. *The Intemperate Rainforest: Nature, Culture, and Power on Canada's West Coast*. Minneapolis: University of Minnesota Press.
Braun, B., and N. Castree, eds. 1998. *Remaking Reality: Nature at the Millennium*. New York: Routledge.
Braun-Fahrlander, C., M. Gassner, L. Grize, U. Neu, F. Sennhauser, H. Varonier, J. Vuille, and B. Wuthrich. 1999. Prevalence of hay fever and allergic sensitisation in farmer's children and their peers living in the same rural community: SCARPOL Team Swiss study on childhood allergy and respiratory symptoms with respect to air pollution. *Clinical and Experimental Allergy* 29:28–34.
Broad, R., and J. Cavanagh. 1993. *Plundering Paradise: The Struggle for the Environment in the Philippines*. Berkeley: University of California Press.
Brockington, D. 2002. *Fortress Conservation: The Preservation of the Mkomazi Game Reserve, Tanzania*. Oxford: James Currey for International African Institute.
Broderick, A. C., R. Frauenstein, F. Glen, G. C. Hays, A. D. Jackson, T. Pelembe, G. R. Ruxton, and B. J. Godley. 2006. Are green turtles globally endangered? *Global Ecology and Biogeography* 15:21–26.
Brody, J. 2006. Trouble in the gut: When antibiotics work too well. *New York Times*, January 24. http://www.nytimes.com/2006/01/24/health/24brod.html (accessed March 10, 2006).
Brookfield, H., and C. Padoch. 1994. Appreciating agrodiversity: A look at the dynamism and diversity of indigenous farming practices. *Environment* 36 (5): 6–11, 37–45.
Brooks, R. 2004. Understanding Organization (Self) beyond Computation. MIT Artificial Intelligence Laboratory. http://www.ai.mit.edu/people/brooks (accessed 2004).
Brosius, J. P. 2006. Seeing communities: Technologies of visualization in conservation. In *Reconsidering Community: The Unintended Consequences of an Intellectual Romance*, edited by G. Creed, 227–54. Santa Fe, NM: SAR Press.
Brosius, J. P., A. L. Tsing, and C. Zerner. 1998. Representing communities: Histories and

politics of community-based natural resource management. *Society and Natural Resources* 11 (2): 157–68.
———, eds. 2005. *Communities and Conservation: Histories and Politics of Community-Based Natural Resource Management*. Walnut Creek, CA: AltaMira Press.
Brown, J., ed. 1989. *Environmental Threats: Perception, Analysis and Management*. New York: Belhaven Press.
Brown, L. 2001. *Eco-Economy: Building an Economy for the Earth*. London: Earthscan; Washington, DC: Earth Policy Institute.
Bryant, R. L. 1996. *Third World Political Ecology: An Introduction*. New York: Routledge.
———. 1997. *The Political Ecology of Forestry in Burma*. Honolulu: University of Hawaii Press.
———. 1998. Power, knowledge and political ecology in the third world: A review. *Progress in Physical Geography* 22 (1): 79–94.
Buckley, A. 1993a. Birckel relieved. *Yukon News*, January 13.
———. 1993b. More input. *Yukon News*, January 15, 1–2.
———. 1996. Wolf plan ignored: YCS. *Yukon News*, April 19, 12.
Budds, J. 2009. Contested H_2o: Science, policy and politics in water resource management in Chile. *Geoforum* 40:418–30.
Butler, J. 1990. *Gender Trouble: Feminism and the Subversion of Identity*. Toronto: Routledge.
Calder, I. 1999. *The Blue Revolution: Land Use and Integrated Water Resources*. London: Earthscan.
Callon, M. 1986. Some elements of a sociology of translation: Domestication of the scallops and fishermen of St. Brieuc Bay. In *Power, Action, and Belief*, edited by J. Law, 196–233. London: Routledge and Kegan Paul.
Callon, M., and B. Latour. 1981. Unscrewing the big Leviathan: How actors macro-structure reality and how sociologists help them to do so. In *Advances in Social Theory and Methodology: Toward an Integration of Micro- and Macro-sociologies*, edited by K. Knorr-Cetina and A. V. Cicourel, 277–303. Boston: Routledge and Kegan Paul.
Callon, M., and J. Law. 1989. On the construction of sociotechnical networks: Content and context revisited. *Knowledge and Society* 8:57–83.
———. 1995. Agency and the hybrid collectif. *South Atlantic Quarterly* 94:481–507.
Callon, M., and V. Rabeharisoa. 2003. Research in the wild and the shaping of new social identities. *Technology in Society* 25:193–204.
Calvert, J., and J. H. Fujimura. 2009. Calculating life? A sociological perspective on systems biology. *EMBO Reports* 10 (August): 546–49.
Campbell, J. M. 1917. Laterite: Its origin, structure and minerals. *Mineralogical Magazine* 17:67–71, 120–28, 171–79, 220–29.
Campbell, L. M. 1998. Use them or lose them? The consumptive use of marine turtle eggs at Ostional, Costa Rica. *Environmental Conservation* 24:305–19.
———. 2000. Human need in rural developing areas: Perceptions of wildlife conservation experts. *Canadian Geographer* 44 (2): 167–81.
———. 2002a. Conservation narratives and the "received wisdom" of ecotourism: Case studies from Costa Rica. *International Journal of Sustainable Development* 5 (3): 300–325.
———. 2002b. Science and sustainable use: Views of conservation experts. *Ecological Applications* 12 (4): 1229–46.

———. 2003. Contemporary culture, use, and conservation of sea turtles. In *The Biology of Sea Turtles*, vol. 2, edited by P. L. Lutz, J. A. Musick, and J. Wyneken, 307–38. Boca Raton, FL: CRC Press.

———. 2007a. Reconciling local conservation practice with global discourse: A political ecology of sea turtle conservation. *Annals of the Association of American Geographers* 97 (2): 313–34.

———. 2007b. Understanding human use of olive ridleys: Implications for conservation. In *Biology and Conservation of Ridley Sea Turtles*, 23–43. Baltimore: Johns Hopkins University Press.

Campbell, L. M., M. H. Godfrey, and O. Drif. 2002. Community based conservation via global legislation? Limitations of the Inter-American Convention for the Conservation of Sea Turtles. *Journal of International Wildlife Law and Policy* 5:121–43.

Campbell, L. M., B. J. Haalboom, and J. Trow. 2007. Sustainability of community-based conservation: Sea turtle egg harvesting in Ostional (Costa Rica) ten years later. *Environmental Conservation* 34 (2): 122–31.

Campbell, L. M., J. J. Silver, N. J. Gray, S. Ranger, A. Broderick, T. Fisher, M. H. Godfrey, et al. 2009. Comanagement of sea turtle fisheries: Biogeography versus geopolitics. *Marine Policy* 33:137–45.

Carney, J. A. 1991. Indigenous soil and water management in Senegambian rice farming systems. *Agriculture and Human Values* 8 (1–2): 37–48.

———. 1992. Peasant women and economic transformation in the Gambia. *Development and Change* 23:67–90.

———. 2001. *Black Rice: The African Origins of Rice Cultivation in the Americas*. Cambridge, MA: Harvard University Press.

Carson, R. 1994. *Silent Spring*. Reprint, New York: Houghton Mifflin Co. Original edition 1962.

Casida, J. E., and G. B. Quistad. 2004. Organophosphate toxicology: Safety aspects of nonacetylcholinesterase secondary targets. *Chemical Research in Toxicology* 17 (8): 983–98.

Castells, M. 1996. *The Information Age: Economy, Society, and Culture*. Oxford: Blackwell.

———. 2000. *The Rise of the Network Society*. Oxford: Blackwell.

Castree, N. 2002. False antithesis? Marxism, nature and actor networks. *Antipode* 34 (1): 111–46.

———. 2005. *Nature*. New York: Routledge.

Castree, N., and B. Braun, eds. 2001. *Social Nature: Theory, Practice, and Politics*. Malden, MA: Blackwell.

Chaloupka, M. 2002. Stochastic simulation modelling of southern Great Barrier Reef green turtle population dynamics. *Ecological Modelling* 148:79–109.

Chaloupka, M., and G. Balazs. 2007. Using Bayesian state-space modelling to assess the recovery and harvest potential of the Hawaiian green sea turtle stock. *Ecological Modelling* 205:93–109.

Chang, J.-M., T.-H. Chen, and T. J. Fang. 2005. Pesticide residue monitoring in marketed fresh vegetables and fruits in central Taiwan (1999–2004) and an introduction to the HACCP system. *Journal of Food and Drug Analysis* 13 (4): 368–76.

channelnewsasia.com. 2008. Officials say Chinese-made dumplings not contaminated in China. http://www.channelnewsasia.com/stories/eastasia/ view/ 331663/1/.html (accessed February 28, 2008).

Chautard, J. 1905. *Étude sur la géographie physique et la géologie du Fouta-Djallon et de ses abords orientaux et occidentaux*. Paris: Henri Jouve.

Cheah Boon Kheng. 1988. *The Peasant Robbers of Kedah, 1900–1929*. Singapore: Oxford University Press.

Chevalier, A. 1906. *La situation agricole de l'ouest africain*. Domfront, France: H. Senen.

———. 1909. Les hauts plateaux du Fouta Djalon. *Annales de Géographie* 18 (99): 253–61.

———. 1928. Sur la dégradation des sols tropicaux causée par les feux de brousse et sur les formations végétales régressives qui en sont la conséquence. *Comptes Rendus de l'Académie des Sciences* 187:84–86.

Chopra, R., and D. Sen. 1991. *Natural Bounty and Artificial Scarcity*. Dehra Dun, India: People's Science Institute.

Chudeau, R. 1921. Le problème du dessechement en Afrique occidentale. *Bulletin du Comité d'Études de l'Afrique Occidentale Française* 4:353–69.

CIDRE (Centro de Información y Documentación para el Desarrollo Regional). 1984. *Monografía de la Provincia Esteban Arce*. Cochabamba, Bolivia: CIDRE.

———. 1985. *Monografía de la Provincia Punata*. Cochabamba, Bolivia: CIDRE.

Clark, J., and J. Murdoch. 1997. Local knowledge and the precarious extension of scientific networks. *Sociologia Ruralis* 37 (1): 38–60.

Clarke, A. E. 2005. *Situational Analysis: Grounded Theory after the Postmodern Turn*. Thousand Oaks, CA: Sage.

Clarke, A. E., and J. H. Fujimura, eds. 1992a. *The Right Tools for the Job: At Work in Twentieth-Century Life Sciences*. Princeton, NJ: Princeton University Press.

———. 1992b. What tools? Which jobs? Why right? In *The Right Tools for the Job: At Work in Twentieth-Century Life Sciences*, edited by A. E. Clarke and J. H. Fujimura, 3–44. Princeton, NJ: Princeton University Press.

Clarke, A. E., and S. L. Star. 2008. The social worlds framework: A theory/methods package. In *The Handbook of Science and Technology Studies*, edited by E. Hackett, O. Amsterdamska, M. Lynch, and J. Wajcman, 113–37. Cambridge, MA: MIT Press.

Collins, H. M., and S. Yearley. 1992. Epistemological chicken. In *Science as Practice and Culture*, edited by A. Pickering, 301–26. Chicago: University of Chicago Press.

Conklin, H. 1954. An ethnoecological approach to shifting agriculture. *Transactions of the New York Academy of Sciences* 77:133–42.

Coppel, C. A. 1983. *Indonesian Chinese in Crisis*. New York: Oxford University Press.

Costanza, R., B. Low, E. Ostrom, and J. Wilson, eds. 2001. *Institutions, Ecosystems, and Sustainability*. Boca Raton, FL: Lewis Publishers.

Cronon, W. 1991. *Nature's Metropolis: Chicago and the Great West*. New York: Norton.

———, ed. 1996. *Uncommon Ground: Rethinking the Human Place in Nature*. New York: Norton.

Crooks, K. R., and M. Sanjayan, eds. 2006. *Connectivity Conservation*. Cambridge: Cambridge University Press.

Crowley, E. L., and S. E. Carter. 2000. Agrarian change and the changing relationships

between toil and soil in Maragoli, western Kenya (1900–1994). *Human Ecology* 28 (3): 383–414.
Csete, M. E., and J. C. Doyle. 2002. Reverse engineering of biological complexity. *Science* 295:1664–69.
Daily Nation. 2007a. Behind-the-scenes foreign efforts to change policies. *Daily Nation*, 24 February.
———. 2007b. New bid to lift hunting ban. *Daily Nation*, 24 February.
Datye, K. R., S. Paranjape, K. J. Joy, and S. Kulkarni. 1997. Krishna Valley Development: Emerging sustainable development options and conflict management. Unpublished paper.
Davidson, J. S. 2008. *From Rebellion to Riots: Collective Violence on Indonesian Borneo*. Madison: University of Wisconsin Press.
Davidson, J. S., and D. Kammen. 2002. Indonesia's unknown wars and the lineages of violence in West Kalimantan. *Indonesia* 73:53–87.
de Chételat, E. 1938. Le modelé latéritique de l'ouest de la Guinée française. *Revue de Géographie Physique et de Géologie Dynamique* 11 (1): 5–120.
de Jong, W., D. Donovan, and K. Abe, eds. 2007. *Extreme Conflict and Tropical Forests*. Dordrecht, Netherlands: Springer.
Delborne, J. 2008. Transgenes and transgressions: Scientific dissent as heterogeneous practice. *Social Studies of Science* 38 (4): 509–41.
Deleuze, G., and F. Guattari. 1987. *A Thousand Plateaus: Capitalism and Schizophrenia*. Minneapolis: University of Minnesota Press.
Demeritt, D. 1996. Social theory and the reconstruction of science and geography. *Transactions of Institute of British Geographers* 21:484–503.
———. 2001. The construction of global warming and the politics of science. *Annals of the Association of American Geographers* 91 (2): 307–37.
Dennis, P., and J. Grey. 1996. *Emergency and Confrontation: Australian Military Operations in Malaya and Borneo, 1950–1966*. Vol. 5 of *The Official History of Australia's Involvement in Southeast Asian Conflicts, 1948–1975*. Sydney: Allen and Unwin / Australian War Memorial.
de Swardt, A. M. J. 1964. Lateritisation and landscape development in parts of Equatorial Africa. *Zeitschrift für Geomorphologie* 8 (3): 313–33.
d'Hoore, J. 1954a. L'accumulation des sesquioxydes libres dans les sols tropicaux. *Publications de l'Institut National pour l'Étude Agronomique du Congo Belge, Série Scientifique* 62:5–132.
———. 1954b. Le facteur humain et l'accumulation de sesquioxydes libres dans les sols tropicaux. In *Deuxième Conférence Interafricaine des Sols*, 1:241–55. Brussels: Direction d'Agriculture, Forêts, et Élevages du Ministère des Colonies.
Diamond, J. M. 2005. *Collapse: How Societies Choose to Fail or Succeed*. New York: Viking.
Dich, J., S. Hoar Zahm, A. Hanberg, and H.–O. Adami. 1997. Pesticides and cancer. *Cancer Causes and Control* 8 (3): 420–43.
Diez-Gonzalez, F., T. R. Callaway, M. G. Kizoulis, and J. B. Russell. 1998. Grain feeding and the dissemination of acid-resistant *Escherichia coli* from cattle. *Science* 281:1666–68.
Dilts, D., and S. Hate. 1996. IPM farmer field schools: Changing paradigms and scaling

up. *Agricultural Research and Extension Network (AGREN) Paper*, 59b:1–4. London: Overseas Development Institute.

Donnelly, M. 1994. *Sea Turtle Mariculture: A Review of Relevant Information for Conservation and Commerce*. Washington, DC: Center for Marine Conservation.

Dove, M. R. 1992. The dialectical history of "jungle" in Pakistan. *Journal of Anthropological Research* 48 (3): 231–53.

———. 1996. Process versus product in Bornean augury: A traditional knowledge system's solution to the problem of knowing. In *Redefining Nature: Ecology, Culture and Domestication*, edited by R. Ellen and K. Fukui, 557–96. Oxford: Berg.

Dowie, M. 1995. *Losing Ground: American Environmentalism at the Close of the Twentieth Century*. Cambridge, MA: MIT Press.

Downey, G. L., and J. Dumit. 1997. Locating and intervening. In *Anthropological Interventions in Emerging Sciences and Technologies*, edited by G. L. Downey, J. Dumit, and S. Traweek, 5–30. Santa Fe, NM: SAR Press.

Doyle, M. W., D. E. Miller, and J. M. Harbor. 1999. Should river restoration be based on classification schemes or process models? Insights from the history of geomorphology. Paper read at American Society of Civil Engineers International Conference on Water Resources Engineering, Seattle, WA.

Doyle, M. W., F. D. Shields, K. F. Boyd, P. B. Skidmore, and D. W. Dominick. 2007. Channel-forming discharge selection in river restoration design. *Journal of Hydraulic Engineering* 133 (7): 831–37.

Drayton, R. 2000. *Nature's Government: Science, Imperial Britain and the "Improvement" of the World*. New Haven: Yale University Press.

Dreher, Rod. 2003. USDA-disapproved: Small farmers and big government. *National Review*, January 27.

Duvall, C. S. 2003. Symbols, not data: Rare trees and vegetation history in Mali. *Geographical Journal* 169 (4): 295–312.

Eckert, K. L., K. A. Bjorndal, F. A. Abreu-Grobois, and M. Donnelly, eds. 1999. *Research and Management Techniques for the Conservation of Sea Turtles*. Publication No. 4. Washington, DC: IUCN/SSC Marine Turtle Specialist Group.

Edwards, P. 2003. Infrastructure and modernity. In *Modernity and Technology*, edited by T. J. Misa, P. Brey, and A. Feenberg, 185–225. Cambridge, MA: MIT Press.

ELI (Environmental Law Institute). 2006. Status Report on Compensatory Mitigation in the United States. Washington, DC: ELI.

Ellen, R., and K. Fukui, eds. 1996. *Redefining Nature: Ecology, Culture and Domestication*. Oxford: Berg.

Endicott, K. 1997. Review: Violence and the dream people. *Journal of Asian Studies* 56 (1): 262–63.

England, P. C., P. Molnar, and F. M. Richter. 2007. Kelvin, Perry and the age of the earth. *American Scientist* 95 (4): 342–49.

Enticott, G. 2003. Lay immunology, local foods and rural identity: Defending unpasteurised milk in England. *Sociologia Ruralis* 43 (3): 257–70.

EPA (Environmental Protection Agency). 2007. Title 40: *Protection of Environment*. Revised July 1,

2007. http://www.access.gpo.gov/nara/cfr/waisidx_07/40cfr180 _07.html (accessed May 14, 2008).

Escobar, A. 1995. *Encountering Development: The Making and Unmaking of the Third World.* Princeton, NJ: Princeton University Press.

———. 1998. Whose knowledge, whose nature? Biodiversity, conservation, and the political ecology of social movements. *Political Ecology* 5:53–82.

———. 1999. After nature: Steps to an anti-essentialist political ecology. *Current Anthropology* 40 (1): 1–30.

———. 2001. Culture sits in places: Reflections on globalization and subaltern strategies of localization. *Political Geography* 20 (2): 139–74.

———. 2004. Actores, redes e novos produtores de conhecimento: Os movimentos sociais e a transição paradigmática nas ciências (Actor networks and new knowledge producers: Social movements and the paradigmatic transition in the sciences). In *Conhecimento prudente para uma vida decente*, edited by B. de Sousa Santos, 639–66. São Paulo: Cortez Editora. Draft English translation at http://www.unc.edu/escobar.

———. 2008. *Territories of Difference: Place, Movements, Life.* Durham, NC: Duke University Press.

Ettema, C. H. 1994. Indigenous soil classifications: What is their structure and function, and how do they compare to scientific soil classifications? University of Georgia. http://www.itc.nl/~rossiter/Docs/Misc/IntroToEthnopedology.pdf (accessed June 20, 2008).

Ezrahi, Y. 1990. *The Descent of Icarus: Science and the Transformation of Contemporary Democracy.* Cambridge, MA: Harvard University Press.

Fairhead, J., and M. Leach. 1996. *Misreading the African Landscape.* Cambridge: Cambridge University Press.

———. 1998. *Reframing Deforestation: Global Analysis and Local Realities.* London: Routledge.

Fairhead, J., and I. Scoones. 2005. Local knowledge and the social shaping of soil investments: Critical perspectives on the assessment of soil degradation in Africa. *Land Use Policy* 22:33–41.

Falconer, J. D. 1911. *The Geology and Geography of Northern Nigeria.* London: Macmillan.

FAO (Food and Agriculture Organization). 2008. Codex Alimentarius: Pesticide Residues in Food. http://www.codexalimentarius.net/mrls/pestdes/jsp/pest_q-e.jsp (accessed May 14, 2008).

Fath, B. D. 2007. Community-level relations and network mutualism. *Ecological Modelling* 208:56–67.

FDA (Food and Drug Administration). 1999. *Pesticide Analytical Manual.* Vol. 1, *Multiresidue Methods.* 3rd ed. Washington, DC: U.S. Department of Health and Human Services.

———. 2004. Pesticide program residue monitoring 2002. Washington, DC: FDA.

———. 2008. FDA Pesticide Program Residue Monitoring, 1993–2006. http://www.fda.gov/Food/FoodSafety/ FoodContaminantsAdulteration/Pesticides/ResidueMonitoringReports/default.htm (accessed August 1, 2008).

Feit, H. 2001. Hunting, nature, and metaphor: Political and discursive strategies in James Bay. In *Indigenous Traditions and Ecology: The Interbeing of Cosmology and Community*, edited by J. Grim, 411–52. Cambridge, MA: Harvard University Press.

Ferber, D. 2004. Microbes made to order. *Science* 303:158–61.
Ferguson, J. 1990. *The Anti-politics Machine: "Development," Depoliticization, and Bureaucratic Power in Lesotho*. Cambridge: Cambridge University Press.
Ferraz, H. B., P. H. F. Bertolucci, J. S. Pereira, J. G. C. Lima, and L. A. F. Andrade. 1988. Chronic exposure to the fungicide maneb may produce symptoms and signs of CNS manganese intoxication. *Neurology* 38:550–53.
Fischer, F. 2000. *Citizens, Experts, and the Environment*. Durham, NC: Duke University Press.
FitzSimmons, M., and D. Goodman. 1998. Incorporating nature: Environmental narratives and the production of food. In *The Production of Nature at the End of the Twentieth Century*, edited by N. Castree and B. Willems-Braun, 194–200. London: Routledge.
Fleming, E. H. 2001. *Swimming against the Tide: Recent Surveys of Exploitation, Trade, and Management of Marine Turtles in the Northern Caribbean*. Washington, DC: TRAFFIC (the Wildlife Trade Monitoring Network).
Fletcher, J. 2005. French cheeses fall victim to import rules. *San Francisco Chronicle*, May 4.
Forman, R. T. T. 1991. Landscape corridors: From theoretical foundations to public policy. In *Nature Conservation 2: The Role of Corridors*, edited by D. A. Saunders and R. J. Hobbs, 71–84. Chipping Norton, Australia: Surrey Beatty and Sons.
Forman, R. T. T., and M. Gordon. 1981. Patches and structural components for a landscape. *BioScience* 31:733–40.
Forsyth, T. 1996. Science, myth and knowledge: Testing Himalayan environmental degradation in northern Thailand. *Geoforum* 27 (3): 375–92.
———. 1998. Mountain myths revisited: Integrating natural and social environmental science. *Mountain Research and Development* 18:107–16.
———. 2003. *Critical Political Ecology: The Politics of Environmental Science*. London: Routledge.
———. 2004. Review: *States of Knowledge: The Co-production of Science and Social Order* edited by S. Jasanoff. *Environment and Planning A* 36:1708–9.
———. 2008. Political ecology and the epistemology of social justice. *Geoforum* 39 (2): 756–64.
Forsyth, T., and A. Walker. 2008. *Forest Guardians, Forest Destroyers: The Politics of Environmental Knowledge in Northern Thailand*. Seattle: University of Washington Press.
Foucault, M. 1977. *Discipline and Punish: The Birth of the Prison*. New York: Pantheon Books.
Freire, P. 1970. *Pedagogy of the Oppressed*. New York: Continuum.
Fujimura, J. H. 1987. Constructing "do-able" problems in cancer research: Articulating alignment. *Social Studies of Science* 17:267–93.
———. 1988. The molecular biological bandwagon in cancer research: Where social worlds meet. *Social Problems* 35 (3): 261–83.
———. 1992. Crafting science: Standardized packages, boundary objects, and "translation." In *Science as Practice and Culture*, edited by A. Pickering, 168–11. Chicago: University of Chicago Press.
———. 1996. *Crafting Science: A Sociohistory of the Quest for the Genetics of Cancer*. Cambridge, MA: Harvard University Press.
———. 2003. Future imaginaries: Genome scientists as socio-cultural entrepreneurs. In

Genetic Nature/Culture: Anthropology and Science beyond the Two Culture Divide, edited by A. Goodman, D. Heath, and S. Lindee, 176–99. Berkeley: University of California Press.

———. 2005. Postgenomic futures: Translations across the machine-nature border in systems biology. *New Genetics and Society* 24 (2): 195–225.

Fuller, S. 1994. Making agency count. *American Behavioral Scientist* 37 (6): 741–53.

———. 2000. *The Governance of Science*. Philadelphia: Open University Press.

Funtowicz, S., and J. Ravetz. 1993. Science for the post-normal age. *Futures*, September, 739–56.

Galt, R. E. 2007. Regulatory risk and farmers' caution with pesticides in Costa Rica. *Transactions of the Institute of British Geographers* 32 (3): 377–94.

———. 2009. It just goes to kill Ticos: National market regulation and the political ecology of farmers' pesticide use in Costa Rica. *Journal of Political Ecology* 16:1–33.

GAO (General Accounting Office). 1986. Better sampling and enforcement needed on imported food. Report GAO/RCED-86-219. Washington, DC: GAO.

García-Barrios, R., and L. García-Barrios. 1990. Environmental and technological degradation in peasant agriculture: A consequence of development in Mexico. *World Development* 18 (11): 1569–85.

Gareau, B. 2005. We have never been human: Agential nature, ant, and Marxist political ecology. *Capitalism Nature Socialism* 16:127–40.

Garrett, G. H. 1892. Sierra Leone and the interior to the upper waters of the Niger. *Proceedings of the Royal Geographical Society* 14 (7): 433–55.

Gautier, E.-F. 1928. *Le Sahara*. Paris: Payot.

———. 1932. Remarques sur la morphologie du Fouta Djallon. *Bulletin de l'Association de Géographes Français* 58:83–87.

Geertz, C. 1963. *Agricultural Involution: The Processes of Ecological Change in Indonesia*. Berkeley: University of California Press.

Geilfus, F. 1995. *From Tree-Haters to Tree-Farmers: Promoting Farm Forestry in the Dominican Republic*. London: Overseas Development Institute.

Gerchman, Y., and R. Weiss. 2004. Teaching bacteria a new language. *Proceedings of the National Academy of Sciences USA* 101 (8): 2221–22.

Gibson, C. C., E. Ostrom, and T. K. Ahn. 2000. The concept of scale and the human dimensions of global change. *Ecological Economics* 32:217–39.

Gichohi, H. 2000. Functional relationships between parks and agricultural areas in East Africa: The case of Nairobi National Park. In *Wildlife Conservation by Sustainable Use*, edited by H. H. T. Prins, J. G. Grootenhuis, and T. Dolan, 141–67. Boston: Kluwer Academic.

Gieryn, T. F. 1995. Boundaries of science. In *Handbook of Science and Technology Studies*, edited by S. Jasanoff, G. E. Markle, J. C. Petersen, and T. Pinch, 393–443. London: Sage.

———. 1999. *Cultural Boundaries of Science: Credibility on the Line*. Chicago: University of Chicago Press.

Gilbert, S. F., and S. Sarkar. 2000. Embracing complexity: Organicism for the 21st century. *Developmental Dynamics* 219:1–9.

Gillilan, S. 1996. Use and misuse of channel classification schemes. *Stream Notes*, October, 2–3.

Glacken, C. J. 1967. *Traces on the Rhodian Shore*. Berkeley: University of California Press.
Godfrey, M. H., J. W. Grahame, S. Webb, C. Manolis, and N. Mrosovsky. 2007. Hawksbill sea turtles: Can phylogenetics inform harvesting? *Molecular Ecology* 16 (17): 3511–13.
Godley, B. J., J. M. Blumenthal, A. C. Broderick, M. S. Coyne, M. H. Godfrey, L. A. Hawkes, and M. J. Witt. 2008. Satellite tracking of sea turtles: Where have we been and where do we go next? *Endangered Species Research* 4:3–22.
Goldman, M. 2003. Partitioned nature, privileged knowledge: Community based conservation in Tanzania. *Development and Change* 34 (5): 833–62.
———. 2005. *Imperial Nature: The World Bank and Struggles for Justice in the Age of Globalization*. New Haven: Yale University Press.
———. 2007. Tracking wildebeest, locating knowledge: Maasai and conservation biology understandings of wildebeest behavior in northern Tanzania. *Environment and Planning D: Society and Space* 25:307–31.
———. 2009. Constructing connectivity? Conservation corridors and conservation politics in East African rangelands. *Annals of the Association of American Geographers* 9 (2): 335–59.
Gottlieb, R. 1993. Reconstructing environmentalism: Complex movements, diverse roots. *Environmental History Review* 17 (4): 1–19.
Goudie, A. S. 1973. *Duricrusts in Tropical and Subtropical Landscapes*. Oxford: Clarendon.
Grandstaff, T. 1980. *Shifting Cultivation in Northern Thailand: Possibilities for Development*. UNU Resource Systems Theory and Methodology Series no. 3. Tokyo: United Nations University.
Gray, N. J., and L. M. Campbell. 2009. Science, policy advocacy, and marine protected areas. *Conservation Biology* 23 (2): 460–68.
Green, P. 2007. The year without toilet paper. *New York Times*, March 22, section F.
Grove, R. H. 1996. *Green Imperialism*. Cambridge: Cambridge University Press.
Guha, R. 2000. *Environmentalism: A Global History*. Oxford: Oxford University Press.
Gupta, Akhil. 1998. *Postcolonial Developments: Agriculture and the Making of Modern India*. Durham, NC: Duke University Press.
Gupta, Anil. 1997. The Honey Bee Network: Linking knowledge-rich grassroots innovations. *Development* 40 (4): 36-40.
Guston, D. H. 2001. Boundary organizations in environmental policy and science: An introduction. *Science Technology and Human Values* 26 (4): 399–408.
Hackett, E., O. Amsterdamska, M. Lynch, and J. Wajcman, eds. 2008. *The Handbook of Science and Technology Studies*. 3rd ed. Cambridge, MA: MIT Press.
Haila, Y., and C. Dyke, eds. 2006. *How Nature Speaks: The Dynamics of the Human Ecological Condition*. Durham, NC: Duke University Press.
Hajer, M. 1995. *The Politics of Environmental Discourse*. Oxford: Clarendon.
Hallowell, I. 1960. Ojibwa ontology, behavior, and worldview. In *Culture in History: Essays in Honor of Paul Radin*, edited by S. Diamond, 19–52. New York: Columbia University Press.
Hallsworth, E. 1987. *Anatomy, Physiology, and Psychology of Erosion*. New York: Wiley.
Hamilton, D. J., P. T. Holland, B. Ohlin, W. J. Murray, A. Ambrus, G. C. De Baptista, and J. Kovacicová. 1997. Optimum use of available residue data in the estimation of dietary intake of pesticides. *Pure and Applied Chemistry* 69 (9): 1373–1410.

Hamilton, L. S., and A. J. Pearce. 1988. Soil and water impacts of deforestation. In *Deforestation: Social Dynamics in Watershed and Mountain Ecosystems*, edited by J. Ives and D. Pitt, 75–98. London: Routledge.

Hammond, D. R. 1997. Toward a science of synthesis: The heritage of general systems theory. Ph.D. dissertation, University of California–Berkeley.

Handelsman, J., and K. Smalla. 2003 Conversations with the silent majority. *Current Opinion in Microbiology* 6:271–73.

Haraway, D. J. 1981–82. The high cost of information in post–World War II evolutionary biology: Ergonomics, semiotics, and the sociobiology of communication systems. *Philosophical Forum* 13 (2–3): 206–37.

———. 1988. Situated knowledges: The science question in feminism as a site of discourse on the privilege of partial perspective. *Feminist Studies* 14 (3): 575–99.

———. 1989. *Primate Visions: Gender, Race, and Nature in the World of Modern Science*. New York: Routledge.

———. 1991. *Simians, Cyborgs and Women: The Reinvention of Nature*. New York: Routledge.

———. 1996. Modest witness: Feminist diffractions in science studies. In *The Disunity of Science: Boundaries, Contexts and Power*, edited by P. Galison and D. Stump, 429–39. Stanford, CA: Stanford University Press.

———. 1997. *Modest_Witness@Second_Millennium*. New York: Routledge.

Harder, B. 2005. Joining the Resistance. *Science News*, January 1, 5.

Harder, E. C. 1952. Examples of bauxite deposits illustrating variations in origin. In *Problems of Clay and Laterite Genesis*, edited by American Institute of Mining and Metallurgical Engineers, 35–64. New York: American Institute of Mining and Metallurgical Engineers.

Harding, S. 1986. *The Science Question in Feminism*. Ithaca, NY: Cornell University Press.

———. 1988. *Is Science Multicultural? Postcolonialisms, Feminisms, and Epistemologies*. Bloomington: Indiana University Press.

———. 2003. A world of sciences. In *From Science and Other Cultures: Issues in Philosophies of Science and Technology*, edited by R. Figueroa and S. Harding, 49–69. New York: Routledge.

Hardy, F. 1935. Some aspects of tropical soils. In *Transactions of the Third International Congress of Soil Science*, 2:150–63. London: Thomas Murby and Co.

Harper, D., and S. El-Swaify. 1988. Sustainable agricultural development in north Thailand: Conservation as a component of success in assistance projects. In *Conservation Farming on Steep Slopes*, edited by W. Moldenhauer and N. Hudson. Ankeny, IA: Soil and Water Conservation Society of America.

Harper, T. N. 1997. The politics of the forest in colonial Malaya. *Modern Asian Studies* 31 (1): 1–29.

Harré, R. 1993. *Laws of Nature*. London: Duckworth.

Harris, L. D., and P. B. Gallagher. 1989. New initiatives for wildlife conservation: The need for movement corridors. In *Preserving Communities and Corridors*, edited by G. Mackintosh, 117–34. Washington, DC: Defenders of Wildlife.

Harris, L. M., and H. D. Hazen. 2006. Power of maps: (Counter) mapping for conservation. *Acme: An International E-Journal for Critical Geographies* 4 (1): 99–130.

Hayden, C. 2003. *When Nature Goes Public: The Making and Unmaking of Bioprospecting in Mexico.* Princeton, NJ: Princeton University Press.

Hayes, R. 1992. *An Experimental Design to Test Wolf Regulation of Ungulates in the Aishihik Area, Southwest Yukon.* Whitehorse, YT: Fish and Wildlife Branch, Yukon Territorial Government.

———. 1996. Numerical and functional responses of wolves and regulation of moose in the Yukon. Master's thesis, Simon Fraser University.

Hayles, N. K. 1999. *How We Became Posthuman: Virtual Bodies in Cybernetics, Literature, and Informatics.* Chicago: University of Chicago Press.

Heppell, S. S., D. T. Crouse, L. B. Crowder, S. P. Epperly, W. Gabriel, T. Henwood, R. Marquez, and N. B. Thompson. 2005. A population model to estimate recovery time, population size, and management impacts on Kemp's ridley sea turtles. *Chelonian Conservation and Biology* 4:767–73.

Hess, G. R., and R. A. Fischer. 2001. Communicating clearly about conservation corridors. *Landscape and Urban Planning* 55 (3): 195–208.

Hesse, M. 1980. *Revolutions and Reconstructions in the Philosophy of Science.* Bloomington: Indiana University Press.

Heynen, N., J. McCarthy, S. Prudham, and P. Robbins, eds. 2007. *Neoliberal Environments: False Promises and Unnatural Consequences.* London: Routledge.

Hilty, J. A., W. Z. Lidicker, A. M. Merenlender, and A. P. Dobson, eds. 2006. *Corridor Ecology: The Science and Practice of Linking Landscapes for Biodiversity Conservation.* Washington, DC: Island Press.

Hindery, D. 2004. Social and environmental impacts of World Bank/IMF–funded economic restructuring in Bolivia: An analysis of ENRON and Shell's hydrocarbons projects. *Singapore Journal of Tropical Geography* 25 (3): 281–303.

Hobbs, R. J. 1992. The role of corridors in conservation: Solution or bandwagon? *Trends in Ecology and Evolution* 7 (11): 389–92.

Holland, P. T., D. Hamilton, B. Ohlin, and M. W. Skidmore. 1994. Effects of storage and processing on pesticide residues in plant products. *Pure and Applied Chemistry* 66 (2): 335–56.

Holland, T. H. 1903. On the constitution, origin and dehydration of laterite. *Geological Magazine* 4 (10): 59–69.

Homer-Dixon, T. F. 1993. *Environmental Scarcity and Global Security.* New York: Foreign Policy Association.

Hood, L. 2002. *My Life and Adventures Integrating Biology and Technology.* A commemorative lecture for the 2002 Kyoto Prize in Advanced Technologies. Institute of Systems Biology. http://www.systemsbiology.org.

Hoogendam, P., and G. Vargas. 1999a. La compleja relación entre municipalización, governabilidad, y la gestión de agua. In *Aguas y Municipios*, edited by P. Hoogendam, 21–35. Cochabamba, Bolivia: PEIRAV / Plural Editores.

———. 1999b. Continuidades y discontinuidades en la gestión de agua después de la implementación del nuevo marco legal. In *Aguas y Municipios*, edited by P. Hoogendam, 151–61. Cochabamba, Bolivia: PEIRAV / Plural Editores.

Horowitz, M. 1991. Victims upstream and down. *Journal of Refugee Studies* 4 (2):164–81.

Houeto, P., G. Bindoula, and J. R. Hoffman. 1995. Ethylenebisdithiocarbamates and ethylenethiourea: Possible human health hazards. *Environmental Health Perspectives* 103 (6): 568–73.

Hough, P. 1998. *The Global Politics of Pesticides: Forging Consensus from Conflicting Interests*. London: Earthscan.

Hubert, H. 1920. Le deséchement progressif en Afrique occidentale. *Bulletin du Comité d'Études de l'Afrique Occidentale Française*, 401–67.

Hughes, T. 1983. *Networks of Power: Electrification of Western Society*. Baltimore: Johns Hopkins University Press.

Humbert, S., M. Margni, R. Charles, O. M. Torres Salazar, A. L. Quirós, and O. Jolliet. 2007. Toxicity assessment of the main pesticides used in Costa Rica. *Agriculture, Ecosystems and Environment* 118:183–90.

Hutton, J., W. M. Adams, and J. C. Murombedzi. 2005. Back to the barriers? Changing narratives in biodiversity conservation. *Forum for Development Studies* 2:341–70.

Ideker, T., T. Gaitski, and L. Hood. 2001. A new approach to decoding life: Systems biology. *Annual Review of Genomics and Human Genetics* 2:343–72.

Igoe, J., and D. Brockington. 1999. *Pastoral Land Tenure and Community Conservation: A Case Study from North-east Tanzania*. Pastoral Land Tenure Series. London: International Institute for Environment and Development.

Ingold, T. 1987. *The Appropriation of Nature: Essays on Human Ecology and Social Relations*. Iowa City: University of Iowa Press.

———. 2000. *The Perception of the Environment: Essays on Livelihood, Dwelling and Skill*. London: Routledge.

Ingram, M. 2007a. Biology and beyond: The science of back to nature farming in the U.S. *Annals of the Association of American Geographers* 97 (2): 298–312.

———. 2007b. Disciplining microbes in the implementation of U.S. federal organic standards. *Environment and Planning A* 39:2866–82.

IOM (Institute of Medicine). 2006. *Ending the War Metaphor: The Changing Agenda for Unraveling the Host-Microbe Relationship Workshop Summary*. Washington, DC: National Academies Press.

Irwin, A. 1995. *Citizen Science: A Study of People, Expertise and Sustainable Development*. New York: Routledge.

Irwin, A., and B. Wynne, eds. 1996. *Misunderstanding Science? The Public Reconstruction of Science and Technology*. Cambridge: Cambridge University Press.

IUCN/SSC MTSG. 1969. *Marine Turtles: Proceedings of the Working Meeting of Marine Turtle Specialists*. Vol. Supplementary Paper no. 20, IUCN Publications New Series. Morges, Switzerland: IUCN.

Ivakhiv, A. 2001. Re-animations: Instinct and civility after the ends of "Man" and "Nature." In *From Virgin Land to Disney World: Nature and Its Discontents in the USA of Yesterday and Today*, edited by B. Herzogenrath. *Critical Studies* 15:7–32

Jaeger, P. 1956. Contribution à l'étude des forêts reliques du Soudan occidental. *Bulletin de l'Institute Française d'Afrique Noire* 18A (4): 993–1053.

Jaffry, A., M. Rangararajan, B. Ekbal, and K. P. Kannan. 1983. Towards a people's science movement. *Economic and Political Weekly* 18 (11): 372–76.

Jama, B., C. A. Palm, R. J. Buresh, A. Niang, C. Gachengo, G. Nziguheba, and B. Amadalo. 2000. *Tithonia diversifolia* as a green manure for soil fertility improvement in western Kenya: A review. *Agroforestry Systems* 49:201–21.

Jantawat, S., ed. 1987. *Proceedings of the International Workshop on Soil Erosion and Its Counter Measures*. Bangkok: Chuan Press.

Jasanoff, S. 1987. Contested boundaries in policy-relevant science. *Social Studies of Science* 17 (2): 195–230.

———. 2003. Technologies of humility: Citizen participation in governing science. *Minerva* 41 (3): 223–44.

———. 2004a. Ordering knowledge, ordering society. In *States of Knowledge*, edited by S. Jasanoff, 13–45. London: Routledge.

———, ed. 2004b. *States of Knowledge: The Co-production of Science and the Social Order*. London: Routledge.

Jasanoff, S., G. E. Markle, J. C. Petersen, and T. Pinch, eds. 1995. *Handbook of Science and Technology Studies*. Thousand Oaks, CA: Sage.

Jasanoff, S., and M. L. Martello, eds. 2004. *Earthly Politics: Local and Global in Environmental Governance, Politics, Science, and the Environment*. Cambridge, MA: MIT Press.

Jeanrenaud, S. 2002, Changing people/nature representations in international conservation discourses. *IDS [Institute for Development Studies] Bulletin* 33:111–22.

Jickling, B. 1994. Does Ostashek have moral authority to govern? *Yukon News*, February 23.

Jordan, W. R. III. 2000. Restoration, community, and wilderness. In *Restoring Nature: Perspectives from the Social Sciences and Humanities*, edited by P. H. Gobster and R. B. Hull, 23–36. Washington, DC: Island Press.

Jungers, M. E. 1949. Séance inaugurale: Discours de M. E. Jungers, Gouverneur Général. *Bulletin Agricole du Congo Belge* 40 (1): 27–30.

Juracek, K. E., and F. A. Fitzpatrick. 2003. Limitations and implications of stream classification. *Journal of the American Water Resources Association* 39 (3): 659–70.

Kareiva, P. 2006. Introduction: Evaluating and quantifying the conservation dividends of connectivity. In *Connectivity Conservation*, edited by K. R. Crooks and M. Sanjayan, 293–95. Cambridge: Cambridge University Press.

Kauffman, S. 2000. *Investigations*. Oxford: Oxford University Press.

Kay, L. 2000. *Who Wrote the Book of Life? A History of the Genetic Code*. Palo Alto, CA: Stanford University Press.

Keil, R. 2003. Progress report: Political ecology. *Urban Geography* 7 (8): 723–38.

Keller, E. F. 1999. The gender/science system; or, Is sex to gender as nature is to science? In *The Science Studies Reader*, edited by M. Biagioli, 234–42. New York: Routledge.

———. 2000. *The Century of the Gene*. Cambridge, MA: Harvard University Press.

———. 2002. *Making Sense of Life: Explaining Biological Development with Models, Metaphors, and Machines*. Cambridge, MA: Harvard University Press.

Kenya. 2007. The Wildlife (Conservation and Management) Bill. Ministry of Tourism and Wildlife, Nairobi.

Kilpelainen, M., E. O. Terho, H. Helenius, and M. Koskenvuo. 2000. Farm environment in childhood prevents the development of allergies. *Clinical and Experimental Allergy* 30: 201–8.

Kim, H., J. Boedicker, J. Choi, and R. Ismagilov. 2008. Defined spatial structure stabilizes a synthetic multispecies bacterial community. *Proceedings of the National Academy of Sciences USA* 105 (47): 18075–76.

Kimball, J. 2005. Agroforestry options for ecological and economic recovery of degraded lands on the forest-savanna ecosystem of Guinea, West Africa. In *Moving Agroforestry into the Mainstream: The 9th North American Agroforestry Conference Proceedings*, edited by K. N. Brooks and P. F. Ffolliott. CD-ROM. St. Paul, MN: Department of Forest Resources, University of Minnesota.

Kinchy, A. J., and D. L. Kleinman. 2003. Organizing credibility: Discursive and organizational orthodoxy on the border of ecology and politics. *Social Studies of Science* 33 (6): 869–96.

Kinyungu, C., A. Kisia, and P. Mwakio. 2006. Millions at stake in park venture. *Standard*, December 28.

Kirsch, S., and D. Mitchell. 2004. The nature of things: Dead labor, nonhuman actors, and the persistence of Marxism. *Antipode* 36:687–705.

Kitano, H. 2001. *Foundations of Systems Biology: An Overview of the Methodologies and Techniques of the Emerging Field of Systems Biology*. Cambridge, MA: MIT Press.

———. 2002a. Computational systems biology. *Nature* 420:206–10.

———. 2002b. Systems biology: A brief overview. *Science* 295:1662–64.

———. 2003. Tumour tactics. *Nature* 426:125.

———. 2004a. Biological robustness. *Nature Reviews Genetics* 5:826–37.

———. 2004b. Cancer as a robust system: Implications for anticancer therapy. *Nature Reviews Cancer* 4:227–35.

Klingensmith, D. 2007. *One Valley and a Thousand: Dams, Nationalism and Development*. New Delhi: Oxford University Press.

Knoll, L. 2005. Origins of the regulation of raw milk cheeses in the United States. Paper written for Harvard Law School. Legal Electronic Document Archive. http://leda.law.harvard.edu/leda/data/702/Knoll05.pdf (accessed April 2007).

Knorr-Cetina, K. 1999. *Epistemic Cultures: How the Sciences Make Knowledge*. Cambridge, MA: Harvard University Press.

Kondolf, G. M. 1995. Geomorphological stream channel classification in aquatic habitat restoration: Uses and limitations. *Aquatic Conservation: Marine and Freshwater Ecosystems* 5:127–41.

———. 2007. River restoration in North America: Meandering channels for all? Paper presented at Association of American Geographers, San Francisco, March 19.

Kondolf, G. M., M. W. Smeltzer, and S. Railsback. 2001. Design and performance of a channel reconstruction project in a coastal California gravel-bed stream. *Environmental Management* 28 (6): 761–76.

Kosek, J. 2006. *Understories: The Political Life of Forests in Northern New Mexico*. Durham, NC: Duke University Press.

Krausman, P. R. 2002. *Introduction to Wildlife Management: The Basics*. Upper Saddle River, NJ: Prentice Hall.
Kuhn, T. 1970. *The Structure of Scientific Revolutions*. 2nd ed. Chicago: University of Chicago Press.
Kumar, K. 1984. People's science and development theory. *Economic and Political Weekly* 19 (28): 1082–84.
Kunstadter, P., E. C. Chapman, and S. Sabhasri. 1978. *Farmers in the Forest: Economic Development and Marginal Agriculture in Northern Thailand*. Honolulu: University of Hawaii Press.
KWS-TWF-KILA (Kenyan Wildlife Service, The Wildlife Foundation, and Kitengela-Ilparakuo Land Owners Association). 2006. Proposed pastoralism and wildlife conservation zone for Isinya-Kitengela Master Plan, 2006–2026. Draft.
Lackey, R. T. 2007. Science, scientists, and policy advocacy. *Conservation Biology* 21 (1): 12–17.
Lacroix, A. 1913. Les latérites de la Guinée et les produits d'altération qui leur sont associés. *Nouvelles Archives du Muséum d'Histoire Naturelle* 5:255–358.
Lahache, M. J. 1907. Le desséchement de l'Afrique française: Est-il démontré? *Société de Géographie et d'Études Coloniales de Marseilles* 31:149–85.
Lakoff, G., and M. Johnson. 1980. *Metaphors We Live By*. Chicago: University of Chicago Press.
Langston, N. 1996. *Forest Dreams and Forest Nightmares: The Paradox of Old Growth in the Inland West*. Seattle: University of Washington Press.
Latour, B. 1987. *Science in Action: How to Follow Scientists and Engineers through Society*. Cambridge, MA: Harvard University Press.
———. 1988. *The Pasteurization of France*. Cambridge, MA: Harvard University Press.
———. 1993. *We Have Never Been Modern*. Cambridge, MA: Harvard University Press.
———. 1994. On technical mediation: Philosophy, sociology, genealogy. *Common Knowledge* 3 (2): 29–64.
———. 1996. *Aramis or the Love of Technology*. Cambridge, MA: Harvard University Press.
———. 2004. *Politics of Nature: How to Bring the Sciences into Democracy*. Cambridge, MA: Harvard University Press.
———. 2005. *Re-assembling the Social: An Introduction to Actor Network Theory*. Oxford: Oxford University Press.
Latour, B., and S. Woolgar 1979. *Laboratory Life: The Construction of Scientific Facts*. Princeton, NJ: Princeton University Press.
Laurie, N., and S. Marvin. 1999. Globalization, neoliberalism, and negotiated development in the Andes: Water projects and regional identity in Cochabamba, Bolivia. *Environment and Planning A* 31:1401–15.
Lave, R. 2008. The Rosgen wars and the shifting construction of scientific expertise. Ph.D. dissertation, University of California–Berkeley.
———. 2009. The controversy over Natural Channel Design: Substantive explanations and potential avenues for resolution. *Journal of the American Water Resources Association* 45 (6): 1519–32.
Law, J. 1987. Technology and heterogeneous engineering: The case of Portugese expansion. In *The Social Construction of Technological Systems: New Directions in the Sociology and History of*

Technology, edited by W. E. Bijker, T. P. Hughes, and T. J. Pinch, 111–34. Cambridge, MA: MIT Press.

———. 1991. Introduction: Monsters, machines and sociotechnical relations. In *A Sociology of Monsters: Essays on Power, Technology and Domination*, edited by J. Law, 1–25. New York: Routledge.

———. 2004. *After Method: Mess in Social Science Research*. New York: Routledge.

Law, J., and J. Hassard, eds. 1999. *Actor Network Theory and After*. Oxford: Blackwell.

Law, J., and A. Mol, eds. 2002. *Complexities: Social Studies of Knowledge Practices*. Durham, NC: Duke University Press.

Leach, M., and R. Mearns, eds. 1996. *The Lie of the Land: Challenging Received Wisdom on the African Environment*. Oxford: James Currey; Portsmouth, NH: Heinemann.

Leary, J. D. 1995. *Violence and the Dream People: The Orang Asli in the Malay Emergency*. Athens: Centre for International Studies, Ohio University.

Leopold, A. 1933. *Game Management*. New York: Scribner's.

———. 1949. *A Sand County Almanac, and Sketches Here and There*. New York: Oxford University Press.

Lernoux, P. 1980. *Cry of the People*. New York: Doubleday.

Leroy, J.-F. 1983. André Aubréville (1897–1982). *Bulletin du Muséum d'Histoire Naturelle, B (Adansonia)* 5 (2): 123–40.

Levins, R., and R. Lewontin. 1985. *The Dialectical Biologist*. Cambridge, MA: Harvard University Press.

Levy, S. 2002. Factors impacting on the problem of antibiotic resistance. *Journal of Antimicrobial Chemotherapy* 49:25–30.

Li, T. M. 2007. *The Will to Improve: Governmentality, Development, and the Practice of Politics*. Durham, NC: Duke University Press.

Lochhead, C. 2009. Crops, ponds destroyed in quest for food safety. *San Francisco Chronicle* Washington Bureau, July 13, A1. http://sfgate.com/cgi-bin/article.cgi?f=/c/a/2009/07/13/MN0218DVJ8.DTL.

Locke, J. 1947. *Two Treatises of Government*. New York: Hafner. Original edition 1690.

Lorimer, J. 2007. Nonhuman charisma. *Environment and Planning D: Society and Space* 25: 911–32.

Lovejoy, T. E., and L. Hannah. 2005. *Climate Change and Biodiversity*. New Haven: Yale University Press.

Lowe, C. 2006. *Wild Profussion: Biodiversity Conservation in an Indonesian Archipelago*. Princeton, NJ: Princeton University Press.

Lulka, D. 2006. Grass or grain? Assessing the nature of the US bison industry. *Sociologica Ruralis* 46 (3): 173–91.

Lynch, B. D. 1996. Marking territory and mapping development: Protected area designation in the Dominican Republic. Paper presented to the International Association for the Study of Common Property, University of California–Berkeley, June 5–8.

Lynch, M. 1982. Technical work and critical inquiry: Investigations in a scientific laboratory. *Social Studies of Science* 12:499–533.

———. 1985. *Art and Artifact in Laboratory Science: A Study of Shop Work and Shop Talk in a Research Laboratory*. London: Routledge and Kegan Paul.
MacArthur, R. H., and E. O. Wilson. 1967. *The Theory of Island Biogeography*. Princeton, NJ: Princeton University Press.
Mackie, J. 1970. *Konfrontasi! The Indonesia-Malaysia Dispute, 1963–1966*. Kuala Lumpur: Oxford University Press for Australian Institute of International Affairs.
Maignien, R. 1958. Le cuirassement des sols en Guinée, Afrique occidentale. *Mémoires du Service Cartographique et Géologique d'Alsace-Lorraine* 16:1–239.
———. 1966. *Review of Research on Laterites*. Paris: UNESCO.
Manu S., and A. Somrang. 1980. *Soil Erosion in Thailand*. Bangkok: Land Development Department.
Marks, T. A. 1996. *Maoist Insurgency since Vietnam*. Ilford, Essex, UK: Frank Cass.
Martin, E. 1987. *The Woman in the Body: A Cultural Analysis of Reproduction*. Boston: Beacon Press.
———. 1991. The egg and the sperm: How science has constructed a romance based on male and female roles. *Signs* 16 (3): 485–501.
———. 1994. *Flexible Bodies: Tracking Immunity in American Culture from the Days of Polio to the Age of AIDS*. Boston: Beacon Press.
———. 1998. Fluid bodies, managed nature. In *Remaking Reality: Nature at the Millennium*, edited by B. Braun and N. Castree, 64–83. London: Routledge.
Martin, G. 2006. Farms may cut habitat renewal over E. coli fears. *San Francisco Chronicle*, December 19.
Martin, J. N., and T. K. Nakayama. 1999. Thinking dialectically about culture and communication. *Communication Theory* 9 (1): 1–25.
Massey, D. 1994. *Space, Place and Gender*. Minneapolis: University of Minnesota Press.
Mast (Maritime Studies). 2005. *Marine turtles as flagships*. Special double issue, 3 (2) and 4 (1).
Matondo, J. I., G. Peter, and K. M. Msibi. 2005. Managing water under climate change for peace and prosperity in Swaziland. *Physics and Chemistry of the Earth* 30 (11–16): 943–49.
Mbaria, J. 2006. Battle for park rages on animal migration corridors. *East African*, December 20.
McAfee, K. 1999. Selling nature to save it? Biodiversity and green developmentalism. *Environment and Planning D: Society and Space* 17:133–54.
McCallan, S. E. A. 1930. Studies on fungicides: 2, Testing protective fungicides in the laboratory. *Cornell Agricultural Experimental Station Memoirs* 128:8–24.
McCandless, R. 1985. *Yukon Wildlife: A Social History*. Edmonton: University of Alberta Press.
McCarthy, J. 2002. First world political ecology: Lessons from the wise use movement. *Environment and Planning A* 34:1281–1302.
McClellan, C. 1975. *My Old People Say: An Ethnographic Survey of Southern Yukon Territory*. 2 vols. Ottawa: National Museum of Man.
McConnell, R., and A. J. Hruska. 1993. An epidemic of pesticide poisoning in Nicaragua: Implications of prevention in developing countries. *American Journal of Public Health* 83 (11): 1559–62.

McGee, H. 2004. *Food & Cooking: An Encyclopedia of Kitchen Science, History and Culture*. London: Hodder and Stroughton.
Mead, P. S., L. Slutsker, V. Dietz , L. F. McCaig, J. S. Bresee, C. Shapiro, P. M. Griffin, and R. V. Tauxe. 1999. Food-related illness and death in the United States. *Emerging Infectious Diseases* 5 (6): 840–42.
Meentemeyer, V. 1989. Geographical perspectives of space, time, and scale. *Landscape Ecology* 3 (3/4): 163–73.
Mehta, L. 2007. Whose scarcity? Whose property? The case of water in western India. *Land Use Policy* 24 (4): 654–63.
Meletis, Z., and L. M. Campbell. 2007. Call it consumption! Re-conceptualizing ecotourism as consumption and consumptive. *Geography Compass* 1 (4): 850–70.
Mellon, M., C. Benbrook, and K. L.Benbrook. 2001. *Hogging It: Estimates of Antimicrobial Abuse in Livestock*. Cambridge, MA: Union of Concerned Scientists. http://www.ucsusa.org/food_and_agriculture/science_and_impacts/impacts_industrial_agriculture/hogging-it-estimates-of.html (accessed March 2010).
Merchant, C. 1980. *The Death of Nature: Women, Ecology, and the Scientific Revolution: A Feminist Reappraisal of the Scientific Revolution*. San Francisco: Harper and Row.
———. 1989. *Ecological Revolutions: Nature, Gender, and Science in New England*. Chapel Hill: University of North Carolina Press.
———. 1992. *Radical Ecology: The Search for a Livable World*. New York: Routledge.
Meyer, W. B., D. Gregory, B. L. Turner II, and P. F. McDowell. 1992. The local-global continuum. In *Geography's Inner Worlds*, edited by R. F. Abler, M. G. Marcus, and J. M. Olson, 255–79. New Brunswick, NJ: Rutgers University Press.
Meylan, A. 1998. Hawksbill turtles still endangered. *Nature* 391:117.
Millar, D. 1993. Farmer experimentation and the cosmovision paradigm. In *Cultivating Knowledge*, edited by W. K. de Boef, K. Amanor, and K. Wellard, 44–50. London: Intermediate Technology Publications.
Miller, J. R., and J. B. Ritter. 1996. An examination of the Rosgen classification of natural rivers. *Catena* 27:295–99.
Miller, P., and N. Rose. 1990. Governing economic life. *Economy and Society* 19 (1): 1–31.
Millstone, E. 1996. Food safety: The ethical dimensions. In *Food Ethics*, edited by B. Mepham, 84–100. New York: Routledge.
Mindell, D., J. Segal, and S. Gerovitch. 2003. Cybernetics and information theory in the United States, France and the Soviet Union. In *Science and Ideology: A Comparative History*, edited by Mark Walker, 66–95. London: Routledge.
Ministère de l'Environnement et de l'Eau. 2000. Burkina Faso: Programme d'action national de lutte contre la désertification. United Nations Committee to Combat Desertification Web site. http:/www.unccd.int/actionprogrammes/africa/nationa l/2000/burkina_faso-fre.pdf (accessed July 8, 2009).
Misiko, M. T. 2007. Fertile ground? Soil fertility management and the African smallholder. Ph.D. thesis, Wageningen University.
Misiko, M. T., P. Tittonell, J. J. Ramisch, P. Richards, and K. E. Giller. 2008. Integrating

new soybean varieties for soil fertility management in smallholder systems through participatory research: Lessons from western Kenya. *Agricultural Systems* 97:1–12.

Mission Tilho. 1910–11. *Documents scientifiques de la Mission Tilho, 1906–1909*. Paris: Imprimérie Nationale.

Mitchell, T. 2002. *Rule of Experts: Egypt, Techno-politics, Modernity*. Berkeley: University of California Press.

MKVDC (Maharashtra Krishna Valley Development Corporation). 2001. *Chikotra Medium Irrigation Project Report*. Kolhapur, India: Maharashtra Krishna Valley Development Corporation, Kolhapur Irrigation Circle, Medium Projects Division. In Marathi.

MLHUD (Ministry of Lands, Housing and Urban Development). 1996. *The National Land Policy*. Dar es Salaam: Government Printer.

MNRT (Ministry of Natural Resources and Tourism). 1998. *The Wildlife Policy of Tanzania*. Dar es Salaam: Government Printer.

Mol, A. 1999 Ontological politics: A word and some questions. In *Actor Network Theory and After*, edited by J. Law and J. Hassard. Oxford: Blackwell.

———. 2002. *The Body Multiple: Ontology in Medical Practice*. Durham, NC: Duke University Press.

Mondry, Z., G. Melia, and M. Haupt. 2006. Stability of Engineered Stream Structures in North Carolina Restoration Projects. North Carolina Ecosystem Enhancement Program. http://www.bae.ncsu.edu/programs/extension/wqg/sri/2006conference/presentations.html (accessed April 28, 2008).

Montes de Oca, I. 1989. *Geografía y recursos naturals de Bolivia*. La Paz: Editorial Educacional del Ministerio de Educación y Cultura.

———. 1992. *Sistemas de riego y agricultura en Bolivia*. La Paz: Ministerio de Asuntos Campesinos y Agropecuarios (MACA), Comité Interinstitucional de Riego.

Moore, D. S. 1993. Contesting terrain in Zimbabwe's eastern highlands: Political ecology, ethnography, and peasant resource struggles. *Economic Geography* 69:380–401.

———. 2005. *Suffering for Territory: Race, Place, and Power in Zimbabwe*. Durham, NC: Duke University Press.

Moore, M. 1989. The fruits and fallacies of neoliberalism: The case of irrigation policy. *World Development* 17:1733–50.

Mortimer, J. A., P. A. Meylan, and M. Donnelly. 2007. Whose turtles are they, anyway? *Molecular Ecology* 16:17–18.

Morus, I. R., ed. 2002. *Bodies/Machines*. Oxford: Berg.

Mosse, D. 2003. *The Rule of Water: Statecraft, Ecology and Collective Action in South India*. New Delhi: Oxford University Press.

Mrosovsky, N. 1997. IUCN's credibility threatened. *Nature* 389:436.

———. 2003. Predicting extinction: Fundamental flaws in IUCN's Red List system, exemplified by the case of sea turtles. http://members.seaturtle.org/mrosovsky/.

MTSG (Marine Turtle Specialist Group). 1995. *A Global Strategy for the Conservation of Marine Turtles*. Gland, Switzerland: IUCN.

Murdoch, J., and J. Clark. 1994. Sustainable knowledge. *Geoforum* 25 (2): 115–32.

Muscat, R. J. 1990. *Thailand and the United States: Development, Security, and Foreign Aid*. New York: Columbia University Press.

Nadasdy, P. 1999. The Politics of TEK: Power and the integration of knowledge. *Arctic Anthropology* 36 (1–2): 1–18.

———. 2003. *Hunters and Bureaucrats: Power, Knowledge and Aboriginal-State Relations in the Southwest Yukon*. Vancouver: University of British Columbia Press.

———. 2005. Transcending the debate over the ecologically noble Indian: Indigenous peoples and environmentalism. *Ethnohistory* 52 (2): 291–331.

———. 2007a. Adaptive co-management and the gospel of resilience. In *Adaptive Co-management: Collaboration, Learning, and Multilevel Governance*, edited by D. Armitage, F. Berkes, and N. Doubleday, 208–27. Vancouver: University of British Columbia Press.

———. 2007b. The gift in the animal: The ontology of hunting and human-animal sociality. *American Ethnologist* 34 (1): 25–43.

———. 2008. Wildlife as renewable resource: Competing conceptions of wildlife, time, and management in the Yukon. In *Timely Assets: The Politics of Resources and Their Temporalities*, edited by E. Ferry and M. Limbert, 75–106. Santa Fe, NM: SAR Press.

Nader, L. 1996. *Naked Science: Anthropological Inquiry into Boundaries, Power, and Knowledge*. New York: Routledge.

Naiman, R. J., H. Decamps, and M. Pollock. 1993. The role of riparian corridors in maintaining regional biodiversity. *Ecological Applications* 3:209–12.

Nataro, J. P., and J. B. Kaper. 1998. Diarrheagenic *Escherichia coli*. *Clinical Microbiology Reviews* 11:142–201.

Natcher, D., S. Davis, and C. Hickey. 2005. Co-management: Managing relationships, not resources. *Human Organization* 64 (3): 240–50.

NRC (National Research Council). 1987. *Regulating Pesticides in Food: The Delaney Paradox*. Washington, DC: National Academy Press.

Nelkin, D., ed. 1992. *Controversy: Politics of Technical Decisions*. London: Sage.

Neumann, R. 1997. Primitive ideas: Protected areas buffer zones and the politics of land in Africa. *Development and Change* 28:559–82.

———. 1998. *Imposing Wilderness: Struggles over Livelihood and Nature Preservation in Africa*. Berkeley: University of California Press.

———. 2004. Moral and discursive geographies in the war for biodiversity in Africa. *Political Geography* 23 (7): 813–37.

———. 2005. *Making Political Ecology*. London: Hodder Arnold.

Nicholson, J. K., and I. D. Wilson. 2003. Understanding global systems biology: Metabonomics and the continuum of metabolism. *Nature Reviews Drug Discovery* 2:668–76.

Nieto Z., Oscar. 2001. *Fichas técnicas de plaguicidas a prohibir o restringir incluidos en el Acuerdo No. 9 de la XVI Reunión del Sector Salud de Centroamérica y República Dominicana (RESSCAD)*. San José, Costa Rica: OPS/OMS.

Nkedianye, D., M. Radeny, P. Kristjanson, and M. Herrero. 2009. Assessing returns to land and changing livelihood strategies in Kitengela. In *Staying Maasai? Livelihoods, Conservation and Development in East African Rangelands*, edited by K. Homewood, P. Trench, and P. Kristjanson, 115–50. London: Springer.

Noble, D. 2002. Modeling the heart: From genes to cells to the whole organ. *Science* 295:1678–82.

———. 2006. *The Music of Life: Biology beyond the Genome*. Oxford: Oxford University Press.

Noss, R. F. 1987. Corridors in real landscapes: A reply to Simberloff and Cox. *Conservation Biology* 1 (2): 159–64.

NRC (National Research Council). 1983. *Risk Assessment in the Federal Government: Managing the Process*. Washington, DC: National Academy Press.

———. 1992. *Restoration of Aquatic Ecosystems: Science, Technology, and Public Policy*. Washington, D.C.: National Academy Press.

Oberthür, T., E. Barrios, S. Cook, H. Usma, and G. Escobar. 2004. Increasing the relevance of scientific information in hillside environments through understanding of local soil management in a small watershed of the Colombian Andes. *Soil Use and Management* 19:1–9.

Odum, H. T. 1994. *Ecological and General Systems: An Introduction to Systems Ecology*. Niwot: University Press of Colorado.

Ogunseitan, O. 2005. *Microbial Diversity: Form and Function in Prokaryotes*. Oxford: Blackwell.

Ollier, C. D. 1991. Laterite profiles, ferricrete and landscape evolution. *Zeitschrift für Geomorphologie* 35 (2): 165–73.

Oporto Castro, H. 1999. *Misicuni: Entre la esperanza y la frustración; La problemática del agua en Cochabamba*. Cochabamba, Bolivia: Centro de Estudios de la Realidad Económica y Social.

Orme, S., and S. Kegley. 2008. PAN pesticide database. Pesticide Action Network, North America. http://www.pesticideinfo.org (accessed May 6, 2008).

Orr, A., B. Mwale, and D. Saiti. 2002. Modelling agricultural performance: Smallholder weed management in southern Malawi. *International Journal of Pest Management* 48 (4): 265–78.

Ortony, A., ed. 1979. *Metaphor and Thought*. New York: Cambridge University Press.

Ostrom, E. 2001. Vulnerability and polycentric governance systems. In newsletter of Center for the Study of Institutions, Population, and Enviromental Change, Workshop in Political Theory and Policy Analysis, Indiana University, Bloomington.

Ostrom, V. 1997. *The Meaning of Democracy and the Vulnerability of Democracies*. Ann Arbor: University of Michigan Press.

Otis, L. 2002. The metaphoric circuit: Organic and technological communication in the 19th century. *Journal of the History of Ideas* 63:105–28.

Oyama, S. 2000a. *Evolution's Eye: A Systems View of the Biology-Culture Divide*. Durham, NC: Duke University Press.

———. 2000b. *The Ontogeny of Information: Developmental Systems and Evolution*. 2nd ed., revised and expanded. Durham, NC: Duke University Press. Original edition 1985.

Oyama, S., P. E. Griffiths, and R. D. Gray, eds. 2001. *Cycles of Contingency: Developmental Systems and Evolution*. Cambridge, MA: MIT Press.

Pahlmann, C. 1990. *Farmers' Perceptions of the Sustainability of Upland Farming Systems of Northern Thailand*. Master of science thesis, University of Canberra.

Paxon, H. 2008. Post-Pasteurian cultures: The microbiopolitics of raw-milk cheese in the United States. *Cultural Anthropology* 23 (1): 15–47.

Peet, R., and M. J. Watts. 1993. Introduction: Development theory and environment in an age of market triumphalism. *Economic Geography* 69 (3): 227–53.

———, eds. 1996. *Liberation Ecologies: Environment, Development and Social Movements*. New York: Routledge.

———, eds. 2004. *Liberation Ecologies, Environment, Development, Social Movements*. 2nd ed. New York: Routledge.

Peluso, N. L. 1992. *Rich Forests, Poor People: Resource Control and Resistance in Java*. Berkeley: University of California Press.

———. 1993. Coercing conservation? The politics of state resource control. *Global Environmental Change* 3 (2): 199–217.

———. 2003. Weapons of the wild: Strategic use of wilderness and violence in West Kalimantan. In *In Search of the Rainforest*, edited by C. Slater, 204–45. Durham, NC: Duke University Press.

Peluso, N. L., and E. Harwell. 2001. Territory, custom, and the cultural politics of ethnic war in West Kalimantan, Indonesia. In *Violent Environments*, edited by N. L. Peluso and M. J. Watts, 83–116. Ithaca, NY: Cornell University Press.

Peluso, N. L., and P. Vandergeest. 2001. Genealogies of forest law and customary rights in Malaysia, Indonesia, and Thailand. *Journal of Asian Studies* 60 (3): 761–812.

Peluso, N. L., and M. J. Watts, eds. 2001. *Violent Environments*. Ithaca, NY: Cornell University Press.

PERG (Political Ecology Research Group). 1979. *A First Report of the Work of the Political Ecology Research Group*. Oxford: PERG.

Perkel, J. 2003. Microbiology vigil: Probing what's out there. *Scientist* 17 (9): 40.

Perkins, H. A. 2007. Ecologies of actor-networks and (non)social labor within the urban political economies of nature. *Geoforum* 38:1152–62.

Perlman, M. 1994. *The Power of Trees: The Reforesting of the Soul*. Dallas, TX: Spring Publications.

Perreault, T. 2005. State restructuring the scale politics of rural water governance in Bolivia. *Environment and Planning A* 37:263–84.

———. 2006. From the *Guerra del agua* to the *Guerra del gas*: Resource governance, neoliberalism, and popular protest in Bolivia. *Antipode* 38 (1): 150–72.

———. 2008. Custom and contradiction: Rural water governance and the politics of *usos y costumbres* in Bolivia's irrigators' movement. *Annals of the Association of American Geographers* 98 (4): 834–54.

Phadke, R. 2002. Assessing water scarcity and watershed development in Maharashtra, India: A case study of the Baliraja Memorial Dam. *Science, Technology, and Human Values* 27 (2): 236–61.

———. 2003. *Some Water for All: People's Science Lessons from the Krishna Valley*. Ph.D. thesis, University of California–Santa Cruz.

Phatharathananunth, S. 2006. *Civil Society and Democratization: Social Movements in Northeast Thailand*. Leifsgade, Denmark: Nordic Institute of Asian Studies Press.

Philo, C., and C. Wilbert, eds. 2000. *Animal Spaces, Beastly Places: New Geographies of Human-Animal Relations*. New York: Routledge.

Pickering, A., ed. 1992. *Science as Practice and Culture*. Chicago: University of Chicago Press.

Pickering, A., and K. Guzik, eds. 2008. *The Mangle in Practice: Science, Society, and Becoming.* Durham, NC: Duke University Press.

Pitot, A. 1953. Feux sauvages, végétation et sols en Afrique Occidentale Française. *Bulletin de l'Institute Française d'Afrique Noire* 15 (4): 1369–83.

Pobéguin, H. 1906. *Essai sur la flore de la Guinée Française.* Paris: Augustin Challamel.

Pollack, A. 2001. Drug makers listen in while bacteria talk. *New York Times*, February 27.

Pollan, M. 2001. *The Botany of Desire: A Plant's-Eye View of the World.* New York: Random House.

Poulgrain, G. 1998. *The Genesis of Konfrontasi: Malaysia, Brunei, Indonesia, 1945–1965.* Bathurst, NSW: Crawford House; London: C. Hurst.

Powledge, T. 2006. Microbes do vital work in human gut. *Scientist.* http://www.the-scientist.com/news/daily/23519/ (accessed April 7, 2007).

Prakash, G. 1999. *Another Reason.* Princeton, NJ: Princeton University Press.

Pratt, A. C. 1995. Putting critical realism to work: The practical implications for geographical research. *Progress in Human Geography* 19 (1): 61–74.

Proctor, J. D. 1998. The social construction of nature: Relativist accusations, pragmatist and critical realist responses. *Annals of the Association of American Geographers* 88 (3): 352–76.

Prudham, S. 2005. *Knock on Wood: Nature as Commodity in Douglas Fir Country.* New York: Routledge.

Putnam, R. D. 2000. *Bowling Alone: The Collapse and Revival of American Community.* New York: Simon and Schuster.

Raloff, J. 1998. Spray guards chicks from infections. Science News Online, March 28. http://www.sciencenews.org/pages/sn_arc98/3_28_98/src1.htm (accessed April 2007).

———. 2005. Antibiotics afield. *Science News*, November 26, 349.

Ralston, G. 1996a. NDP will continue wolf kill. *Yukon News*, December 18.

———. 1996b. To kill or not to kill, that is the question. *Yukon News*, November 27.

Ramchandani, M., A. Manges, C. DebRoy, S. Smith, J. Johnson, and L. Riley. 2005. Possible animal origin of human-associated, multidrug-resistant, uropathogenic *Escherichia coli*. *Clinical Infectious Diseases* 40:251–57.http://www.journals.uchicago.edu/CID/journal/issues/v40n2/34442/brief/.

Ramisch, J. J., M. T. Misiko, I. E. Ekise, and J. B. Mukalama. 2006. Strengthening "folk ecology": Community-based learning for soil fertility management, western Kenya. *International Journal of Agricultural Sustainability* 4 (2): 154–68.

Rangachari, R., N. Sengupta, R. Iyer, P. Baneri, and S. Singh. 2000. *Large Dams: India's Experience.* Capetown: World Commission on Dams.

Rangan, H. 2000. *Of Myth and Movements: Rewriting Chipko into Himalayan History.* London: Verso.

Rapoport, A. 1986. *General System Theory.* Cambridge, MA: Abacus Press.

Raynolds, L. 1994. The re-structuring of third world agro-exports: Changing production relations in the Dominican Republic. In *The Global Re-structuring of Agro-food Systems*, edited by P. McMichael, 214–37. Ithaca, NY: Cornell University Press.

Reid, R., and S. Traweek. 2000. *Doing Science + Culture.* New York: Routledge.

Reid, R. S., H. Gichohi, M. Y. Said, D. Nkedianye, J. O. Ogutu, M. Kshatriya, P. Kristjanson, et al. 2008. Fragmentation of a peri-urban savanna, Athi-Kaputiei plains, Kenya. In

Fragmentation in Semi-arid and Arid Landscapes: Consequences for Human and Natural Systems, edited by K. A. Galvin, R. S. Reid, R. H. Behnke, and N. T. Hobbs, 195–224. Dordrecht, Netherlands: Springer.

Reid, R. S., M. Rainy, J. Oguto, R. L. Kruska, M. McCartney, M. Nyabenge, K. Kimani, et al. 2003. *People, Wildlife and Livestock in the Mara Ecosystem: The Mara Count 2002*. Nairobi: International Livestock Research Institute.

Remedios, F. 2003. *Legitimizing Scientific Knowledge: An Introduction to Steve Fuller's Social Epistemology*. Lanham, MD: Lexington Books.

Renner, G. T. Jr. 1926. A famine zone in Africa: The Sudan. *Geographical Review* 16 (4): 583–96.

Richards, P. 1985. *Indigenous Agricultural Revolution*. Boulder, CO: Westview Press.

———. 1989. Agriculture as a performance. In *Farmer First: Farmer Innovation and Agricultural Research*, edited by R. Chambers, A. Pacey, and L. A. Thrupp, 39–43. London: Intermediate Technology Publications.

———. 1993. Cultivation: Knowledge or performance. In *An Anthropological Critique of Development: The Growth of Ignorance*, edited by M. Hobart, 61–78. London: Routledge.

Richardson, P., A. Broderick, L. M. Campbell, B. Godley, and S. Ranger. 2006. Marine turtle fisheries in the UK Overseas Territories of the Caribbean: Domestic legislation and the requirements of multilateral agreements. *Journal of International Wildlife Law and Policy* 9:223–46.

Ricourt, M. 2000. From Mama Tingo to globalization: The Dominican women peasant movement. *Women's Studies Review* 9:1–10.

Robbins, P. 1998. Paper forests: Imagining and deploying exogenous ecologies in arid India. *Geoforum* 29 (1): 69–86.

———. 2001. Fixed categories in a portable landscape: The causes and consequences of land-cover categorization. *Environment and Planning A* 33:161–79.

———. 2004. *Political Ecology: A Critical Introduction*. London: Blackwell.

———. 2007. *Lawn People: How Grasses, Weeds, and Chemicals Make Us Who We Are*. Philadelphia: Temple University Press.

Roberts, R. 1995. Taking nature-culture hybrids seriously in agricultural geography. *Environment and Planning D: Society and Space* 27:673–75.

Robertson, M. M. 2006. The nature that capital can see: Science, state, and market in the commodification of ecosystem services. *Environment and Planning D: Society and Space* 24:367–87.

Rocheleau, D. 1991. Gender, ecology and the science of survival: Stories and lessons from Kenya. *Agriculture and Human Values* 8 (1): 156–64.

———. 1995. Maps, numbers, text and context: Mixing methods in feminist political ecology. *Professional Geographer* 47 (4): 458–67.

———. 2005. Political landscapes and ecologies of Zambrana-Chacuey: The legacy of Mama Tingo. In *Women and the Politics of Place*, edited by W. Harcourt and A. Escobar, 72–85. Bloomfield, CT: Kumarian.

———. 2008. Political ecology in the key of policy: From chains of explanation to webs of relation. *Geoforum* 39:716–27.

Rocheleau, D., and L. Ross. 1995. Trees as tools, trees as text: Struggles over resources in Zambrana-Chacuey, Dominican Republic. *Antipode* 27:407–28.

Rocheleau, D., and R. Roth. 2007. Rooted networks, relational webs and powers of connection: Rethinking human and political ecologies. *Geoforum* 38:433–37.

Rocheleau, D., B. Thomas-Slayter, and E. Wangari. 1996. *Feminist Political Ecology: Global Issues and Local Experiences*. New York: Zed Books.

Roe, E. 1991. Development narratives or making the best of blueprint development. *World Development* 19 (4): 287–300.

Roper, B. B., J. M. Buffington, E. Archer, C. Moyer, and M. Ward. 2008. The role of observer variation in determining Rosgen stream types in northeastern Oregon mountain streams. *Journal of the American Water Resources Association* 44 (2): 417–27.

Rosgen, D. L. 1994. A classification of natural rivers. *Catena* 22 (3): 169–99.

———. 1996. *Applied River Morphology*. 2nd ed. Pagosa Springs, CO: Wildland Hydrology.

———. 2006. *Watershed Assessment of River Stability and Sediment Supply*. Fort Collins, CO: Wildland Hydrology.

———. 2007. The Rosgen geomorphic approach for natural channel design. In *Stream Restoration Design. National Engineering Handbook Part 654*, edited by J. Bernard, J. Fripp, and K. Robinson. Des Moines, IA: U.S. Department of Agriculture National Resources Conservation Service.

Rouget, M., R. M. Cowling, A. T. Lombard, A. T. Knight, and G. I. H. Kerley. 2006. Designing large-scale conservation corridors for pattern and process. *Conservation Biology* 20 (2): 549–61.

Russell, E. 2001. *War and Nature: Fighting Humans and Insects with Chemicals from World War I to Silent Spring*. New York: Cambridge University Press.

Russell, P. E. 2005. A century of fungicide evolution. *Journal of Agricultural Science* 143 (1): 11–25.

Sayre, N. F. 2008. The genesis, history, and limits of carrying capacity. *Annals of the Association of American Geographers* 98:120–34.

Scaëtta, H. 1937. Variations du climat pléistocene en Afrique centrale. *Annales de Géographie* 46:164–71.

———. 1938. La notion de cycle dans l'évolution du sol tropical d'après des recherches en Afrique centrale et en Afrique occidentale. *Comptes Rendus de l'Académie des Sciences* 206:1222–24.

———. 1941. L'évolution des sols et de la végétation dans la zone des latérites en Afrique occidentale. *Comptes Rendus de l'Académie des Sciences* 212:169–71.

Schlosser, E. 2006. Has politics contaminated the food supply? *New York Times*, December 11, Op Ed.

Schmitz, A., A. O. Fall, and S. Rouchiche. 1996. *Contrôle et utilisation du feu en zones arides et subhumides africaines*. Rome: Food and Agriculture Organization.

Schön, Donald. 1979. Generative metaphor: A perspective on problem-setting in social policy. In *Metaphor and Thought*, edited by A. Ortony, 137–63. Cambridge: Cambridge University Press.

Schonck de Goldfiem, J. 1936. L'influence de l'homme sur les variations de la flore en Guinée. *Revue Scientifique* 74 (2): 48–53.
Schreinemachers, D. M., J. P. Creason, and V. F. Garry. 1999. Cancer mortality in agricultural regions of Minnesota. *Environmental Health Perspectives* 107 (3): 205–11.
Schroeder, R. A. 1999a. Geographies of environmental intervention in Africa. *Progress in Human Geography* 23 (3): 359–78.
———. 1999b. *Shady Practices: Agroforestry and Gender Politics in the Gambia*. Berkeley: University of California Press.
———. 2008. Environmental justice and the market: The politics of sharing wildlife revenues in Tanzania. *Society and Natural Resources* 21 (7): 583–96.
Science Studies Centre. 2004. Actor Network Resource: An Annotated Bibliography. http://www.lancs.ac.uk/fass/centres/css/ant/antres.htm (accessed August 8, 2009).
Sclove, R. 1995. *Democracy and Technology*. New York: Guilford Press.
Scott, Colin. 1996. Science for the west, myth for the rest? The case of James Bay Cree knowledge production. In *Naked Science: Anthropological Inquiry into Boundaries, Power, and Knowledge*, edited by L. Nader, 69–86. New York: Routledge.
Scott, J. C. 1998. *Seeing like a State: How Certain Schemes to Improve the Human Condition Have Failed*. New Haven: Yale University Press.
Searle, J. 1995. *The Construction of Social Reality*. New York: Free Press.
Seminoff, J. A., and K. Shanker. 2008. Marine turtles and IUCN Red Listing: A review of the process, the pitfalls, and novel assessment approaches. *Journal of Experimental Marine Biology and Ecology* 356 (1–2): 52–68.
Shantz, H. L., and C. F. Marbut. 1923. *The Vegetation and Soils of Africa*. New York: American Geographical Society.
Shepard, H. H. 1951. *The Chemistry and Action of Insecticides*. New York: McGraw-Hill.
Shields, F. D., and R. R. Copeland. 2006. Empirical and analytical approaches for stream channel design. Paper presented at Federal Interagency Sediment Conference, April 2–6, Reno, NV.
Shields, F. D., R. R. Copeland, P. C. Klingeman, M. W. Doyle, and A. Simon. 2003. Design for stream restoration. *Journal of Hydraulic Engineering* 129 (3): 575–84.
———. 2009. Chapter 9. Stream Restoration. In *Manual 54 Update*. Reston, Virginia: American Society of Civil Engineers.
Shrader-Frechette, K. 1996. Throwing out the bathwater of positivism, keeping the baby of objectivity: Relativism and advocacy in conservation biology. *Conservation Biology* 10:912–14.
Sikana, P. 1993. Mismatched models: How farmers and scientists see soils. *Institute for Low External Input Agriculture (ILEIA) Newsletter* 9 (1): 15–16.
Sillans, R. 1958. *Les savanes de l'Afrique centrale*. Paris: Lechevalier.
Simberloff, D., J. A. Farr, J. Cox, and D. W. Mehlman. 1992. Movement corridors: Conservation bargains or poor investments? *Conservation Biology* 6 (4): 493–504.
Simon, A., M. W. Doyle, G. M. Kondolf, F. D. Shields, B. Rhoads, G. Grant, F. A. Fitzpatrick, K. E. Juracek, M. McPhillips, and J. MacBroom. 2005. How well do the Rosgen classification and associated natural channel design methods integrate

and quantify fluvial processes and channel response? Paper presented at the American Society of Civil Engineers Environmental and Water Resources Institute, Anchorage, AL.

Simon, A., M. W. Doyle, G. M. Kondolf, F. D. Shields, B. Rhoads, and M. McPhillips. 2007. Critical evaluation of how the Rosgen classification and associated natural channel design methods fail to integrate and quantify fluvial processes and channel response. *Journal of the American Water Resources Association* 43 (5): 1–15.

Sioh, M. 2004. An ecology of postcoloniality: Disciplining nature and society in Malaya, 1984–1957. *Journal of Historical Geography* 30:729–46.

Sivaramakrishnan, K. 1997. A limited forest conservancy in southwest Bengal, 1864–1912. *Journal of Asian Studies* 56 (1): 75–112

Sivaramakrishnan, K., and A. Agrawal. 2000. *Agrarian Environments: Resources, Representation, and Rule in India*. Durham, NC: Duke University Press.

Slate, L. O., F. D. Shields, J. S. Schwartz, D. D. Carpenter, and G. E. Freeman. 2007. Engineering design standards and liability for stream channel restoration. *Journal of Hydraulic Engineering* 133 (10): 1099–1102.

Slater, C. 1995. Amazonia as edenic narrative. In *Uncommon Ground: Toward Reinventing Nature*, edited by W. Cronon, 114–31. New York: Norton.

Smith, K. 1982. Shiyatah and his wolf helper. In *Nindäl Kwädindär/I'm Going to Tell You a Story*. Recorded by J. Cruikshank. Whitehorse, YT: Council for Yukon Indians and Government of Yukon.

Smith, K. E., J. M. Besser, C. W. Hedberg, F. T. Leano, J. B. Bender, J. H. Wicklund, B. P. Johnson, K. A. Moore, and M. T. Osterholm. 1999. Quinolone-resistant *Campylobacter jejuni* infections in Minnesota, 1992–1998. *New England Journal of Medicine* 340:1525–32.

Smith, S. M., and K. L. Prestegaard. 2005. Hydraulic performance of a morphology-based stream channel design. *Water Resources Research* 41 (W11413).

Snapp, S. 1999. Mother and baby trials: A novel trial design being tried out in Malawi. In *Target: The Newsletter of the Soil Fertility Research Network for Maize-Based Cropping Systems in Malawi and Zimbabwe* 17 (January): 8.

Snover, M. L., and A. A. Hohn. 2004. Validation and interpretation of annual skeletal marks in loggerhead (*Caretta caretta*) and Kemp's ridley (*Lepidochelys kempii*) sea turtles. *Fisheries Bulletin* 102:682–92.

Somers-Heidhues, M. 2003. *Golddiggers, Farmers, and Traders in the "Chinese District" of West Kalimantan, Indonesia*. Ithaca, NY: Cornell Southeast Asia Program.

SOPPECOM (Society for Promoting Participative Ecosystem Management). 2002. A framework for restructuring of water sector in sustainable, equitable and participatory lines. Unpublished paper, Pune.

Soulé, M. E., and M. E. Gilpin. 1991. The theory of wildlife corridor capability. In *Nature Conservation 2: The Roles of Corridors*, edited by D. A. Saunders and R. J. Hobbs, 3–8. Chipping Norton, Australia: Surrey Beatty and Sons.

Soulé, M. E., and J. Terborgh, eds. 1999. *Continental Conservation: Scientific Foundations of Regional Reserve Networks*. Washington, DC: Island Press.

Spence, C. 1998. Fertility control and the ecological consequences of managing northern wolf populations. Master's thesis, University of Toronto.

Spiegel-Rosing, I., and D. J. de Solla Price, eds. 1977. *Handbook of Science, Technology, and Society*. Thousand Oaks, CA: Sage.

Stan, H.-J. 2000. Pesticide residue analysis in foodstuffs applying capillary gas chromatography with mass spectrometric detection: State-of-the-art use of modified DFG-multimethod S19 and automated data evaluation. *Journal of Chromatography A* 892 (1–2): 347–77.

Star, S. L., ed. 1995. *Ecologies of Knowledge: Work and Politics in Science and Technology*. Albany: State University of New York Press.

Star, S. L., and J. R. Griesemer. 1989. Institutional ecology, "translation," and boundary objects: Amateurs and professionals in Berkeley's Museum of Vertebrate Zoology, 1907–39. *Social Studies of Science* 19:387–420.

Stebbing, E. P. 1935. The encroaching Sahara: The threat to the West African colonies. *Geographical Journal* 85:506–24.

Steinmuller, N., L. Demma, J. Bender, M. Eidson, and F. Angulo. 2006. Outbreaks of enteric disease associated with animal contact: Not just a foodborne problem anymore. *Clinical Infectious Diseases* 43:1596–1602.

Stott, P., and S. Sullivan, eds. 2000. *Political Ecology: Science, Myth and Power*. London: Arnold.

Strathern, M. 1991. *Partial Connections*. Savage, MD: Rowman and Littlefield.

———. 1992. *After Nature: English Kinship in the Late Twentieth Century*. Cambridge: Cambridge University Press.

Stubbs, R. 1989. *Hearts and Minds in Guerrilla Warfare: The Malayan Emergency, 1948–1960*. Singapore: Oxford University Press.

Suchman, L. 2001. Thinking with cyborgs: Feminist reflections on sociomaterialities. Paper presented to the Feminist Science Studies seminar, Lancaster University, February 5.

Sumberg, J., C. Okali, and D. Reece. 2003. Agricultural research in the face of diversity, local knowledge and the participation imperative: Theoretical considerations. *Agricultural Systems* 76:739–53.

Sundberg, J. 2003. Strategies for authenticity and space in the Maya Biosphere Reserve, Petén, Guatemala. In *Political Ecology: An Integrative Approach to Geography and Environment-Development Studies*, edited by K. S. Zimmerer and T. J. Bassett, 50–69. New York: Guilford Press.

Svanes, C., D. Jarvis, S. Chinn, and P. Burney. 1999. Childhood environment and adult atopy: Results from the European Community Respiratory Health Survey. *Journal of Allergy and Clinical Immunology* 103:415–20.

Swerdlow, J., and A. Johnson. 2002. Living with microbes. *Wilson Quarterly* 26 (2): 42–59.

Swift, J. 1996. Desertification: Narratives, winners and losers. In *The Lie of the Land*, edited by M. Leach and R. Mearns, 73–90. Oxford: James Currey.

Swyngedouw, E. 1999. Modernity and hybridity: Nature, *regeneracionismo*, and the production of the Spanish waterscape, 1890–1930. *Annals of the Association of American Geographers* 89 (3): 443–65.

Swyngedouw, E., and N. Heynen. 2003. Urban political ecology, justice and the politics of scale. *Antipode* 35 (5): 898–918.

Takacs, David. 1996. *The Idea of Biodiversity*. Baltimore: Johns Hopkins University Press.

Tapp, N. 1989. *Sovereignty and Rebellion: The White Hmong of Northern Thailand*. Singapore: Oxford University Press.

Tardy, Y. 1992. Diversity and terminology of lateritic profiles. In *Weathering, Soils and Paleosols*, edited by I. P. Martini and W. Chesworth, 379–405. London: Elsevier.

Tardy, Y., and C. Roquin. 1992. Geochemistry and evolution of lateritic landscapes. In *Weathering, Soils and Paleosols*, edited by I. P. Martini and W. Chesworth, 407–43. London: Elsevier.

Taylor, P. 2001. Anticholinesterase agents. In *Goodman and Gilman's The Pharmacological Basis of Therapeutics*, 10th ed., edited by J. G. Hardman and L. E. Limbird, 175–91. New York: McGraw-Hill.

Taylor, P. J. 2005. *Unruly Complexity: Ecology, Interpretation, Engagement*. Chicago: University of Chicago Press.

Taylor, P. J., and R. García-Barrios. 1995. The social analysis of ecological change: From systems to intersecting processes. *Social Science Information* 34 (1): 5–30.

Terborgh, J. 1991. *Requiem for Nature*. Washington, DC: Island Press.

TEWG (Turtle Expert Working Group). 2007. An assessment of the leatherback turtle population in the Atlantic Ocean. NOAA Technical Memorandum NMFS-SEFSC-555.

Thailand Development Research Institute. 1987. *Thailand Natural Resource Profile*. Bangkok: Thailand Development Research Institute.

Theberge, J. 1993. Letter to the editor. *Whitehorse Star*, February 3.

Thitirojanawat, P., and S. Chareonsuk. 2000. *Verification of the Universal Soil Loss Equation Application USLE on Forest Area*. Royal Forest Department, Bangkok. http://www.forest.go.th/Research/English/abstracts_water/wse64.html.

Thomas, W. G., J. Van Dusen Lewis, and J. Dorsey. 2003. Guinea agricultural sector assessment. USAID/Guinea Rural and Agricultural Incomes with a Sustainable Environment (RAISE) program, contract PCE-1-00-99-00001-00.

Thomas, W. I., and D. S. Thomas. 1929. *The Child in America*. 2nd ed. New York: Alfred Knopf.

Thompson, H. N. 1911. The forests of southern Nigeria. *Journal of the Royal African Society* 10 (38): 121–45.

Thompson, M., M. Warburton, and T. Hatley. 1986. *Uncertainty on a Himalayan Scale*. London: Ethnographica.

Thrift, N., S. Harrison, and S. Pile, eds. 2004. *Patterned Ground: Entanglements of Nature and Culture*. London: Reaktion Books.

Thrupp, L. A. 1988. Pesticides and policies: Approaches to pest-control dilemmas in Nicaragua and Costa Rica. *Latin American Perspectives* 15 (4): 37–70.

Tilho, J. 1928. Variations et disparition possible du Tchad. *Annales de Géographie* 37 (207): 238–60.

Tilman, D., K. G. Cassman, P. A. Matson, R. Naylor, and S. Polasky. 2002. Agricultural sustainability and intensive production practices. *Nature* 418:671–77.

Timofeyev, B. V., A. D. Dembele, and A. Danioko. 1988. Fertility of ferralitic concretionary degraded African savanna soils. *Soviet Soil Science* 20 (2): 23–28.

Todes, D. 1989. *Darwin without Malthus: The Struggle for Existence in Russian Evolutionary Thought*. New York: Oxford University Press.

Traweek, S. 1988. *Beamtimes and Lifetimes: The World of High Energy Physicists*. Cambridge, MA: Harvard University Press.

———. 1992. Border crossings: Narrative strategies in science studies and among physicists in Tsukuba Science City, Japan. In *Science as Practice and Culture*, edited by A. Pickering, 429–65. Chicago: University of Chicago Press.

Trendall, A. F. 1962. The formation of "apparent peneplains" by a process of combined laterisation and surface wash. *Zeitschrift für Geomorphologie* 6:183–97.

Troëng, S., and E. Rankin. 2005. Long-term conservation efforts contribute to positive green turtle *Chelonia mydas* nesting trend at Tortuguero, Costa Rica. *Biological Conservation* 121:111–16.

Tsing, A. L. 2004. *Friction: An Ethnography of Global Connection*. Durham, NC: Duke University Press.

Turkelboom, F. 1999. On-farm diagnosis of steepland erosion in northern Thailand: Integrating spatial scales with household strategies. Ph.D. thesis, Katholieke Universiteit Leuven, Belgium.

Turkelboom, F., J. Poesen, and G. Trébuil. 2008. The multiple land degradation effects caused by land-use intensification in tropical steeplands: A catchment study from northern Thailand. *Catena* 75 (1): 102–16.

Turnbull, D. 2000. *Masons, Tricksters and Cartographers: Comparative Studies in the Sociology of Scientific and Indigenous Knowledge*. Amsterdam: Harwood Academic.

Turner, M. D. 1993. Overstocking the range: A critical analysis of the environmental science of Sahelian pastoralism. *Economic Geography* 69 (4): 402–21.

———. 1998. The interaction of grazing history with rainfall and its influence on annual rangeland dynamics in the Sahel. In *Nature's Geography: New Lessons for Conservation in Developing Countries*, edited by K. Zimmerer and K. Young, 237–61. Madison: University of Wisconsin Press.

———. 2003. Methodological reflections on the use of remote sensing and geographic information science in human ecological research. *Human Ecology* 31 (2): 255–79.

———. 2004. Political ecology and the moral dimensions of resource conflicts: The case of farmer-herder conflicts in the Sahel. *Political Geography* 23 (7): 863–89.

Twidale, C. 1982. *Granite Landforms*. Amsterdam: Elsevier.

Uhlig, H., ed. 1984. *Spontaneous and Planned Land Settlement in Southeast Asia*. Giessener Geographische Schriften, vol. 58. Hamburg: Institute of Asian Affairs.

USDA (United States Department of Agriculture). 1961. *A Universal Equation for Predicting Rainfall-Erosion Losses*. USDA-ARS special report, U.S. Department of Agricultural Research Service.

van den Bogaard, A. E., R. Willems, N. London, J. Top, and E. E. Stobberingh. 2002. Antibiotic resistance of faecal enterococci in poultry, poultry farmers and poultry slaughterers. *Journal of Antimicrobial Chemotherapy* 49:497–505.

Vandergeest, P. 2003. Racialization and citizenship in Thai forest politics. *Society and Natural Resources* 16 (1): 19–37.

Vandergeest, P., and N. L. Peluso. 1995. Territorialization and state power in Thailand. *Theory and Society* 24:385–426.

———. 2006a. Empires of forestry: Professional forestry and state power in Southeast Asia, part 1. *Environment and History* 12:31–64.

———. 2006b. Empires of forestry: Professional forestry and state power in Southeast Asia, part 2. *Environment and History* 12:359–93.

Varma, R. 2001. People's science movements and science wars? *Economic and Political Weekly* 36 (52): 4796–4802.

Vera Varela, R. 1995. *Misicuni: La frustración de un pueblo?* Cochabamba, Bolivia: author.

von Meyer, W. C. 1977. *A Study of Liver and Thyroid Cancer Mortality as Related to Areas of Use of Ethylene Bisdithiocarbamate Fungicides.* Philadelphia: Rohm and Haas Co.

Wade, R. 1988. *Village Republics: Economics Condition for Collective Action.* Cambridge: Cambridge University Press.

———. 1990. On the technical case of irrigation hoarding behavior. In *Social, Economic and Institutional Issues in Third World Irrigation Management*, edited by R. K. Sampath and R. A. Young. Boulder, CO: Westview Press.

Wainwright, J., and M. M. Robertson. 2003. Territorialization, science, and the colonial state: The case of Highway 55 in Minneapolis, USA. *Cultural Geographies* 10:196–217.

Walker, A. 1983. In mountain and ulu: A comparative history of development strategies for ethnic minorities in Thailand and Malaysia. *Contemporary Southeast Asia* 4:451–85.

Walker, P. A. 2005. Political ecology: Where is the ecology? *Progress in Human Geography* 29 (1): 72–83.

Walley, C. J. 2004. *Rough Waters: Nature and Development in an East African Marine Park.* Princeton, NJ: Princeton University Press.

Ward, F. A., J. F. Booker, and A. M. Michelsen. 2006. Integrated economic, hydrologic, and institutional analysis of policy responses to mitigate drought impacts in Río Grande Basin. *Journal of Water Resources Planning and Management* 132 (6): 488–502.

Wargo, J. 1998. *Our Children's Toxic Legacy: How Science and Law Fail to Protect Us from Pesticides.* 2nd ed. New Haven: Yale University Press.

Watanabe, H., P. Sahunalu, and C. Khemnark. 1988. Combination of trees and crops in the tuangya method as applied in Thailand. *Agroforestry Systems* 6:169–77.

Waterton, C., and B. Wynne. 2004. Knowledge and political order in the European Environment Agency. In *States of Knowledge: The Co-production of Science and the Social Order*, edited by S. Jasanoff, 87–108. London: Routledge.

Watnick, P., and R. Kolter. 2000. Biofilm, city of microbes. *Journal of Bacteriology* 182 (10): 2675–79.

Watson-Verran, H., and D. Turnbull. 1995. Science and other indigenous knowledge systems. In *Handbook of Science and Technology Studies*, edited by S. Jasanoff, G. E. Markle, J. C. Peterson, and T. Pinch. Thousand Oaks, CA: Sage.

Watts, M. J. 1983. *Silent Violence: Food, Famine and Peasantry in Northern Nigeria.* Berkeley: University of California Press.

———. 2000. Political ecology. In *A Companion to Economic Geography*, edited by E. S. Sheppard and T. Barnes, 254–74. Malden, MA: Blackwell.

Watts, M. J., and J. McCarthy. 1997. Nature as artifice, nature as artifact. In *Geographies of Economies*, edited by R. Lee and J. Wills. London: Edward Arnold.

Watts, M. J., and R. Peet. 1993. Environment and development. Special double issue of *Economic Geography* 69 (3–4): 227–448. Republished with some changes as *Liberation Ecologies: Environment, Development, and Social Movements*, edited by R. Peet and M. J. Watts. London: Routledge, 1996.

Webb, G. J. W., and E. Carrillo. 2000. Risk of extinction and categories of endangerment: Perspectives from long-lived reptiles. *Population Ecology* 41:11–17.

Weins, J. A. 2006. Introduction: Connectivity research; What are the issues? In *Connectivity Conservation*, edited by K. R. Crooks and M. Sanjayan, 23–27. Cambridge: Cambridge University Press.

Weiss, P. 1955. Beauty and the beast: Life and the rule of order. *Scientific Monthly* 81:286–99.

Westfall, R. S. 1978. *The Construction of Modern Science: Mechanisms and Mechanics*. Cambridge: Cambridge University Press.

Whatmore, S. 2002. *Hybrid Geographies: Natures, Cultures, Spaces*. London: Sage.

———. 2006. Materialist returns: Practising cultural geographies in and for a more-than-human world. *Cultural Geographies* 13 (4): 600–610.

Whatmore, S., and L. Thorne. 1997. Nourishing networks: Alternative geographies of food. In *Globalising Food: Agrarian Questions and Global Restructuring*, edited by D. Goodman and M. Watts, 287–304. London: Routledge.

———. 2000. Elephants on the move: Spatial formations of wildlife exchange. *Environment and Planning D: Society and Space* 18:185–203.

Whitaker, R., D. Grogan, and J. Taylor. 2003. Geographic barriers isolate endemic populations of hyperthermophilic Archaea. *Science* 301:976–78.

White, R. 1983. *The Roots of Dependency: Subsistence, Environment, and Social Change among the Choctaws, Pawnees, and Navajos*. Lincoln: University of Nebraska Press.

Wilbanks, T. J., and R. W. Kates. 1999. Global change in local places: How scale matters. *Climatic Change* 43:601–28.

Wilcox, B. A., and D. D. Murphy. 1985. Conservation strategy: The effects of fragmentation on extinction. *American Naturalist* 125:879–87.

Williams, R. 1980. Ideas of nature. In *Problems in Materialism and Culture*, 67–85. London: Verso.

Winklerprins, A. M. G. A., and N. Barrera-Bassols. 2004. Latin American ethnopedology: A vision of its past, present and future. *Agriculture and Human Values* 21:35–52.

Winner, L. 1985. Do artifacts have politics? In *The Social Shaping of Technology*, edited by D. MacKenzie and J. Wacsman. Philadelphia: Open University Press.

Wolch, J., and J. Emel, eds. 1998. *Animal Geographies: Place, Politics and Identity in the Nature-Culture Borderlands*. London: Verso.

Wolford, W. W. 2001. *This Land Is Ours Now: Social Mobilization and the Struggle for Agrarian Reform in Brazil*. Ph.D. dissertation, University of California–Berkeley.

Wolman, M. G., and R. Gerson. 1978. Relative scales of time and effectiveness of climate in watershed geomorphology. *Earth Surface Processes* 3:189–208.

Wolman, M. G., and J. P. Miller. 1960. Magnitude and frequency of forces in geomorphic processes. *Journal of Geology* 68:57–74.

Worster, D. 1993. *The Wealth of Nature*. New York: Oxford University Press.

Wright, Angus. 1990. *The Death of Ramón González: The Modern Agricultural Dilemma*. Austin: University of Texas Press.

Wu, M.-L., J.-F. Deng, W.-J. Tsai, J. Ger, S.-S. Wong, and H.-P. Li. 2001. Food poisoning due to methamidophos-contaminated vegetables. *Clinical Toxicology* 39 (4): 333–36.

Wyatt-Smith, J. 1947. Save the belukar. *Malayan Forester* 11:24–26.

———. 1949. Regrowth in cleared areas. *Malayan Forester* 13:83–86.

Wynne, B. 1991. Knowledges in context. *Science, Technology, and Human Values* 16 (1): 111–21.

———. 1992. Misunderstood misunderstanding: Social identities and public uptake of science. *Public Understanding of Science* 1 (3): 281–304.

YCS (Yukon Conservation Society). 1994. Yukon Conservation Society position paper on the wolf kill. *YCS Quarterly*, December, 8–11.

Yearley, S. 1996. *Sociology, Environmentalism, Globalization*. London: Sage.

YFWB (Yukon Fish and Wildlife Branch). 1992. Designing an experiment for large mammal recovery in the Aishihik area, Yukon Territory. Minutes of technical meeting, Department of Renewable Resources, Yukon Territorial Government, Whitehorse.

Yukon News. 1993. Wolf kill lacks respect. *Yukon News*, December 3.

Zerner, C., ed. 2000. *People, Plants, and Justice: The Politics of Nature Conservation*. New York: Columbia University Press.

Zimmerer, K. S. 1995. The origins of Andean irrigation. *Nature* 378:481–83.

———. 1996. *Changing Fortunes: Biodiversity and Peasant Livelihood in the Peruvian Andes*. Berkeley: University of California Press.

———. 2000a. Rescaling irrigation in Latin America: The cultural images and political ecology of water resources. *Ecumene* 7 (2): 150–75.

———. 2000b. The reworking of conservation geographies: Nonequilibrium landscapes and nature-society hybrids. *Annals of the Association of American Geographers* 90 (2): 356–69.

———. 2009. Community-based resource management, environmental conservation, and farmer-and-food movements, 1985–present. In *Beyond Neoliberalism in Latin America? Societies and Politics at the Crossroads*, edited by J. Burdick, P. Oxhorn, and K. M. Roberts, 157–74. New York: Palgrave Macmillan.

———. 2010. Woodlands and agrobiodiversity in irrigation landscapes amidst global change: Bolivia, 1990–2002. *Professional Geographer* 62 (3): 335–56.

———. N.d. The conservation boom, land use transitions, and environmental governance in Latin America, 1985–2008 (Mexico, Costa Rica, Brazil, Peru, Bolivia). Manuscript.

Zimmerer, K. S., and T. J. Bassett, eds. 2003. *Political Ecology: An Integrative Approach to Geography and Environment-Development Studies*. New York: Guilford Press.

Zimmerer, K. S., R. E. Galt, and M. V. Buck. 2004. Globalization and the multi-spatial trends in the coverage of protected-area conservation (1980–2000). *Ambio* 33 (8): 520–29.

CONTRIBUTORS

Lisa M. Campbell
Nicholas School of the Environment, Marine Lab
Duke University
135 Duke Marine Lab Road
Beaufort, NC 28516 USA

Chris Duvall
Department of Geography
University of New Mexico
Albuquerque, NM 87131 USA

Tim Forsyth
Development Studies Institute
London School of Economics and Political Science
London WC2A 2AE UK

Joan H. Fujimura
Department of Sociology and the Program in Science and Technology Studies
University of Wisconsin
Madison, WI 53706 USA

Ryan E. Galt
Human and Community Development
University of California
Davis, CA 95616 USA

Mara J. Goldman
Department of Geography
University of Colorado
Boulder, CO 80309 USA

Mrill Ingram
Department of Geography
University of Arizona
Tucson, AZ 85721 USA

Rebecca Lave
Department of Geography
Indiana University
Bloomington, IN 47405 USA

Paul Nadasdy
Department of Anthropology and the American Indian Studies Program
Cornell University
Ithaca, NY 14853 USA

Nancy Lee Peluso
Department of Environmental Science, Policy, and Management
University of California
Berkeley, CA 94720 USA

Roopali Phadke
Department of Environmental Studies
Macalester College
St. Paul, MN 55105 USA

Joshua J. Ramisch
School of International Development and Global Studies
University of Ottawa
Ottawa, Ontario
Canada K1N 6N5

Dianne Rocheleau
Department of Geography
Clark University
Worcester, MA 01610 USA

Peter J. Taylor
Program in Critical and Creative Thinking
University of Massachusetts
Boston, MA 02125 USA

Matthew D. Turner
Department of Geography
University of Wisconsin
Madison, WI 53706 USA

Peter Vandergeest
Department of Geography
York University
Toronto, Ontario
Canada M3J 1P3

Karl S. Zimmerer
Department of Geography
The Pennsylvania State University
University Park, PA 16802 USA

INDEX